Introduction to Algebraic Geometry

Algebraic geometry has a reputation for being difficult and inaccessible, even among mathematicians! This must be overcome. The subject is central to pure mathematics, and applications in fields like physics, computer science, statistics, engineering, and computational biology are increasingly important. This book is based on courses given at Rice University and the Chinese University of Hong Kong, introducing algebraic geometry to a diverse audience consisting of advanced undergraduate and beginning graduate students in mathematics, as well as researchers in related fields.

For readers with a grasp of linear algebra and elementary abstract algebra, the book covers the fundamental ideas and techniques of the subject and places these in a wider mathematical context. However, a full understanding of algebraic geometry requires a good knowledge of guiding classical examples, and this book offers numerous exercises fleshing out the theory. It introduces Gröbner bases early on and offers algorithms for almost every technique described. Both students of mathematics and researchers in related areas benefit from the emphasis on computational methods and concrete examples.

Brendan Hassett is Professor of Mathematics at Rice University, Houston.

Introduction to Algebraic Geometry

Brendan Hassett
Department of Mathematics, Rice University, Houston

CAMBRIDGE UNIVERSITY PRESS
Cambridge, New York, Melbourne, Madrid, Cape Town, Singapore, São Paulo, Delhi

Cambridge University Press
The Edinburgh Building, Cambridge CB2 8RU, UK

Published in the United States of America by Cambridge University Press, New York

www.cambridge.org
Information on this title: www.cambridge.org/9780521691413

© Brendan Hassett 2007

This publication is in copyright. Subject to statutory exception
and to the provisions of relevant collective licensing agreements,
no reproduction of any part may take place without
the written permission of Cambridge University Press.

First published 2007
Reprinted 2008

Printed in the United Kingdom at the University Press, Cambridge

A catalog record for this publication is available from the British Library

ISBN 978-0-521-87094-8 hardback
ISBN 978-0-521-69141-3 paperback

Cambridge University Press has no responsibility for the persistence or accuracy
of URLs for external or third-party internet websites referred to in this publication,
and does not guarantee that any content on such websites is, or will remain,
accurate or appropriate.

To Eileen and William

Contents

Preface — xi

1 Guiding problems — 1
1.1 Implicitization — 1
1.2 Ideal membership — 4
1.3 Interpolation — 5
1.4 Exercises — 8

2 Division algorithm and Gröbner bases — 11
2.1 Monomial orders — 11
2.2 Gröbner bases and the division algorithm — 13
2.3 Normal forms — 16
2.4 Existence and chain conditions — 19
2.5 Buchberger's Criterion — 22
2.6 Syzygies — 26
2.7 Exercises — 29

3 Affine varieties — 33
3.1 Ideals and varieties — 33
3.2 Closed sets and the Zariski topology — 38
3.3 Coordinate rings and morphisms — 39
3.4 Rational maps — 43
3.5 Resolving rational maps — 46
3.6 Rational and unirational varieties — 50
3.7 Exercises — 53

4 Elimination — 57
4.1 Projections and graphs — 57
4.2 Images of rational maps — 61
4.3 Secant varieties, joins, and scrolls — 65
4.4 Exercises — 68

5 Resultants — 73
- 5.1 Common roots of univariate polynomials — 73
- 5.2 The resultant as a function of the roots — 80
- 5.3 Resultants and elimination theory — 82
- 5.4 Remarks on higher-dimensional resultants — 84
- 5.5 Exercises — 87

6 Irreducible varieties — 89
- 6.1 Existence of the decomposition — 90
- 6.2 Irreducibility and domains — 91
- 6.3 Dominant morphisms — 92
- 6.4 Algorithms for intersections of ideals — 94
- 6.5 Domains and field extensions — 96
- 6.6 Exercises — 98

7 Nullstellensatz — 101
- 7.1 Statement of the Nullstellensatz — 102
- 7.2 Classification of maximal ideals — 103
- 7.3 Transcendence bases — 104
- 7.4 Integral elements — 106
- 7.5 Proof of Nullstellensatz I — 108
- 7.6 Applications — 109
- 7.7 Dimension — 111
- 7.8 Exercises — 112

8 Primary decomposition — 116
- 8.1 Irreducible ideals — 116
- 8.2 Quotient ideals — 118
- 8.3 Primary ideals — 119
- 8.4 Uniqueness of primary decomposition — 122
- 8.5 An application to rational maps — 128
- 8.6 Exercises — 131

9 Projective geometry — 134
- 9.1 Introduction to projective space — 134
- 9.2 Homogenization and dehomogenization — 137
- 9.3 Projective varieties — 140
- 9.4 Equations for projective varieties — 141
- 9.5 Projective Nullstellensatz — 144
- 9.6 Morphisms of projective varieties — 145
- 9.7 Products — 154
- 9.8 Abstract varieties — 156
- 9.9 Exercises — 162

10 Projective elimination theory — 169
- 10.1 Homogeneous equations revisited — 170
- 10.2 Projective elimination ideals — 171
- 10.3 Computing the projective elimination ideal — 174
- 10.4 Images of projective varieties are closed — 175
- 10.5 Further elimination results — 176
- 10.6 Exercises — 177

11 Parametrizing linear subspaces — 181
- 11.1 Dual projective spaces — 181
- 11.2 Tangent spaces and dual varieties — 182
- 11.3 Grassmannians: Abstract approach — 187
- 11.4 Exterior algebra — 191
- 11.5 Grassmannians as projective varieties — 197
- 11.6 Equations for the Grassmannian — 199
- 11.7 Exercises — 202

12 Hilbert polynomials and the Bezout Theorem — 207
- 12.1 Hilbert functions defined — 207
- 12.2 Hilbert polynomials and algorithms — 211
- 12.3 Intersection multiplicities — 215
- 12.4 Bezout Theorem — 219
- 12.5 Interpolation problems revisited — 225
- 12.6 Classification of projective varieties — 229
- 12.7 Exercises — 231

Appendix A Notions from abstract algebra — 235
- A.1 Rings and homomorphisms — 235
- A.2 Constructing new rings from old — 236
- A.3 Modules — 238
- A.4 Prime and maximal ideals — 239
- A.5 Factorization of polynomials — 240
- A.6 Field extensions — 242
- A.7 Exercises — 244

Bibliography — 246
Index — 249

Preface

This book is an introduction to algebraic geometry, based on courses given at Rice University and the Institute of Mathematical Sciences of the Chinese University of Hong Kong from 2001 to 2006. The audience for these lectures was quite diverse, ranging from second-year undergraduate students to senior professors in fields like geometric modeling or differential geometry. Thus the algebraic prerequisites are kept to a minimum: a good working knowledge of linear algebra is crucial, along with some familiarity with basic concepts from abstract algebra. A semester of formal training in abstract algebra is more than enough, provided it touches on rings, ideals, and factorization. In practice, motivated students managed to learn the necessary algebra as they went along.

There are two overlapping and intertwining paths to understanding algebraic geometry. The first leads through sheaf theory, cohomology, derived functors and categories, and abstract commutative algebra – and these are just the prerequisites! We will not take this path. Rather, we will focus on specific examples and limit the formalism to what we need for these examples. Indeed, we will emphasize the strand of the formalism most useful for computations: We introduce *Gröbner bases* early on and develop algorithms for almost every technique we describe. The development of algebraic geometry since the mid 1990s vindicates this approach. The term 'Groebner' occurs in 1053 Math Reviews from 1995 to 2004, with most of these occurring in the last five years. The development of computers fast enough to do significant symbolic computations has had a profound influence on research in the field.

A word about what this book will *not* do: We develop computational techniques as a means to the end of learning algebraic geometry. However, we will not dwell on the technical questions of computability that might interest a computer scientist. We will also not spend time introducing the syntax of any particular computer algebra system. However, it is necessary that the reader be willing to carry out involved computations using elementary algebra, preferably with the help of a computer algebra system such as *Maple*, *Macaulay II*, or *Singular*.

Our broader goal is to display the core techniques of algebraic geometry in their natural habitat. These are developed systematically, with the necessary commutative algebra integrated with the geometry. Classical topics like resultants and elimination theory, are discussed in parallel with affine varieties, morphisms, and rational maps. Important examples of projective varieties (Grassmannians, Veronese varieties, Segre varieties) are emphasized, along with the matrix and exterior algebra needed to write down their defining equations.

It must be said that this book is not a comprehensive introduction to all of algebraic geometry. Shafarevich's book [37, 38] comes closest to this ideal; it addresses many important issues we leave untouched. Most other standard texts develop the material from a specific point of view, e.g., sheaf cohomology and schemes (Hartshorne [19]), classical geometry (Harris [17]), complex algebraic differential geometry (Griffiths and Harris [14]), or algebraic curves (Fulton [11]).

Acknowledgments

I am grateful to Ron Goldman, Donghoon Hyeon, Frank Jones, Sándor Kovács, Manuel Ladra, Dajiang Liu, Miles Reid, Ryan Scott, Burt Totaro, Yuri Tschinkel, and Fei Xu for helpful suggestions and corrections. I am indebted to Bradley Duesler for his comments on drafts of the text. My research has been supported by the Alfred P. Sloan Foundation and the National Science Foundation (DMS 0554491, 0134259, and 0196187).

The treatment of topics in this book owes a great deal to my teachers and the fine textbooks they have written: Serge Lang [27], Donal O'Shea [8], Joe Harris [17], David Eisenbud [9], and William Fulton [11]. My first exposure to algebraic geometry was through drafts of [8]; it has had a profound influence on how I teach the subject.

1 Guiding problems

Let k denote a field and $k[x_1, x_2, \ldots, x_n]$ the polynomials in $x_1, x_2, \ldots, x_n$ with coefficients in k. We often refer to k as the *base field*. A nonzero polynomial

$$f = \sum_{\alpha_1, \ldots, \alpha_n} c_{\alpha_1 \ldots \alpha_n} x_1^{\alpha_1} \ldots x_n^{\alpha_n}, \quad c_{\alpha_1 \ldots \alpha_n} \in k,$$

has degree d if $c_{\alpha_1 \ldots \alpha_n} = 0$ when $\alpha_1 + \cdots + \alpha_n > d$ and $c_{\alpha_1 \ldots \alpha_n} \neq 0$ for some index with $\alpha_1 + \cdots + \alpha_n = d$. It is *homogeneous* if $c_{\alpha_1 \ldots \alpha_n} = 0$ whenever $\alpha_1 + \cdots + \alpha_n < d$. We will sometimes use multiindex notation

$$f = \sum_{\alpha} c_\alpha x^\alpha$$

where $\alpha = (\alpha_1, \ldots, \alpha_n)$, $c_\alpha = c_{\alpha_1 \ldots \alpha_n}$, $x^\alpha = x_1^{\alpha_1} \ldots x_n^{\alpha_n}$, and $|\alpha| = \alpha_1 + \cdots + \alpha_n$.

1.1 Implicitization

Definition 1.1 *Affine space* of dimension n over k is defined

$$\mathbb{A}^n(k) = \{(a_1, a_2, \ldots, a_n) : a_i \in k\}.$$

For $k = \mathbb{R}$ this is just the ubiquitous $\mathbb{R}^n$. Why don't we use the notation k^n for affine space? We write $\mathbb{A}^n(k)$ when we want to emphasize the geometric nature of k^n rather than its algebraic properties (e.g., as a vector space). Indeed, when our discussion does not involve the base field in an essential way we drop it from the notation, writing $\mathbb{A}^n$.

We shall study maps between affine spaces, but not just any maps are allowed in algebraic geometry. We consider only maps given by polynomials:

Definition 1.2 A *morphism* of affine spaces

$$\phi : \mathbb{A}^n(k) \to \mathbb{A}^m(k)$$

is a map given by a polynomial rule

$$(x_1, x_2, \ldots, x_n) \mapsto (\phi_1(x_1, \ldots, x_n), \ldots, \phi_m(x_1, \ldots, x_n)),$$

with the $\phi_i \in k[x_1, \ldots, x_n]$.

Remark 1.3 This makes a tacit reference to the base field k, in that the polynomials ϕ_i have coefficients in k. If we want to make this explicit, we say that the morphism is *defined over k*.

Example 1.4 An affine-linear transformation is a morphism: given an $m \times n$ matrix $A = (a_{ij})$ and an $m \times 1$ matrix $b = (b_i)$ with entries in k, we define

$$\phi_{A,b} : \mathbb{A}^n(k) \to \mathbb{A}^m(k)$$
$$\begin{pmatrix} x_1 \\ \vdots \\ x_n \end{pmatrix} \mapsto \begin{pmatrix} a_{11}x_1 + \cdots + a_{1n}x_n + b_1 \\ \vdots \\ a_{m1}x_1 + \cdots + a_{mn}x_n + b_m \end{pmatrix}.$$

Example 1.5 Consider

$$\mathbb{A}^1(\mathbb{R}) \to \mathbb{A}^2(\mathbb{R})$$

given by the rule

$$t \mapsto (t, t^2).$$

If y_1 and y_2 are the corresponding coordinates on $\mathbb{R}^2$ then the image is the parabola $\{(y_1, y_2) : y_2 = y_1^2\}$. More generally, consider the morphism

$$\phi : \mathbb{A}^1(k) \to \mathbb{A}^m(k)$$
$$t \mapsto (t, t^2, t^3, \ldots, t^m).$$

Can we visualize the image of ϕ in $\mathbb{A}^m(k)$? Just as for the parabola, we write down polynomial equations for this locus. Fix coordinates $y_1, \ldots, y_m$ on $\mathbb{A}^m(k)$ so that ϕ is given by $y_i \mapsto t^i$. We find the equations

$$y_i y_j = y_{i+j} \quad 1 \leq i < j \leq m$$
$$y_i y_j = y_k y_l \quad i + j = k + l$$

corresponding to the relations $t^i t^j = t^{i+j}$ and $t^i t^j = t^k t^l$ respectively.

The polynomial equations describing the image of our morphism are an *implicit* description of this locus. Here the sense of 'implicit' is the same as the 'implicit function theorem' from calculus. We can consider the general question:

Problem 1.6 (Implicitization) Write down the polynomial equations satisfied by the image of a morphism.

1.1.1 A special case: linear transformations

Elementary row operations from linear algebra solve Problem 1.6 in the case where ϕ is a linear transformation.

Suppose ϕ is given by the rule

$$\mathbb{A}^2(\mathbb{Q}) \to \mathbb{A}^3(\mathbb{Q})$$
$$(x_1, x_2) \mapsto (x_1 + x_2, x_1 - x_2, x_1 + 2x_2)$$

and assign coordinates y_1, y_2, y_3 to affine three-space. From this, we extract the system

$$y_1 = x_1 + x_2$$
$$y_2 = x_1 - x_2$$
$$y_3 = x_1 + 2x_2,$$

or equivalently,

$$\begin{aligned} x_1 + x_2 - y_1 &= 0 \\ x_1 - x_2 \quad\quad - y_2 &= 0 \\ x_1 + 2x_2 \quad\quad\quad - y_3 &= 0, \end{aligned}$$

which in turn are equivalent to

$$\begin{aligned} x_1 + x_2 - y_1 &= 0 \\ -2x_2 + y_1 - y_2 &= 0 \\ x_2 + y_1 \quad\quad - y_3 &= 0, \end{aligned}$$

and

$$\begin{aligned} x_1 + x_2 - y_1 &= 0 \\ -2x_2 + y_1 - y_2 &= 0 \\ +\tfrac{3}{2} y_1 - \tfrac{1}{2} y_2 - y_3 &= 0. \end{aligned}$$

Thus the image of our morphism is given by

$$3y_1 - y_2 - 2y_3 = 0.$$

Our key tool for solving Problem 1.6 in general – Buchberger's Algorithm – will contain elementary row operations as a special case.

Moral 1: To solve Problem 1.6, choosing an order on the variables is very useful.

1.1.2 A converse to implicitization?

The implicitization problem seeks equations for the image of a morphism

$$\phi : \mathbb{A}^n(k) \to \mathbb{A}^m(k).$$

We will eventually show that this admits an algorithmic solution, at least when the base field is algebraically closed. However, there is a natural converse to this question which is much deeper.

GUIDING PROBLEMS

Definition 1.7 A *hypersurface* of degree d is the locus

$$V(f) := \{(a_1, \ldots, a_m) \in \mathbb{A}^m(k) : f(a_1, \ldots, a_m) = 0\} \subset \mathbb{A}^m(k),$$

where f is a polynomial of degree d.

A *regular parametrization* of a hypersurface $V(f) \subset \mathbb{A}^m(\mathbb{C})$ is a morphism

$$\phi : \mathbb{A}^n(\mathbb{C}) \to \mathbb{A}^m(\mathbb{C})$$

such that

1. the image of ϕ is contained in the hypersurface, i.e., $f \circ \phi = 0$;
2. the image of ϕ is not contained in any other hypersurface, i.e., for any $h \in \mathbb{C}[y_1, \ldots, y_m]$ with $h \circ \phi = 0$ we have $f | h$.

Problem 1.8 Which hypersurfaces admit regular parametrizations?

Example 1.9 Here are some cases where parametrizations exist:

1. hypersurfaces of degree one (see Exercise 1.5);
2. the curve $V(f) \subset \mathbb{A}^2$, $f = y_1^2 - y_2^3$, has parametrization (cf. Exercise 1.8)

$$\phi : \mathbb{A}^1(\mathbb{C}) \to \mathbb{A}^2(\mathbb{C})$$
$$t \mapsto (t^3, t^2)$$

3. if $f = y_0^2 + y_1^2 - y_2^2$ then $V(f)$ has a parametrization

$$\phi(s, t) = (2st, s^2 - t^2, s^2 + t^2);$$

4. if $f = y_0^3 + y_1^3 + y_2^3 + y_3^3$ then $V(f)$ has parametrization

$$y_0 = (u_2 + u_1)u_3^2 + (u_2^2 + 2u_1^2)u_3 - u_2^3 + u_1u_2^2 - 2u_1^2u_2 - u_1^3$$
$$y_1 = u_3^3 - (u_2 + u_1)u_3^2 + (u_2^2 + 2u_1^2)u_3 + u_1u_2^2 - 2u_1^2u_2 + u_1^3$$
$$y_2 = -u_3^3 + (u_2 + u_1)u_3^2 - (u_2^2 + 2u_1^2)u_3 + 2u_1u_2^2 - u_1^2u_2 + 2u_1^3$$
$$y_3 = (u_2 - 2u_1)u_3^2 + (u_1^2 - u_2^2)u_3 + u_2^3 - u_1u_2^2 + 2u_1^2u_2 - 2u_1^3.$$

The form here is due to Noam Elkies.

We will come back to these questions when we discuss unirationality and rational maps in Chapter 3.

1.2 Ideal membership

Our second guiding problem is algebraic in nature.

Problem 1.10 (Ideal Membership Problem) Given $f_1, \ldots, f_r \in k[x_1, \ldots, x_n]$, determine whether $g \in k[x_1, \ldots, x_n]$ belongs to the ideal $\langle f_1, \ldots, f_r \rangle$.

Example 1.11 Consider the ideal

$$I = \langle y_2 - y_1^2, y_3 - y_1 y_2 \rangle \subset k[y_1, y_2, y_3]$$

and the polynomial $g = y_1 y_3 - y_2^2$ (cf. Example 1.5 and the following discussion). Then $g \in I$ because

$$y_1 y_3 - y_2^2 = y_1(y_3 - y_1 y_2) + y_2(y_1^2 - y_2).$$

Again, whenever the f_i and g are all linear, elementary row reductions give a solution to Problem 1.10. However, there is one further case where we already know how to solve the problem. The Euclidean Algorithm yields a procedure to decide whether a polynomial $g \in k[t]$ is contained in a given ideal $I \subset k[t]$. By Theorem A.9, each ideal $I \subset k[t]$ can be expressed $I = \langle f \rangle$ for some $f \in k[t]$. Therefore $g \in I$ if and only if f divides g.

Example 1.12 Check whether $t^5 + t^3 + 1 \in \langle t^3 + 1 \rangle$:

$$\begin{array}{r}
t^2 + 1 \\
t^3 + 1 \overline{\smash{\big)} t^5 + t^3 + 1} \\
\underline{t^5 + t^2 } \\
+ t^3 - t^2 + 1 \\
\underline{+ t^3 + 1} \\
-t^2
\end{array}$$

thus $q = t^2 + 1$ and $r = -t^2$. We conclude $t^5 + t^3 + 1 \notin \langle t^3 + 1 \rangle$:

Moral 2: In solving Problem 1.10, keeping track of *degrees* of polynomials is crucial.

1.3 Interpolation

Let $P_{n,d} \subset k[x_1, \ldots, x_n]$ denote the vector subspace of polynomials of degree $\leq d$. The monomials

$$x^\alpha = x_1^{\alpha_1} \ldots x_n^{\alpha_n}, \quad \alpha_1 + \cdots + \alpha_n \leq d$$

form a basis for $P_{n,d}$, so we have (see Exercise 1.4)

$$\dim P_{n,d} = \binom{n+d}{n}.$$

Problem 1.13 (Simple Interpolation Problem) Given distinct points

$$p_1, \ldots, p_N \in \mathbb{A}^n(k)$$

what is the dimension of the vector space $I_d(p_1, \ldots, p_N)$ of polynomials of degree $\leq d$ vanishing at each of the points?

Here is some common terminology used in these questions:

Definition 1.14 Given $S \subset \mathbb{A}^n(k)$, the number of conditions imposed by S on polynomials of degree $\leq d$ is defined

$$C_d(S) := \dim P_{n,d} - \dim I_d(S).$$

S is said to *impose independent conditions on* $P_{n,d}$ if

$$C_d(S) = |S|.$$

It *fails to impose independent conditions* otherwise.

Another formulation of the Simple Interpolation Problem is:

When do N points in $\mathbb{A}^n(k)$ fail to impose independent conditions on polynomials of degree $\leq d$?

In analyzing examples, it is useful to keep in mind that affine linear transformations do not affect the number conditions imposed on $P_{n,d}$:

Proposition 1.15 *Let $S \subset \mathbb{A}^n(k)$ and consider an invertible affine-linear transformation $\phi : \mathbb{A}^n(k) \to \mathbb{A}^n(k)$. Then $C_d(S) = C_d(\phi(S))$ for each d.*

Proof By Exercise 1.11, ϕ induces an invertible linear transformation $\phi^* : P_{n,d} \to P_{n,d}$ with $\phi^*(f(x_1, \ldots, x_n)) = (f \circ \phi)(x_1, \ldots, x_n)$. Thus $(\phi^* f)(p) = 0$ for each $p \in S$ if and only if $f(q) = 0$ for each $q \in \phi(S)$. In particular, $\phi^*(I_d(\phi(S))) = I_d(S)$ so these spaces have the same dimension. $\square$

1.3.1 Some examples
Let $S = \{p_1, p_2, p_3\} \subset \mathbb{A}^n(k)$ be collinear with $n > 1$ or $S = \{p_1, p_2, p_3, p_4\} \subset \mathbb{A}^n(k)$ coplanar with $n > 2$. Then S fails to impose independent conditions on polynomials of degree ≤ 1.

Let $S = \{p_1, p_2, p_3, p_4, p_5, p_6\} \subset \mathbb{A}^2(\mathbb{R})$ lie on the unit circle

$$x_1^2 + x_2^2 = 1.$$

Then S fails to impose independent conditions on polynomials of degree ≤ 2; indeed, $C_2(S) = 5 < 6$.

When does a set of four points $\{p_1, p_2, p_3, p_4\} \subset \mathbb{A}^2(k)$ fail to impose independent conditions on quadrics ($d = 2$)? Assume that three of the points are non-collinear, e.g., p_1, p_2, p_3. After translating suitably we may assume $p_1 = (0, 0)$, and after a further linear change of coordinates we may assume $p_2 = (1, 0)$ and $p_3 = (0, 1)$. (Proposition 1.15 allows us to change coordinates without affecting the number of conditions imposed.) If $p_4 = (a_1, a_2)$ then the conditions on

$$c_{00} + c_{10}x_1 + c_{01}x_2 + c_{20}x_1^2 + c_{11}x_1x_2 + c_{02}x_2^2 \in P_{2,2}$$

1.3 INTERPOLATION

take the form

$$c_{00} = 0 \quad (p_1)$$
$$c_{00} + c_{10} + c_{20} = 0 \quad (p_2)$$
$$c_{00} + c_{01} + c_{02} = 0 \quad (p_3)$$
$$c_{00} + c_{10}a_1 + c_{01}a_2 + c_{20}a_1^2 + c_{11}a_1 a_2 + c_{02}a_2^2 = 0. \quad (p_4)$$

If these are not independent, the matrix

$$\begin{pmatrix} 1 & 0 & 0 & 0 & 0 & 0 \\ 0 & 1 & 0 & 1 & 0 & 0 \\ 0 & 0 & 1 & 0 & 0 & 1 \\ 0 & 0 & 0 & a_1^2 - a_1 & a_1 a_2 & a_2^2 - a_2 \end{pmatrix}$$

has rank 3. This can only happen if

$$a_1^2 - a_1 = a_1 a_2 = a_2^2 - a_2 = 0,$$

which means $p_4 \in \{(0,0), (1,0), (0,1)\} = \{p_1, p_2, p_3\}$, a contradiction. Thus we have shown:

Proposition 1.16 *Four distinct points in the plane fail to impose independent conditions on quadrics only if they are all collinear.*

Here are some sample results:

Proposition 1.17 *Any N points in the affine line $\mathbb{A}^1(k)$ impose independent conditions on $P_{1,d}$ for $d \geq N - 1$.*
Assume k is infinite. For each $N \leq \binom{n+d}{d}$, there exist N points in $\mathbb{A}^n(k)$ imposing independent conditions on $P_{n,d}$.

Proof For the first statement, suppose that $f \in k[x_1]$ is a polynomial vanishing at

$$p_1, \ldots, p_N \in \mathbb{A}^1(k).$$

The Euclidean Algorithm implies that f is divisible by $x - p_j$ for each $j = 1, \ldots, N$. Consequently, it is also divisible by the product $(x_1 - p_1) \ldots (x_1 - p_N)$ (see Exercise A.13). Moreover, if $f \neq 0$ we have a unique expression

$$f = q(x_1 - p_1) \ldots (x_1 - p_N), \quad q \in P_{1, d-N}.$$

The polynomials of this form (along with 0) form a vector space of dimension $d - N + 1$, so

$$C_d(p_1, \ldots, p_N) = \min(N, d+1).$$

The second statement is established by producing a sequence of points $p_1, \ldots, p_{\binom{n+d}{d}}$ such that

$$I_d(p_1, \ldots, p_j) \supsetneq I_d(p_1, \ldots, p_{j+1})$$

for each $j < \binom{n+d}{d}$. The argument proceeds by induction. Given $p_1, \ldots, p_j$, linear algebra gives a nonzero $f \in P_{n,d}$ with $f(p_1) = \ldots = f(p_j) = 0$. It suffices to find some $p_{j+1} \in \mathbb{A}^n(k)$ such that $f(p_{j+1}) \neq 0$, which follows from the fact (Exercise 1.9) that every nonzero polynomial over an infinite field takes a nonzero value somewhere in $\mathbb{A}^n(k)$. $\square$

1.4 Exercises

1.1 Consider the linear morphism

$$\phi : \mathbb{A}^3(\mathbb{R}) \to \mathbb{A}^4(\mathbb{R})$$
$$(t_1, t_2, t_3) \mapsto (3t_1 + t_3, t_2 + 4t_3, t_1 + t_2 + t_3, t_1 - t_2 - t_3).$$

Describe image(ϕ) as the locus where a linear polynomial vanishes.

1.2 Decide whether $g = t^3 + t^2 - 2$ is contained in the ideal

$$\langle t^3 - 1, t^5 - 1 \rangle \subset \mathbb{Q}[t].$$

If so, produce $h_1, h_2 \in \mathbb{Q}[t]$ such that

$$g = h_1(t^3 - 1) + h_2(t^5 - 1).$$

1.3 Consider the ideal

$$I = \langle y_2 - y_1^2, y_3 - y_1 y_2, \ldots, y_m - y_1 y_{m-1} \rangle \subset k[y_1, \ldots, y_m].$$

Show this contains all the polynomials $y_{i+j} - y_i y_j$ and $y_i y_j - y_k y_l$ where $i + j = k + l$ (cf. Example 1.5.)

1.4 Show that the dimension of the vector space of polynomials of degree $\leq d$ in n variables is equal to the binomial coefficient

$$\binom{n+d}{d} = \frac{(n+d)!}{d! \, n!}.$$

Compute the dimension of the vector space of homogeneous polyonomials of degreee d in $n + 1$ variables.

1.5 Given

$$f = c_1 x_1 + c_2 x_2 + \cdots + c_n x_n + c_0$$

with $c_i \neq 0$ for some $i > 0$, exhibit a morphism

$$\phi : \mathbb{A}^{n-1} \to \mathbb{A}^n$$

such that image(ϕ) = $V(f)$ and ϕ is one-to-one.

1.4 EXERCISES

1.6 Let $A = (a_{ij})$ be an $m \times n$ matrix with entries in k and $b = (b_1, \ldots, b_n) \in k^n$. For each $i = 1, \ldots, m$, set
$$f_i = a_{i1}x_1 + \cdots + a_{in}x_n \in k[x_1, \ldots, x_n]$$
and $g = b_1x_1 + \cdots + b_nx_n$. Show that $g \in \langle f_1, \ldots, f_m \rangle$ if and only if b is contained in the span of the rows of A.

1.7 Let k be an infinite field. Consider the morphism
$$j : \mathbb{A}^3(k) \to \mathbb{A}^6(k)$$
$$(u, v, w) \mapsto (u^2, uv, v^2, vw, w^2, uw).$$

Let $a_{11}, a_{12}, a_{22}, a_{23}, a_{33}$, and a_{13} be the corresponding coordinates on $\mathbb{A}^6(k)$ and
$$A = \begin{pmatrix} a_{11} & a_{12} & a_{13} \\ a_{12} & a_{22} & a_{23} \\ a_{13} & a_{23} & a_{33} \end{pmatrix}$$
the symmetric matrix with these entries.

(a) Show that the image of j satisfies the equations given by the two-by-two minors of A.

(b) Compute the dimension of the vector space V in
$$R = k[a_{11}, a_{12}, a_{22}, a_{23}, a_{33}, a_{13}]$$
spanned by these two-by-two minors.

(c) Show that every homogeneous polynomial of degree 2 in R vanishing on the image of j is contained in V. *Hint:* Degree-2 polynomials in R yield degree-4 polynomials in $k[u, v, w]$. Count dimensions!

1.8 Show that the parametrization given for the curve $V(f) \subset \mathbb{A}^2(\mathbb{C})$, $f = x_1^2 - x_2^3$ satisfies the required properties.

1.9 Let k be an infinite field. Suppose that $f \in k[x_1, \ldots, x_n]$ is nonzero. Show there exists $a = (a_1, \ldots, a_n) \in \mathbb{A}^n(k)$ with $f(a_1, \ldots, a_n) \neq 0$.

1.10 Let $S \subset \mathbb{A}^n(k)$ be a finite nonempty subset and let $k[S]$ denote the ring of k-valued functions on S. Show that the linear transformation
$$P_{n,d} \to k[S]$$
$$f \mapsto f|S$$
is surjective if and only if S imposes independent conditions on polynomials of degree d.

1.11 Let $\phi : \mathbb{A}^n(k) \to \mathbb{A}^m(k)$ be an affine linear transformation given by the matrix formula $\phi(x) = Ax + b$ (see Example 1.4). Consider the map induced by composition of polynomials
$$\phi^* : k[y_1, \ldots, y_m] \to k[x_1, \ldots, x_n]$$
$$P(y) \mapsto P(Ax + b).$$

Show that
(a) ϕ^* takes polynomials of degree $\leq d$ to polynomials of degree $\leq d$;
(b) ϕ^* is a k-algebra homomorphism;
(c) if the matrix A is invertible then so is ϕ^*.

Moreover, in case (c) the induced linear transformation $\phi^* : P_{n,d} \to P_{n,d}$ is also invertible.

1.12 Consider five distinct points in $\mathbb{A}^2(\mathbb{R})$ that fail to impose independent conditions on $P_{2,3}$. Show that these points are collinear, preferably by concrete linear algebra.

1.13 Show that $d + 1$ distinct points

$$p_1, \ldots, p_{d+1} \in \mathbb{A}^n(\mathbb{Q})$$

always impose independent conditions on polynomials in $P_{n,d}$.

1.14 Let ℓ_1, ℓ_2, ℓ_3 be arbitrary lines in $\mathbb{A}^3(\mathbb{Q})$. (By definition, a line $\ell \subset \mathbb{A}^3$ is the locus where two consistent independent linear equations are simultaneously satisfied, e.g., $x_1 + x_2 + x_3 - 1 = x_1 - x_2 + 2x_3 - 4 = 0$.) Show there exists a nonzero polynomial $f \in P_{3,2}$ such that f vanishes on ℓ_1, ℓ_2, and ℓ_3.

Optional Challenge: Assume that ℓ_1, ℓ_2, and ℓ_3 are pairwise skew. Show that f is unique up to scalar.

2 Division algorithm and Gröbner bases

In this chapter we solve the Ideal Membership Problem for polynomial ideals. The key tool is *Gröbner bases*: producing a Gröbner basis for a polynomial ideal is analogous to putting a system of linear equations in row echelon form. Once we have a Gröbner basis, a multivariate division algorithm can be applied to decide whether a given polynomial sits in our ideal. We also discuss normal forms for polynomials modulo ideals.

The existence of a Gröbner base can be deduced from nonconstructive arguments, but actually finding one can be challenging computationally. *Buchberger's Algorithm* gives a general solution. The proof that it works requires a systematic understanding of the 'cancellations' among polynomials, which are usually called *syzygies*.

2.1 Monomial orders

As we have seen, in order to do calculations we need a system for ordering the terms of a polynomial. For polynomials in one variable, the natural order is by degree, i.e.,

$$x_1^\alpha > x_1^\beta \quad \text{if } \alpha > \beta.$$

However, for linear polynomials in many variables, we have seen that the order is essentially arbitrary.

We first fix terminology. Given a polynomial

$$\sum c_{\alpha_1\ldots\alpha_n} x_1^{\alpha_1} \ldots x_n^{\alpha_n}$$

each $c_{\alpha_1\ldots\alpha_n} x_1^{\alpha_1} \ldots x_n^{\alpha_n}$ is a *term*. A polynomial of the form

$$x^\alpha = x_1^{\alpha_1} \ldots x_n^{\alpha_n}$$

is called a *monomial*.

Definition 2.1   A *monomial order* $>$ on $k[x_1, \ldots, x_n]$ is a total order on monomials satisfying the following:

1. **Multiplicative property**   If $x^\alpha > x^\beta$ then $x^\alpha x^\gamma > x^\beta x^\gamma$ (for any α, β, γ).
2. **Well ordering**   An arbitrary set of monomials

$$\{x^\alpha\}_{\alpha \in A}$$

has a least element.

The stipulation that $>$ be a total order means that any monomials x^α and x^β are comparable in the order, i.e., either $x^\alpha > x^\beta$, $x^\alpha < x^\beta$, or $x^\alpha = x^\beta$.

Remark 2.2   The well-ordering condition is equivalent to the requirement that any decreasing sequence of monomials

$$x^{\alpha(1)} > x^{\alpha(2)} > x^{\alpha(3)} > \ldots$$

eventually terminates.

We give some basic examples of monomial orders:

Example 2.3 (Pure lexicographic order)   This is basically the order on words in a dictionary. We have $x^\alpha >_{\text{lex}} x^\beta$ if the first nonzero entry of $(\alpha_1 - \beta_1, \alpha_2 - \beta_2, \ldots, \alpha_n - \beta_n)$ is positive. For example, we have

$$x_1 >_{\text{lex}} x_2^3 >_{\text{lex}} x_2 x_3 >_{\text{lex}} x_3^{100}.$$

We prove this is a monomial order. Any two monomials are comparable: given x^α and x^β, either some $\alpha_j - \beta_j \neq 0$ (in which case $x^\alpha >_{\text{lex}} x^\beta$ or $x^\alpha <_{\text{lex}} x^\beta$) or $\alpha_j = \beta_j$ for each j (and $x^\alpha = x^\beta$). For the multiplicative condition, it suffices to observe that for any $\gamma = (\gamma_1, \ldots, \gamma_n)$ we have

$$(\alpha_j + \gamma_j) - (\beta_j + \gamma_j) = \alpha_j - \beta_j,$$

so $x^\alpha x^\gamma >_{\text{lex}} x^\beta x^\gamma$ if and only if $x^\alpha >_{\text{lex}} x^\beta$.

Finally, given any set of monomials $\{x^\alpha\}_{\alpha \in A}$, we extract the smallest element. Consider the descending sequence of subsets

$$A = A_0 \supset A_1 \supset A_2 \supset \ldots \supset A_n$$

defined recursively by

$$A_j = \{\alpha \in A_{j-1} : \alpha_j \text{ is minimal}\}.$$

Each element of A_j is smaller (with respect to $>_{\text{lex}}$) than all the elements of $A \setminus A_j$. On the other hand, A_n has a unique element, which is therefore the minimal element in A.

Definition 2.4   Fix a monomial order on $k[x_1, \ldots, x_n]$ and consider a nonzero polynomial

$$f = \sum_\alpha c_\alpha x^\alpha.$$

The *leading monomial* of f (denoted LM(f)) is the largest monomial x^α such that $c_\alpha \neq 0$. The *leading term* of f (denoted LT(f)) is the corresponding term $c_\alpha x^\alpha$.

For instance, in lexicographic order the polynomial $f = 5x_1x_2 + 7x_2^5 + 19x_3^{17}$ has leading monomial LM(f) $= x_1x_2$ and leading term LT(f) $= 5x_1x_2$. One nonintuitive aspect of lexicographic order is that the degree of the terms is not paramount: the smallest degree term could be the leading one. We can remedy this easily:

Example 2.5 (Graded lexicographic order) $x^\alpha >_{\text{grlex}} x^\beta$ if $\deg(x^\alpha) > \deg(x^\beta)$ or $\deg(x^\alpha) = \deg(x^\beta)$ and $x^\alpha >_{\text{lex}} x^\beta$.

Example 2.6 (Graded reverse lexicographic order) $x^\alpha >_{\text{grelex}} x^\beta$ if $\deg(x^\alpha) > \deg(x^\beta)$ or $\deg(x^\alpha) = \deg(x^\beta)$ and the *last* nonzero $\alpha_j - \beta_j < 0$. (Yes, this inequality goes the right way!) Note that $x_1x_2x_4 >_{\text{grlex}} x_1x_3^2$ but $x_1x_2x_4 <_{\text{grelex}} x_1x_3^2$. Generally, this is more efficient than lexicographic order (see Exercise 2.10).

2.2 Gröbner bases and the division algorithm

Algorithm 2.7 (Division procedure)   Fix a monomial order $>$ on $k[x_1, \ldots, x_n]$ and nonzero polynomials $f_1, \ldots, f_r \in k[x_1, \ldots, x_n]$. Given $g \in k[x_1, \ldots, x_n]$, we want to determine whether $g \in \langle f_1, \ldots, f_r \rangle$:

Step 0   Put $g_0 = g$. If there exists no f_j with LM(f_j)|$LM(g_0)$ then we STOP. Otherwise, pick such an f_{j_0} and cancel leading terms by putting

$$g_1 = g_0 - f_{j_0}\text{LT}(g_0)/\text{LT}(f_{j_0}).$$

...

Step i   Given g_i, if there exists no f_j with LM(f_j)|LM(g_i) then we STOP. Otherwise, pick such an f_{j_i} and cancel leading terms by putting

$$g_{i+1} = g_i - f_{j_i}\text{LT}(g_i)/\text{LT}(f_{j_i}). \tag{2.1}$$

As we are cancelling leading terms at each stage, we have

$$\text{LM}(g) = \text{LM}(g_0) > \text{LM}(g_1) > \ldots > \text{LM}(g_i) > \text{LM}(g_{i+1}) > \ldots.$$

By the well-ordering property of the monomial order, such a chain of decreasing monomials must eventually terminate. If this procedure does not stop, then we must have $g_N = 0$ for some N. Back-substituting using Equation 2.1, we obtain

$$g = \sum_{i=0}^{N-1} f_{j_i} \mathrm{LT}(g_i)/\mathrm{LT}(f_{j_i}) = \sum_{j=1}^{r} \left(\sum_{j_i=j} \mathrm{LT}(g_i)/\mathrm{LT}(f_{j_i}) \right) f_j = \sum_{j=1}^{r} h_j f_j,$$

where the last sum is obtained by regrouping terms.

Unfortunately, this procedure often stops prematurely. Even when $g \in \langle f_1, \ldots, f_r \rangle$, it may happen that $\mathrm{LM}(g)$ is not divisible by any $\mathrm{LM}(f_j)$.

Example 2.8

1. Let $f_1 = x + 1$, $f_2 = x$ and $g = 1$. We certainly have $g \in \langle f_1, f_2 \rangle$ but $\mathrm{LM}(g)$ is not divisible by $\mathrm{LM}(f_1)$ or $\mathrm{LM}(f_2)$, so the procedure stops at the initial step.
2. If $f_1 = x + 2y + 1$, $f_2 = x - y - 5$, and $g = y + 2$ then we have the same problem. Linear algebra presents a solution: Our system of equations corresponds to the augmented matrix

$$\begin{pmatrix} 1 & 2 & | & 1 \\ 1 & -1 & | & -5 \end{pmatrix}.$$

Put this matrix in 'row echelon form' by subtracting the first row from the second

$$\begin{pmatrix} 1 & 2 & | & 1 \\ 0 & -3 & | & -6 \end{pmatrix},$$

which corresponds to the new set of generators $f_1, \tilde{f}_2 = -3y - 6$. Our division algorithm works fine for these new generators.

To understand better why this breakdown occurs, we make the following definitions:

Definition 2.9 A *monomial ideal* $J \subset k[x_1, \ldots, x_n]$ is an ideal generated by a collection of monomials $\{x^\alpha\}_{\alpha \in A}$.

The main example is the ideal of leading terms of an arbitrary ideal $I \subset k[x_1, \ldots, x_n]$.

Definition 2.10 Fix a monomial order $>$ and let $I \subset k[x_1, \ldots, x_n]$ be an ideal. The *ideal of leading terms* is defined

$$\mathrm{LT}(I) := \langle \mathrm{LT}(g) : g \in I \rangle.$$

By convention, $\mathrm{LT}(\langle 0 \rangle) = \langle 0 \rangle$.

2.2 GRÖBNER BASES AND THE DIVISION ALGORITHM

Definition 2.11 Fix a monomial order $>$ and let $I \subset k[x_1, \ldots, x_n]$ be an ideal. A *Gröbner basis* for I is a collection of nonzero polynomials

$$\{f_1, \ldots, f_r\} \subset I$$

such that $\mathrm{LT}(f_1), \ldots, \mathrm{LT}(f_r)$ generate $\mathrm{LT}(I)$.

Nothing in the definition says that a Gröbner basis actually generates I! We prove this *a posteriori*.

Remark 2.12 Every generator for a principal ideal is a Gröbner basis.

Proposition 2.13 Let $I \subset k[x_1, \ldots, x_n]$ be an ideal and $f_1, \ldots, f_r$ a Gröbner basis for I. The Division Algorithm terminates in a finite number of steps, with either $g_i = 0$ or $\mathrm{LT}(g_i)$ not divisible by any of the leading terms $\mathrm{LT}(f_j)$.

1. *In the first case, it returns a representation*

$$g = h_1 f_1 + \cdots + h_r f_r \quad h_j \in k[x_1, \ldots, x_n],$$

and $g \in I$.

2. *In the second case, we obtain an expression*

$$g = h_1 f_1 + \cdots + h_r f_r + g_i \quad \mathrm{LT}(g_i) \notin \langle \mathrm{LT}(f_1), \ldots, \mathrm{LT}(f_r) \rangle,$$

hence $g \notin I$.

The proposition immediately implies the following corollary.

Corollary 2.14 Fix a monomial order $>$. Let $I \subset k[x_1, \ldots, x_n]$ be an ideal and $f_1, \ldots, f_r$ a Gröbner basis for I. Then $I = \langle f_1, \ldots, f_r \rangle$.

The proof of the proposition will use the following lemma.

Lemma 2.15 (Key lemma) Let $I = \langle x^\alpha \rangle_{\alpha \in A}$ be a monomial ideal. Then every monomial in I is a multiple of some x^α.

Proof of lemma Let x^β be a monomial in I. Then we can write

$$x^\beta = \sum_i x^{\alpha(i)} w_i,$$

where the w_i are polynomials. In particular, x^β appears in the right-hand side, is a monomial of $x^{\alpha(i)} w_i$ for some i, and thus is divisible by $x^{\alpha(i)}$. $\square$

Proof of proposition We have already shown that we obtain a representation

$$g = h_1 f_1 + \cdots + h_r f_r$$

unless the algorithm stops. We need to show the algorithm terminates with $g_i = 0$ for some i whenever $g \in I$.

If $g \in I$ then the intermediate $g_i \in I$ as well. We now use the definition of a Gröbner basis: If, for some i, the leading term $\mathrm{LT}(g_i)$ is not divisible by $\mathrm{LT}(f_j)$ for any j, then

$$\mathrm{LT}(g_i) \notin \langle \mathrm{LT}(f_1), \ldots, \mathrm{LT}(f_r)\rangle$$

by Lemma 2.15. It follows that $g_i \notin I$; the formula relating g and g_i guarantees that $g \notin I$. □

2.3 Normal forms

Theorem 2.16 Fix a monomial order $>$ on $k[x_1, \ldots, x_n]$ and an ideal $I \subset k[x_1, \ldots, x_n]$. Then each $g \in k[x_1, \ldots, x_n]$ has a unique expression

$$g \equiv \sum_{x^\alpha \notin \mathrm{LT}(I)} c_\alpha x^\alpha \pmod{I},$$

where $c_\alpha \in k$ and all but a finite number are zero. The expression $\sum_\alpha c_\alpha x^\alpha$ is called the normal form *of g modulo I.*

Equivalently, the monomials $\{x^\alpha : x^\alpha \notin \mathrm{LT}(I)\}$ form a k-vector-space basis for the quotient $k[x_1, \ldots, x_n]/I$.

Corollary 2.17 Fix a monomial order $>$ on $k[x_1, \ldots, x_n]$, an ideal $I \subset k[x_1, \ldots, x_n]$, and Gröbner basis $f_1, \ldots, f_r$ for I. Then each $g \in k[x_1, \ldots, x_n]$ has a unique expression

$$g \equiv \sum c_\alpha x^\alpha \pmod{I},$$

where $\mathrm{LM}(f_j)$ does not divide x^α for any j or α.

Proof of theorem: We first establish existence: the proof is essentially an induction on $\mathrm{LM}(g)$. Suppose the result is false, and consider the nonempty set

$$\{\mathrm{LM}(g) : g \text{ does not admit a normal form}\}.$$

One of the defining properties of monomial orders guarantees that this set has a least element x^β; choose g such that $\mathrm{LT}(g) = x^\beta$.

Suppose $x^\beta \in \mathrm{LT}(I)$. Choose $h \in I$ with $\mathrm{LT}(h) = x^\beta$ and consider $\tilde{g} = g - h$. Note that $\mathrm{LM}(\tilde{g}) < \mathrm{LM}(g)$ and $\tilde{g} \equiv g \pmod{I}$. By the minimality of g, we obtain a

normal form
$$\tilde{g} \equiv \sum_{x^\alpha \notin \mathrm{LT}(I)} c_\alpha x^\alpha \pmod{I}.$$

But this is also a normal form for g, a contradiction.

Now suppose $x^\beta \notin \mathrm{LT}(I)$. Consider $\tilde{g} = g - x^\beta$ so that $\mathrm{LM}(\tilde{g}) < \mathrm{LM}(g)$. By minimality, we have a normal form

$$\tilde{g} \equiv \sum_{x^\alpha \notin \mathrm{LT}(I)} c_\alpha x^\alpha \pmod{I}.$$

But then we have

$$g \equiv x^\beta + \sum_{x^\alpha \notin \mathrm{LT}(I)} c_\alpha x^\alpha \pmod{I},$$

i.e., a normal form for g, which is a contradiction.

Now for uniqueness: Suppose we have

$$g \equiv \sum_\alpha c_\alpha x^\alpha \equiv \sum_\alpha \tilde{c}_\alpha x^\alpha \pmod{I}$$

with $c_\alpha \neq \tilde{c}_\alpha$ for some α. It follows that

$$h := \sum_\alpha (c_\alpha - \tilde{c}_\alpha) x^\alpha \in I, \quad h \neq 0,$$

and $\mathrm{LT}(h) = (c_\alpha - \tilde{c}_\alpha) x^\alpha$ for some α. We have $x^\alpha \in \mathrm{LT}(I)$, a contradiction. $\square$

Example 2.18 Choose > such that

$$x_1 > x_2 > \ldots > x_n.$$

Let $0 \subsetneq I \subsetneq k[x_1, \ldots, x_n]$ be an ideal generated by linear forms

$$g_i = \sum_{j=1}^n a_{ij} x_j + a_{i0}, \quad a_{ij} \in k.$$

It is an exercise to show that $\langle g_i \rangle$ admits a Gröbner basis of the form:

$$f_1 = \sum_{j \geq \ell(1)} b_{1j} x_j + b_{10}, \quad b_{1\ell(1)} \neq 0$$
$$f_2 = \sum_{j \geq \ell(2)} b_{2j} x_j + b_{20}, \quad b_{2\ell(2)} \neq 0$$
$$\vdots$$
$$f_r = \sum_{j \geq \ell(r)} b_{rj} x_j + b_{r0}, \quad b_{r\ell(r)} \neq 0$$

where $\ell(1) < \ell(2) < \ldots < \ell(r)$. The numbers $\ell(1), \ldots, \ell(r)$ are positions of the pivots of the row echelon form of the matrix

$$\begin{pmatrix} a_{11} & \ldots & a_{1n} & a_{10} \\ a_{21} & \ldots & a_{2n} & a_{20} \\ \vdots & \ldots & \vdots & \vdots \end{pmatrix}.$$

We may write

$$\{1, 2, \ldots, n\} = \{\ell(1), \ldots, \ell(r)\} \cup \{m(1), \ldots, m(n-r)\}.$$

Theorem 2.16 says that for each $g \in k[x_1, \ldots, x_n]$ there exists a unique $P \in k[x_{m(1)}, \ldots, x_{m(n-r)}]$ with $g \equiv P \pmod{I}$. Its proof implies that if g is linear we can write

$$g \equiv c_1 x_{m(1)} + \cdots + c_{n-r} x_{m(n-r)} + c_0 \pmod{I}$$

for unique $c_0, c_1, \ldots, c_{n-r} \in k$.

Algorithm 2.19   *Fix a monomial order $>$ on $k[x_1, \ldots, x_n]$, a nonzero ideal $I \subset k[x_1, \ldots, x_n]$, and a Gröbner basis $f_1, \ldots, f_r$ for I. Given a nonzero element $g \in k[x_1, \ldots, x_n]$, we find the normal form of $g \pmod{I}$ as follows:*

Step 0   *Put $g_0 = g$: If each monomial appearing in g_0 is not divisible by any $\mathrm{LM}(f_j)$ then g_0 is already a normal form. Otherwise, let $c_{\beta(0)} x^{\beta(0)}$ be the largest term in g_0 divisible by some $\mathrm{LM}(f_j)$, say $\mathrm{LM}(f_{j_0})$. Set*

$$g_1 = g_0 - c_{\beta(0)} x^{\beta(0)} f_{j_0} / \mathrm{LT}(f_{j_0})$$

so that $g_1 \equiv g_0 \pmod{I}$....

Step i   *Given g_i, if each monomial appearing in g_i is not divisible by any $\mathrm{LM}(f_j)$ then g_i is already a normal form. Otherwise, let $c_{\beta(i)} x^{\beta(i)}$ be the largest term in g_i divisible by some $\mathrm{LM}(f_j)$, say $\mathrm{LM}(f_{j_i})$. Set*

$$g_{i+1} = g_i - c_{\beta(i)} x^{\beta(i)} f_{j_i} / \mathrm{LT}(f_{j_i})$$

so that $g_{i+1} \equiv g_i \pmod{I}$.

Proposition 2.20   *The algorithm terminates in a finite number of steps, with either $g_i = 0$ or g_i in normal form.*

Proof   In passing from g_i to g_{i+1}, we replace the largest term of g_i appearing in $\mathrm{LT}(I)$ with a sum of terms of lower degrees. Thus we have

$$x^{\beta(0)} > x^{\beta(1)} > \ldots > x^{\beta(i)} > x^{\beta(i+1)} > \ldots.$$

2.4 EXISTENCE AND CHAIN CONDITIONS

However, one of the defining properties of a monomial order is that every descending sequence of monomials eventually terminates, so the algorithm must terminate as well. □

2.4 Existence and chain conditions

We have not yet established that Gröbner bases *exist*, or even that each ideal of $k[x_1, \ldots, x_n]$ is finitely generated. In this section, we shall prove the following theorem.

Theorem 2.21 (Existence Theorem) *Fix a monomial order $>$ and an arbitrary nonzero ideal $I \subset k[x_1, \ldots, x_n]$. Then I admits a finite Gröbner basis for the prescribed order.*

We obtain the following result, named in honor of David Hilbert (1862–1943), who pioneered the use of nonconstructive arguments in algebraic geometry and invariant theory at the end of the nineteenth century:

Corollary 2.22 (Hilbert Basis Theorem) *Every polynomial ideal is finitely generated.*

It suffices to show that $\mathrm{LT}(I)$ is finitely generated. Indeed, if $f_1, \ldots, f_r \in I$ are chosen such that
$$\mathrm{LT}(I) = \langle \mathrm{LT}(f_1), \ldots, \mathrm{LT}(f_r) \rangle$$
then Corollary 2.14 implies
$$I = \langle f_1, \ldots, f_r \rangle.$$
Thus the proof of the Existence Theorem is reduced to the case of monomial ideals:

Proposition 2.23 (Dickson's Lemma) *Every monomial ideal in a polynomial ring over a field is generated by a finite collection of monomials.*

Proof Let $J \subset k[x_1, \ldots, x_n]$ be a monomial ideal; we want to find a finite number of monomials $\{x^{\alpha(1)}, \ldots, x^{\alpha(s)}\} \in J$ generating J. The proof is by induction on n, the number of variables. The case $n = 1$ mirrors the proof in Appendix A that every ideal in $k[x_1]$ is principal: If x_1^α is the monomial of minimal degree in J and $x_1^\beta \in J$, then $\alpha \leq \beta$ and $x_1^\alpha | x_1^\beta$.

For the inductive step, we assume the result is valid for $k[x_1, \ldots, x_n]$ and deduce it for $k[x_1, \ldots, x_n, y]$. Consider the following set of auxillary monomial ideals $J_m \subset k[x_1, \ldots, x_n]$:
$$J_m = \langle x^\alpha \in k[x_1, \ldots, x_n] : x^\alpha y^m \in J \rangle.$$

DIVISION ALGORITHM AND GRÖBNER BASES

Note that we have an ascending chain of ideals:

$$J_0 \subset J_1 \subset J_2 \ldots$$

The following result will prove useful:

Proposition 2.24 (Noether's Proposition) *Let R be a ring. Then the following conditions are equivalent:*

1. *every ideal $I \subset R$ is finitely generated;*
2. *every ascending chain of ideals*

$$I_0 \subset I_1 \subset I_2 \subset \ldots$$

terminates, i.e., $I_N = I_{N+1}$ for sufficiently large N.

Then we say the ring R is Noetherian.

This terminology pays homage to Emmy Noether (1882–1935), who pioneered abstract approaches to finiteness conditions and primary decomposition. [33]

Proof of Proposition 2.24 Suppose every ideal is finitely generated. Consider

$$I_\infty = \cup_n I_n,$$

which is also an ideal (see Exercise 2.13). Pick generators $g_1, \ldots, g_r \in I_\infty$; each $g_i \in I_{n_i}$ for some n_i. If $N = \max(n_1, \ldots, n_r)$ then $I_\infty = I_N$.

Conversely, suppose every ascending chain terminates. Let I be an ideal and write

$$I = \langle f_\alpha \rangle_{\alpha \in A}.$$

If I is not generated by a finite number of α then we may construct an infinite sequence $f_{\alpha(1)}, f_{\alpha(2)}, \ldots$ with

$$I_r := \langle f_{\alpha(1)}, \ldots, f_{\alpha(r)} \rangle \subsetneq I_{r+1} := \langle f_{\alpha(1)}, \ldots, f_{\alpha(r+1)} \rangle$$

for each r, violating the ascending chain condition. $\square$

Remark 2.25 The same statement applies to $S = k[x_1, \ldots, x_n]$ with the ideals restricted to *monomial* ideals. Note that every monomial ideal with a finite set of generators has a finite set of monomial generators.

Completion of Proposition 2.23 The sequence of monomial ideals $J_m \subset k[x_1, \ldots, x_n]$ terminates at some J_N. Therefore, there is a finite sequence of

2.4 EXISTENCE AND CHAIN CONDITIONS

monomials:

$$\langle x^{\alpha(0,1)}, \ldots, x^{\alpha(0,n_0)} \rangle = J_0$$
$$\langle x^{\alpha(1,1)}, \ldots, x^{\alpha(1,n_1)} \rangle = J_1$$
$$\vdots$$
$$\langle x^{\alpha(N,1)}, \ldots, x^{\alpha(N,n_N)} \rangle = J_N$$

generating each of the J_m for $m \geq N$. The ideal J is therefore generated by the terms $x^{\alpha(m,j)} y^m$ for $m = 0, \ldots, N$. □

Essentially the same argument proves the following more general theorem:

Theorem 2.26 *Let R be a Noetherian ring. Then $R[y]$ is also Noetherian.*

Proof Given $J \subset R[y]$, consider

$$J_m = \{a_m \in R : a_m y^m + a_{m-1} y^{m-1} + \cdots + a_0 \in J \text{ for some } a_0, \ldots, a_{m-1} \in R\},$$

i.e., the leading terms of degree m polynomials in J. We leave it to the reader to check that J_m is an ideal. Again we have an ascending sequence

$$J_0 \subset J_1 \subset J_2 \ldots$$

so our Noetherian assumption implies the sequence terminates at J_N. Thus we can find $a_{ij} \in R$ with

$$\langle a_{0,1}, \ldots, a_{0,n_0} \rangle = J_0$$
$$\langle a_{1,1}, \ldots, a_{1,n_1} \rangle = J_1$$
$$\vdots$$
$$\langle a_{N,1}, \ldots, a_{N,n_N} \rangle = J_N$$

Choose polynomials $f_{ij} \in J$ with leading terms $a_{ij} y^i$. We claim these generate J. The proof is by induction on the degree in y. Indeed, given $f \in J$ we have

$$f = b_d y^d + \text{lower-order terms}$$

with $b_d \in J_d$. There exist $h_{ij} \in R$ such that

$$b_d = \sum_{i \leq d} h_{ij} a_{ij}.$$

The difference $g := f - \sum h_{ij} f_{ij} y^{d-i}$ has degree $d - 1$ and is contained in J. Hence $g \in \langle f_{ij} \rangle$ by induction and $f \in \langle f_{ij} \rangle$ as well. □

2.5 Buchberger's Criterion

In this section, we give an algorithm for finding a Gröbner basis for an ideal in $k[x_1, \ldots, x_n]$. We first study how a set of generators for an ideal might *fail* to be a Gröbner basis. Consider

$$I = \langle f_1, \ldots, f_r \rangle$$

and assume that

$$h = f_1 h_1 + \cdots + f_r h_r \tag{2.2}$$

has leading term not contained in $J = \langle \text{LM}(f_1), \ldots, \text{LM}(f_r) \rangle$. Consider the monomial

$$x^\delta = \max_j \{\text{LM}(f_j h_j)\} = \max_j \{\text{LM}(f_j)\text{LM}(h_j)\},$$

which is contained in J. Therefore, the occurences of x^δ in (2.2) necessarily cancel, and some smaller monomial takes up the mantle of being the leading term.

We will describe precisely how such cancellation might occur:

Definition 2.27 The *least common multiple* of monomials x^α and x^β is defined

$$\text{LCM}(x^\alpha, x^\beta) = x_1^{\max(\alpha_1, \beta_1)} \ldots x_n^{\max(\alpha_n, \beta_n)}.$$

Fix a monomial order on $k[x_1, \ldots, x_n]$. Let f_1 and f_2 be polynomials in $k[x_1, \ldots, x_n]$ and set

$$x^{\gamma(12)} = \text{LCM}(\text{LM}(f_1), \text{LM}(f_2)).$$

The S-polynomial $S(f_1, f_2)$ is defined

$$S(f_1, f_2) := \left(x^{\gamma(12)}/\text{LT}(f_1)\right) f_1 - \left(x^{\gamma(12)}/\text{LT}(f_2)\right) f_2.$$

The S-polynomial is constructed to ensure the sort of cancellation alluded to above: we have

$$x^{\gamma(12)} = \max_{i=1,2} \left\{\text{LM}\left(f_i x^{\gamma(12)}/\text{LT}(f_i)\right)\right\}$$

but the $x^{\gamma(1,2)}$ terms cancel; in particular,

$$\text{LM}(S(f_1, f_2)) < \text{LCM}(\text{LM}(f_1), \text{LM}(f_2)).$$

For example, using lexicographic order and

$$f_1 = 2x_1 x_2 - x_3^2, \quad f_2 = 3x_1^2 - x_3,$$

2.5 BUCHBERGER'S CRITERION

the S-polynomial is

$$S(f_1, f_2) = \frac{x_1^2 x_2}{2x_1 x_2}(2x_1 x_2 - x_3^2) - \frac{x_1^2 x_2}{3x_1^2}(3x_1^2 - x_3) = -1/2 x_1 x_3^2 + 1/3 x_2 x_3.$$

Our goal is to show that all cancellations can be expressed in terms of S-polynomials. Later on, we will put this on a more systematic footing using syzygies. Now we will prove the following:

Theorem 2.28 (Buchberger's Criterion) *Fix a monomial order and polynomials $f_1, \ldots, f_r$ in $k[x_1, \ldots, x_n]$. The following are equivalent:*

1. *$f_1, \ldots, f_r$ form a Gröbner basis for $\langle f_1, \ldots, f_r \rangle$.*
2. *Each S-polynomial $S(f_i, f_j)$ gives remainder zero on application of the division algorithm.*

Proof ($\Rightarrow$) Each S-polynomial is contained in the ideal $I = \langle f_1, \ldots, f_r \rangle$. If we have a Gröbner basis, the division algorithm terminates with a representation

$$S(f_i, f_j) = \sum_{l=1}^{r} h(ij)_l f_l \quad \mathrm{LM}(S(f_i, f_j)) \geq \mathrm{LM}(h(ij)_l f_l) \tag{2.3}$$

In particular, $S(f_i, f_j)$ has remainder zero.

($\Leftarrow$) Suppose that each S-polynomial gives remainder zero; for each i, j we have an expression in the form (2.3). If the f_i do not form a Gröbner basis, some $h \in I$ does not have leading term in $\langle \mathrm{LM}(f_1), \ldots, \mathrm{LM}(f_r) \rangle$. Choose a representation as in (2.2)

$$h = h_1 f_1 + \cdots + h_r f_r$$

satisfying the following minimality assumptions:

1. $x^\delta := \max_j \{\mathrm{LM}(f_j h_j)\}$ is minimal;
2. the number of indices j realizing the maximum (i.e., $\mathrm{LM}(f_j h_j) = x^\delta$) is minimal.

After reordering the f_j, we may assume that

$$\mathrm{LM}(f_1 h_1) = \mathrm{LM}(f_2 h_2) = \cdots = \mathrm{LM}(f_m h_m) = x^\delta$$

but $\mathrm{LM}(f_j h_j) < x^\delta$ for $j > m$. Note that $m \geq 2$ because the x^δ term cancels in (2.2).
For $i = 1, j = 2$, (2.3) takes the form

$$S(f_1, f_2) = \sum_{l=1}^{r} h(12)_l f_l, \quad \mathrm{LM}(S(f_1, f_2)) \geq \mathrm{LM}(h(12)_l f_l).$$

Write out the S-polynomial using the definition

$$\left(x^{\gamma(12)}/\mathrm{LT}(f_1)\right) f_1 - \left(x^{\gamma(12)}/\mathrm{LT}(f_2)\right) f_2 - \sum_{l=1}^{r} h(12)_l f_l = 0 \quad x^{\gamma(12)} > \mathrm{LM}(h(12)_l f_l).$$

(2.4)

Since $\text{LM}(f_i h_i) = x^\delta$, $i = 1, 2$, we know $x^{\gamma(1,2)} | x^\delta$ and $\mu x^{\gamma(12)} = \text{LT}(f_1)\text{LT}(h_1)$ for some monomial μ. We subtract $\mu \times (2.4)$ from (2.2), to get a new expression

$$h = \tilde{h}_1 f_1 + \tilde{h}_2 f_2 + \cdots + \tilde{h}_r f_r$$

such that $x^\delta \geq (\text{LM}(f_j \tilde{h}_j))$, with strict inequality for $j > m$ and $j = 1$. This contradicts the minimality assumption for (2.2). $\square$

Corollary 2.29 (Buchberger's Algorithm) *Fix a monomial order and polynomials $f_1, \ldots, f_r \in k[x_1, \ldots, x_n]$. A Gröbner basis for $\langle f_1, \ldots, f_r \rangle$ is obtained by iterating the following procedure:*

For each i, j apply the division algorithm to the S-polynomials to get expressions

$$S(f_i, f_j) = \sum_{l=1}^{r} h(ij)_l f_l + r(ij), \quad \text{LM}(S(f_i, f_j)) \geq \text{LM}(h(ij)_l f_l)$$

where each $\text{LM}(r(ij))$ is not divisible by any of the $\text{LM}(f_l)$. If all the remainders $r(ij) = 0$ then $f_1, \ldots, f_r$ are already a Gröbner basis. Otherwise, let $f_{r+1}, \ldots, f_{r+s}$ denote the nonzero $r(ij)$ and adjoin these to get a new set of generators

$$\{f_1, \ldots, f_r, f_{r+1}, \ldots, f_{r+s}\}.$$

Proof Write $I = \langle f_1, \ldots, f_r \rangle$, $S_1 = \{f_1, \ldots, f_r\}$, and $J_1 = \langle \text{LM}(f_1), \ldots, \text{LM}(f_r) \rangle$. If $J_1 = \text{LT}(I)$ then we are done. Otherwise, at least one of the remainders is nonzero by the Buchberger criterion. Consider $S_2 = \{f_1, \ldots, f_r, f_{r+1}, \ldots, f_{r+s}\}$ and let J_2 denote the ideal generated by leading terms of these polynomials. Iterating, we obtain an ascending chain of monomial ideals

$$J_1 \subsetneq J_2 \subsetneq J_3 \ldots \subset \text{LT}(I)$$

and subsets

$$S_1 \subsetneq S_2 \subsetneq S_3 \ldots \subset I.$$

As long as $J_m \subsetneq \text{LT}(I)$, Buchberger's criterion guarantees that $J_m \subsetneq J_{m+1}$.

The chain terminates at some J_N because $k[x_1, \ldots, x_n]$ is Noetherian. Since $J_N = J_{N+1} = \cdots$, we conclude that $J_N = \text{LT}(I)$ and S_N is a Gröbner basis for I. $\square$

2.5.1 An example

We compute a Gröbner basis of

$$I = \langle f_1, f_2 \rangle = \langle x_1^2 - x_2, x_1^3 - x_3 \rangle$$

with respect to lexicographic order.

The first S-polynomial is

$$S(f_1, f_2) = x_1 f_1 - f_2 = x_1(x_1^2 - x_2) - (x_1^3 - x_3) = -x_1 x_2 + x_3;$$

its leading term is not contained in

$$\langle \mathrm{LM}(f_1), \mathrm{LM}(f_2) \rangle = \langle x_1^2 \rangle.$$

Therefore, we must add

$$f_3 = x_1 x_2 - x_3$$

to the Gröbner basis.

The next S-polynomial is

$$S(f_1, f_3) = x_2 f_1 - x_1 f_3 = x_1 x_3 - x_2^2;$$

its leading term is not contained in

$$\langle \mathrm{LM}(f_1), \mathrm{LM}(f_2), \mathrm{LM}(f_3) \rangle = \langle x_1^2, x_1 x_2 \rangle.$$

Therefore, we must add

$$f_4 = x_1 x_3 - x_2^2$$

to the Gröbner basis.

We have:

$$\begin{aligned}
S(f_2, f_3) &= x_2 f_2 - x_1^2 f_3 = x_1^2 x_3 - x_2 x_3 = x_3 f_1, \\
S(f_1, f_4) &= x_3 f_1 - x_1 f_4 = x_1 x_2^2 - x_2 x_3 = x_2 f_3, \\
S(f_2, f_4) &= x_3 f_2 - x_1^2 f_4 = x_1^2 x_2^2 - x_3^2 = (x_1 x_2 + x_3) f_3, \\
S(f_3, f_4) &= x_3 f_3 - x_2 f_4 = x_2^3 - x_3^2.
\end{aligned}$$

The last has leading term not contained in

$$\langle \mathrm{LM}(f_1), \ldots, \mathrm{LM}(f_4) \rangle = \langle x_1^2, x_1 x_2, x_1 x_3 \rangle.$$

Therefore, we must add

$$f_5 = x_2^3 - x_3^2$$

to the Gröbner basis.

Adding this new generator necessitates computing the S-polynomials involving f_5:

$$\begin{aligned}
S(f_1, f_5) &= x_2^3 f_1 - x_1^2 f_5 = -x_2^4 + x_1^2 x_3^2 = (x_1 x_3 + x_2^2) f_4, \\
S(f_2, f_5) &= x_2^3 f_2 - x_1^3 f_5 = -x_2^3 x_3 + x_1^3 x_3^2 = x_1^2 x_3 f_3 + x_2 x_3 f_1, \\
S(f_3, f_5) &= x_2^2 f_3 - x_1 f_5 = -x_2^2 x_3 + x_1 x_3^2 = x_3 f_4, \\
S(f_4, f_5) &= x_2^3 f_4 - x_1 x_3 f_5 = x_1 x_3^3 - x_2^5 = x_3^2 f_4 - x_2^2 f_5.
\end{aligned}$$

Buchberger's criterion implies $\{f_1, f_2, f_3, f_4, f_5\}$ is a Gröbner basis.

Remark 2.30 Note that

$$\mathrm{LM}(f_2) \in \langle \mathrm{LM}(f_1), \mathrm{LM}(f_3), \mathrm{LM}(f_4), \mathrm{LM}(f_5) \rangle$$

so that f_2 is redundant and can be removed from the minimal Gröbner basis.

The division algorithm applied to the S-polynomials for f_1, f_3, f_4, f_5 gives the following relations

$$0 = S(f_1, f_3) - f_4 = x_2 f_1 - x_1 f_3 - f_4,$$
$$0 = S(f_1, f_4) - x_2 f_3 = x_3 f_1 - x_1 f_4 - x_2 f_3,$$
$$0 = S(f_1, f_5) - \left(x_1 x_3 + x_2^2\right) f_4 = x_2^3 f_1 - x_1^2 f_5 - \left(x_1 x_3 + x_2^2\right) f_4,$$
$$0 = S(f_3, f_4) - f_5 = x_3 f_3 - x_2 f_4 - f_5,$$
$$0 = S(f_3, f_5) - x_3 f_4 = x_2^2 f_3 - x_1 f_5 - x_3 f_4,$$
$$0 = S(f_4, f_5) - x_3^2 f_4 + x_2^2 f_5 = \left(x_2^3 - x_3^3\right) f_4 - \left(x_1 x_3 - x_2^2\right) f_5.$$

2.6 Syzygies

We now formalize the notion of cancellations of leading terms of polynomials, and give an important example of how modules arise in algebraic geometry.

According to the Webster Third International Unabridged Dictionary, a *syzygy* is

> the nearly straight-line configuration of three celestial bodies (as the sun, moon, and earth during a solar or lunar eclipse) in a gravitational system.

Just as the sun or moon is obscured during an eclipse, leading terms of polynomials are obscured by syzygies. The original Greek term συζυγία refers to a yoke, conjunction, or copulation.

Definition 2.31 Let $f_1, \ldots, f_r \in k[x_1, \ldots, x_n]$. A *syzygy* among the f_j is a relation

$$h_1 f_1 + h_2 f_2 + \cdots + h_r f_r = 0$$

where $(h_1, \ldots, h_r) \in k[x_1, \ldots, x_n]^r$. The set of all such relations is denoted

$$\mathrm{Syz}(f_1, \ldots, f_r) \subset k[x_1, \ldots, x_n]^r.$$

It is easy to check the following property of syzygies:

Proposition 2.32 $\mathrm{Syz}(f_1, \ldots, f_r)$ *is a* $k[x_1, \ldots, x_n]$-*submodule of* $k[x_1, \ldots, x_n]^r$.

2.6 SYZYGIES

Example 2.33 Given a finite set of monomials

$$\{\mu_j = x^{\alpha(j)}\}_{j=1,\ldots,r}$$

with $x^{\gamma(i,j)} = \text{LCM}(x^{\alpha(i)}, x^{\alpha(j)})$. The syzygies among the μ_j of the form

$$x^{\gamma(i,j)-\alpha(i)}\mu_i - x^{\gamma(i,j)-\alpha(j)}\mu_j$$

generate $\text{Syz}(\mu_1, \ldots, \mu_r)$. The proof is the same as the inductive argument for the Buchberger criterion, i.e., that all cancellations among leading terms are explained by S-polynomials (see Exercise 2.16).

Theorem 2.34 Let $f_1, \ldots, f_r$ be a Gröbner basis with respect to some monomial order on $k[x_1, \ldots, x_n]$. Consider the relations

$$\left(x^{\gamma(ij)}/\text{LT}(f_i)\right)f_i - \left(x^{\gamma(ij)}/\text{LT}(f_j)\right)f_j - \sum_l h(ij)_l f_l = 0$$

obtained by applying the division algorithm to the S-polynomials

$$S(f_i, f_j) = \sum_l h(ij)_l f_l, \quad \text{LM}(S(f_i, f_j)) \geq \text{LM}(h(ij)_l f_l).$$

These generate $\text{Syz}(f_1, \ldots, f_r)$ as a $k[x_1, \ldots, x_n]$-module.

Corollary 2.35 (Generalized Hilbert Basis Theorem) *The module of syzygies among a set of polynomials is finitely generated.*

Proof The proof is essentially contained in our proof of Buchberger's criterion. Suppose we have a syzygy

$$f_1 h_1 + f_2 h_2 + \cdots + f_r h_r = 0. \tag{2.5}$$

The proof proceeds by induction on

$$x^\delta = \max_j(\text{LM}(f_j h_j))$$

and the number of indices j realizing this maximum. After reordering, we may assume that this set of indices is $\{1, 2, \ldots, m\}$ with $m \geq 2$. Consider the syzygy associated to f_1 and f_2:

$$\left(x^{\gamma(12)}/\text{LT}(f_1)\right)f_1 - \left(x^{\gamma(12)}/\text{LT}(f_2)\right)f_2 - \sum_l h(1,2)_l f_l = 0. \tag{2.6}$$

Choose the monomial μ such that $\mu x^{\gamma(12)} = \text{LT}(f_1)\text{LT}(h_1)$. Subtract $\mu \times (2.6)$ from (2.5) to get a new relation

$$f_1 \tilde{h}_1 + f_2 \tilde{h}_2 + \cdots + f_r \tilde{h}_r = 0$$

with fewer indices realizing x^δ or with

$$\max_j(\mathrm{LM}(f_j\tilde{h}_j)) < x^\delta.$$

□

This can be placed in a much more general context:

Theorem 2.36 *Let R be Noetherian and $M \subset R^n$ an R submodule. Then M is finitely generated.*

Proof The basic strategy of the proof is induction on n; the $n = 1$ case is immediate because R is Noetherian. We record some elements of the argument which might be useful in other contexts:

Lemma 2.37 *Let $M_1 \subset M$ be R-modules such that M_1 and M/M_1 are both finitely generated. Then M is also finitely generated.*

Given generators $m_1, \ldots, m_s \in M_1$ and elements $m_{s+1}, \ldots, m_{s+t} \in M$ with images generating M/M_1, $m_1, \ldots, m_{s+t}$ generate M. Indeed, for an arbitrary element $m \in M$, first choose $r_{s+1}, \ldots, r_{s+t} \in R$ such that

$$m - r_{s+1}m_{s+1} - \cdots - r_{s+t}m_{s+t} \to 0 \in M/M_1.$$

But this difference is also in M_1, so we can write

$$m - r_{s+1}m_{s+1} - \cdots - r_{s+t}m_{s+t} = r_1m_1 + \cdots + r_sm_s$$

for some $r_1, \ldots, r_s \in R$.

Iterating the previous argument gives the following result.

Lemma 2.38 *Suppose there exists a sequence of R-submodules*

$$0 = M_0 \subset M_1 \subset M_2 \ldots \subset M_n = M$$

such that each M_i/M_{i-1} is finitely generated. Then M is finitely generated.

To prove the theorem, consider the sequence of modules

$$0 \subset R^1 \subset R^2 \subset \ldots \subset R^{n-1} \subset R^n$$

where

$$R^j = \{(r_1, r_2, \ldots, r_j, \underbrace{0, \ldots, 0}_{n-j \text{ times}}) : r_1, \ldots, r_j \in R\}.$$

Each $R^j \subset R^n$ is a submodule and $R^j/R^{j-1} \simeq R$. We have an induced sequence

$$0 \subset M_1 \subset M_2 \subset \ldots \subset M_{n-1} \subset M_n = M$$

where $M_j = M \cap R^j$. Each M_j is a submodule of M. One of the standard isomorphism theorems of group theory gives

$$M_j/M_{j-1} = (M \cap R^j)/(M \cap R^{j-1}) \hookrightarrow R^j/R^{j-1} = R.$$

In particular, we can regard each quotient as an ideal. Applying the lemmas above and the case $n = 1$ gives the result. □

Remark 2.39 The 1966 thesis [5] of Bruno Buchberger, supervised by Wolfgang Gröbner (1899–1980), introduces Gröbner bases and the Buchberger algorithm. The extension to syzygies is due to F.O. Schreyer [36]; this is presented in [9, ch. 15].

2.7 Exercises

2.1 Prove the assertion in Remark 2.2.
2.2 Prove the assertion in Remark 2.12.
2.3 Let $I = \langle x^{\alpha(1)}, \ldots, x^{\alpha(r)} \rangle \subset k[x_1, \ldots, x_n]$ be a monomial ideal. Given a polynomial

$$f = \sum_\beta c_\beta x^\beta \in I, \quad c_\beta \neq 0,$$

show that each x^β is divisible by some $x^{\alpha(j)}$, $j = 1, \ldots, r$.

2.4 Let $k[t_1, \ldots, t_n]$ and fix real numbers $w_1, \ldots, w_n$, the *weights* corresponding to the variables, i.e., $w(t_j) = w_j$. Given a monomial t^a with exponent $a = (a_1, \ldots, a_n)$, its weight is defined

$$w(t^a) = w_1 a_1 + \cdots + w_n a_n.$$

We order the monomials by weight: $t^a > t^b$ if and only if $w(t^a) > w(t^b)$. This is called a *weight order*.
(a) Take $n = 2$, $w_1 = 3$, and $w_2 = 7$. Is the weight order a monomial order?
(b) Take $n = 2$, $w_1 = 1$, and $w_2 = \pi$. Show that the weight order is a monomial order.
(c) Give necessary and sufficient conditions on the weights for the weight order to be a monomial order.
(d) Can lexicographic order be defined as a weight order, in the sense defined above? (However, see [8] where a more general notion of weight order is defined.)

2.5 Show there is a unique monomial order on $\mathbb{C}[x]$.
2.6 (a) Give an example of a monomial ideal $I \subset \mathbb{C}[x, y]$ with a minimal set of generators consisting of five elements.
(b) Is there any bound on the number of generators of a monomial ideal in $\mathbb{C}[x, y]$? Prove your answer!
2.7 Show that

$$\langle x_1 - x_2^{37}, x_1 - x_2^{38} \rangle$$

is not a Gröbner basis with respect to lexicographic order.

2.8 Consider the ideal $I \subset \mathbb{C}[x, y]$

$$\langle x^{2n} - y^{3n}, n = 1, 2, 3, 4 \ldots \rangle.$$

Find a finite set of generators for I and show they actually generate the full ideal.

2.9 Consider the polynomial

$$f = x^4 + x^2 y^2 + y^3 - x^3 \in \mathbb{C}[x, y]$$

and the ideal

$$I = \langle f, \partial f/\partial x, \partial f/\partial y \rangle.$$

Compute the dimension of $\mathbb{C}[x, y]/I$ as a complex vector space. Determine whether $x^5 \equiv y^5 \pmod{I}$.

2.10 Using the Buchberger Algorithm, compute Gröbner bases for

$$\langle x_3 - x_1^5, x_2 - x_1^3 \rangle$$

with respect to both lexicographic order and graded reverse lexicographic order. Include all the relevant S-polynomial calculations. Which computation takes more effort?

Compute the normal form of $x_1 x_2 x_3$ with respect to each Gröbner basis.

2.11 Fix a monomial order $<$ on $k[x_1, \ldots, x_n]$ and a nonzero ideal $I \subset k[x_1, \ldots, x_n]$. A *reduced Gröbner basis* for I is a Gröbner basis $\{f_1, \ldots, f_r\}$ with following additional properties:

(1) $\mathrm{LT}(f_j) = \mathrm{LM}(f_j)$ for each j, i.e., the leading coefficient of f_j equals one;
(2) for each i and j with $i \neq j$, no term of f_i is divisible by $\mathrm{LM}(f_j)$.

Show that I admits a unique reduced Gröbner basis.

2.12 Given $\phi_1, \ldots, \phi_m \in k[x_1, \ldots, x_n]$, consider the k-algebra homomorphism

$$\begin{aligned} \psi : k[x_1, \ldots, x_n, y_1, \ldots, y_m] &\to k[x_1, \ldots, x_n] \\ y_j &\mapsto \phi_j \\ x_i &\mapsto x_i. \end{aligned}$$

Show that

$$\ker(\psi) = \langle y_1 - \phi_1, y_2 - \phi_2, \ldots, y_m - \phi_m \rangle.$$

Hint: Check that the generators of

$$I = \langle y_1 - \phi_1, y_2 - \phi_2, \ldots, y_m - \phi_m \rangle \subset \ker(\psi)$$

are a Gröbner basis under lexicographic order with

$$y_1 > y_2 > \ldots > y_m > x_1 > \ldots > x_n.$$

Conclude that normal forms modulo I correspond to polynomials in $k[x_1, \ldots, x_n]$. The inclusion

$$j : k[x_1, \ldots, x_n] \hookrightarrow k[x_1, \ldots, x_n, y_1, \ldots, y_m]$$

is a right inverse for ψ, i.e., $\psi \circ j$ is the identity. Hence $\ker(\psi) \cap k[x_1, \ldots, x_n] = 0$ and the quotient

$$k[x_1, \ldots, x_n, y_1, \ldots, y_m]/I \twoheadrightarrow k[x_1, \ldots, x_n, y_1, \ldots, y_m]/\ker(\psi)$$

is injective.

2.13 Consider an ascending chain of ideals

$$I_0 \subset I_1 \subset I_2 \subset \ldots$$

in a ring R. Show that $I_\infty = \cup_n I_n$ is also an ideal.

2.14 Show that the ring of polynomials in an infinite number of variables

$$k[x_1, x_2, x_3, \ldots, x_n, \ldots]$$

does *not* satisfy the ascending chain condition.

2.15 Let $f_1, f_2 \in k[x_1, \ldots, x_n]$ be polynomials with no common irreducible factors. Show that

$$\mathrm{Syz}(f_1, f_2) = (f_2, -f_1)k[x_1, \ldots, x_n] \subset k[x_1, \ldots, x_n]^2.$$

What happens if f_1 and f_2 have common irreducible factors?

2.16 Let $\mu_j = x_j$ for $j = 1, \ldots, n$. Verify explicitly that

$$\mathrm{Syz}(\mu_1, \ldots, \mu_n) \subset k[x_1, \ldots, x_n]^n$$

is generated by

$$(0, \ldots, 0, \underbrace{-x_j}_{\text{ith place}}, 0, \ldots, 0, \underbrace{x_i}_{\text{jth place}}, 0, \ldots).$$

2.17 Consider the matrix

$$A = \begin{pmatrix} a_{11} & a_{12} & a_{13} \\ a_{21} & a_{22} & a_{23} \end{pmatrix}$$

and the ideal

$$I \subset k[a_{11}, a_{12}, a_{13}, a_{21}, a_{22}, a_{23}]$$

generated by the 2×2 minors of A, i.e.,

$$g_1 = a_{12}a_{23} - a_{13}a_{22}, \quad g_2 = -a_{11}a_{23} + a_{13}a_{21}, \quad g_3 = a_{11}a_{22} - a_{12}a_{21}.$$

(a) Do the minors form a Gröbner basis with respect to lexicographic order $a_{11} > a_{12} > a_{13} > a_{21} > a_{22} > a_{23}$?
(b) Compute a set of generators for the syzygies among g_1, g_2, and g_3.

2.18 Commutative algebra challenge problem:

A ring R satisfies the *descending chain condition* if any descending sequence of ideals in R

$$I_1 \supset I_2 \supset I_3 \ldots$$

stabilizes, i.e., $I_N = I_{N+1}$ for large N. Such a ring is said to be *Artinian*.
(a) Show that $\mathbb{Z}/n\mathbb{Z}$ and $k[t]/\langle t^n \rangle$ are Artinian.
(b) Show that $\mathbb{Z}$ and $k[t]$ are not Artinian.
(c) Show that an Artinian ring has a finite number of maximal ideals. *Hint:* Consider chains

$$\mathfrak{m}_1 \supset \mathfrak{m}_1 \mathfrak{m}_2 \supset \mathfrak{m}_1 \mathfrak{m}_2 \mathfrak{m}_3 \ldots.$$

(d) Show that every prime ideal in an Artinian ring is maximal.
(e) Show that an Artinian ring R finitely generated over a field k satisfies $\dim_k(R) < \infty$. (To say that R is finitely generated over k means that R is a quotient of a polynomial ring $k[x_1, \ldots, x_n]$.)

3 Affine varieties

In this chapter, we introduce algebraic varieties and various kinds of maps between them. Our main goal is to develop a working dictionary between geometric concepts and algebraic techniques. The geometric formulations are dictated by the algebraic structures in the background, e.g., we only consider maps that can be expressed using polynomials.

For researchers in the field, geometric intuition and algebraic formalism are (or ought to be) mutually reinforcing. Most of the algebra is developed with a view toward geometric applications, rather than for its own sake. Often, complex algebraic manipulations are more transparent *in vivo* than *in vitro*.

For much of the rest of this book, the base field is assumed to be infinite. Finite fields exhibit pathologies that complicate their use in algebraic geometry (cf. Definition 3.26). For instance, the polynomial function $x^p - x$ vanishes at every point of $\mathbb{F}_p = \mathbb{Z}/p\mathbb{Z}$, where p is a prime integer. However, it does not vanish over the field extension $\mathbb{F}_{p^2}$, so we cannot identify $x^p - x$ with 0.

3.1 Ideals and varieties

Definition 3.1 Given $S \subset \mathbb{A}^n(k)$, the *ideal of polynomials vanishing on S* is defined

$$I(S) = \{f \in k[x_1, \ldots, x_n] : f(s) = 0 \text{ for each } s \in S\}.$$

This is an ideal: if f_1 and f_2 both vanish on S then so does $f_1 + f_2$. If f vanishes on S and g is arbitrary, then gf also vanishes on S.

Example 3.2

1. $S = \mathbb{A}^n(\mathbb{R})$ then $I(S) = \langle 0 \rangle$;
2. $S = \{(a_1, \ldots, a_n)\}$ then $I(S) = \langle x_1 - a_1, \ldots, x_n - a_n \rangle$;

3. $S = \{(1, 1), (2, 3)\} \subset \mathbb{A}^2(\mathbb{Q})$ then

$$I(S) = \langle (x - 1)(y - 3), (x - 1)(x - 2), (y - 1)(x - 2), (y - 1)(y - 3)\rangle;$$

4. $S = \mathbb{N} \subset \mathbb{A}^1(\mathbb{C})$ then $I(S) = \{0\}$, because every nonconstant polynomial $f \in \mathbb{C}[x]$ has at most $\deg(f)$ distinct roots (see Exercise A.13);

5. $S = \{(x, y) : x^2 + y^2 = 1, x \neq 0\} \subset \mathbb{A}^2(\mathbb{R})$ has ideal $I(S) = \langle x^2 + y^2 - 1\rangle$.

Definition 3.3 An *affine variety* is the locus where a collection of polynomial equations is satisfied, i.e., given $F = \{f_j\}_{j \in J} \subset k[x_1, \ldots, x_n]$ we define

$$V(F) = \{a \in \mathbb{A}^n(k) : f_j(a) = 0 \text{ for each } j \in J\} \subset \mathbb{A}^n(k).$$

These polynomials are said to *define* the variety.

The structure of an algebraic variety on a given set depends on the choice of base field. For example, the set of rational numbers $\mathbb{Q}$ is a variety when it is regarded as a subset of $\mathbb{A}^1(\mathbb{Q})$, but not when it is regarded as a subset of $\mathbb{A}^1(\mathbb{R})$ or $\mathbb{A}^1(\mathbb{C})$. When we want to put particular emphasis on the ground field, we will say that V is an affine variety *defined over k*. This means the defining polynomials have coefficients in k.

In what follows, B is a (possibly infinite) index set:

Proposition 3.4 *For each $\beta \in B$, let $F_\beta \subset k[x_1, \ldots, x_n]$ denote a collection of polynomials. Then we have*

$$V(\cup_{\beta \in B} F_\beta) = \cap_{\beta \in B} V(F_\beta).$$

Proof

$$\begin{aligned} V(\cup_{\beta \in B} F_\beta) &= \{a \in \mathbb{A}^n(k) : f(a) = 0 \text{ for each } f \in \cup_{\beta \in B} F_\beta\} \\ &= \cap_{\beta \in B} \{a \in \mathbb{A}^n(k) : f(a) = 0 \text{ for each } f \in F_\beta\} \\ &= \cap_{\beta \in B} V(F_\beta). \end{aligned}$$
□

Thus as we add new polynomials to our collection, the corresponding variety gets smaller:

Proposition 3.5 *For each collection of polynomials $F = \{f_j\}_{j \in J} \subset k[x_1, \ldots, x_n]$ and each subset $F' \subset F$ we have $V(F') \supset V(F)$.*

We have the analogous statements for ideals, which are left to the reader:

Proposition 3.6 *For each $\beta \in B$, let $S_\beta \subset \mathbb{A}^n(k)$ denote a subset. Then we have*

$$I(\cup_{\beta \in B} S_\beta) = \cap_{\beta \in B} I(S_\beta).$$

3.1 IDEALS AND VARIETIES

Proposition 3.7   *For any subsets $S' \subset S \subset \mathbb{A}^n(k)$ we have $I(S') \supset I(S)$.*

The variety defined by a collection of polynomials only depends on the ideal they define:

Proposition 3.8   *Given $F = \{f_j\}_{j \in J} \subset k[x_1, \ldots, x_n]$ generating an ideal $I = \langle f_j \rangle_{j \in J}$, we have $V(F) = V(I)$.*

Proof   Proposition 3.5 guarantees $V(F) \supset V(I)$. Conversely, for $v \in V(F)$ we have $f_j(v) = 0$ for each $j \in J$. For each $g = \sum_{i=1}^{N} h_i f_{j_i} \in I$ we have $g(v) = \sum_{i=1}^{N} h_i f_{j_i}(v) = 0$, so $v \in V(I)$. □

Definition 3.9   Given a ring R and a collection of ideals $\{I_\beta\}_{\beta \in B}$ of R, we define the *sum* to be

$$\sum_{\beta \in B} I_\beta = \{f_1 + \ldots + f_r : f_j \in I_{\beta_j} \text{ for some } \beta_j\},$$

i.e., all finite sums of elements each taken from one of the I_β.

We leave it to the reader to check (see Exercise 3.4) that this is the smallest ideal containing $\cup_{\beta \in B} I_\beta$.

Proposition 3.10   *For any collection of ideals $\{I_\beta\}_{\beta \in B}$ in $k[x_1, \ldots, x_n]$, we have*

$$V\left(\sum_{\beta \in B} I_\beta\right) = \cap_{\beta \in B} V(I_\beta).$$

Proof   This follows from Proposition 3.4 and the identity

$$V\left(\sum_{\beta \in B} I_\beta\right) = V(\cup_{\beta \in B} I_\beta)$$

of Proposition 3.8. □

Definition 3.11   Given a ring R and ideals $I_1, I_2 \subset R$, the *product* $I_1 I_2$ is the ideal generated by products $f_1 f_2$ with $f_1 \in I_1$ and $f_2 \in I_2$.

Proposition 3.12   *For any ideals $I_1, I_2 \subset k[x_1, \ldots, x_n]$, we have*

$$V(I_1 \cap I_2) = V(I_1 I_2) = V(I_1) \cup V(I_2).$$

Proof   We have inclusions

$$I_1 I_2 \subset I_1 \cap I_2 \subset I_1, I_2$$

so Proposition 3.5 yields

$$V(I_1 I_2) \supset V(I_1 \cap I_2) \supset V(I_1), V(I_2).$$

It remains to show that $V(I_1 I_2) \subset V(I_1) \cup V(I_2)$. Suppose that $v \in V(I_1 I_2)$ but $v \notin V(I_1)$. Then for some $f \in I_1$ we have $f(v) \neq 0$. However, for each $g \in I_2$, all the products $(fg)(v) = 0$. Thus $g(v) = 0$ and $v \in V(I_2)$. $\square$

We describe the behavior of varieties under set-theoretic operations:

Proposition 3.13 *An arbitrary intersection of varieties $\cap_{\beta \in B} V_\beta$ is a variety. A finite union of varieties $\cup_{i=1}^N V_i$ is a variety.*

Proof For the intersection part, write each $V_\beta = V(I_\beta)$ for some ideal I_β. Then Proposition 3.10 gives the result. To show that a finite union of varieties is a variety, we apply Proposition 3.12 successively. $\square$

The above properties and the Hilbert Basis Theorem from Chapter 2 together imply the following.

Proposition 3.14 *Every variety can be defined as the locus where a finite number of polynomials vanish.*

We can take products of affine varieties:

Definition 3.15 Consider affine varieties $V \subset \mathbb{A}^n(k)$ and $W \subset \mathbb{A}^m(k)$. The *product variety*

$$V \times W \subset \mathbb{A}^{n+m}(k)$$

is defined as the set

$$\{(a_1, \ldots, a_n, b_1, \ldots, b_m) : (a_1, \ldots, a_n) \in V, (b_1, \ldots, b_m) \in W\}.$$

The *projections*

$$\pi_1 : V \times W \to V, \quad \pi_2 : V \times W \to W$$

are defined

$$\pi_1(x_1, \ldots, x_n, y_1, \ldots, y_m) = (x_1, \ldots, x_n), \quad \pi_2(x_1, \ldots, x_n, y_1, \ldots, y_m) = (y_1, \ldots, y_m).$$

3.1 IDEALS AND VARIETIES

Proposition 3.16 *Consider ideals $I_1 \subset k[x_1, \ldots, x_n]$ and $I_2 \subset k[y_1, \ldots, y_m]$ and the corresponding varieties $V(I_1) \subset \mathbb{A}^n(k)$ and $V(I_2) \subset \mathbb{A}^m(k)$. Let*

$$J = I_1 k[x_1, \ldots, x_n, y_1, \ldots, y_m] + I_2 k[x_1, \ldots, x_n, y_1, \ldots, y_m],$$

i.e., the ideal in $k[x_1, \ldots, x_n, y_1, \ldots, y_m]$ generated by I_1 and I_2. Then $V(I_1) \times V(I_2) = V(J)$.

Proof We start with some notation: Given a map $\phi : X \to Y$ and a subset $Z \subset Y$, we write

$$\phi^{-1}(Z) = \{x \in X : \phi(x) \in Z\}.$$

Consider the projection morphisms

$$\Pi_1 : \mathbb{A}^{m+n} \to \mathbb{A}^n$$
$$(x_1, \ldots, x_n, y_1, \ldots, y_m) \mapsto (x_1, \ldots, x_n),$$
$$\Pi_2 : \mathbb{A}^{m+n} \to \mathbb{A}^m$$
$$(x_1, \ldots, x_n, y_1, \ldots, y_m) \mapsto (y_1, \ldots, y_m).$$

Since

$$\Pi_1^{-1}(V(I_1)) = \{(a_1, \ldots, a_n, b_1, \ldots, b_m) : f(a_1, \ldots, a_n) = 0 \text{ for each } f \in I_1\}$$

it follows that $\Pi_1^{-1}(V(I_1))$ is the variety in $\mathbb{A}^{m+n}$ defined by $I_1 \subset k[x_1, \ldots, x_n] \subset k[x_1, \ldots, x_n, y_1, \ldots, y_m]$. Proposition 3.6 then implies

$$\Pi_1^{-1}(V(I_1)) = V(I_1 k[x_1, \ldots, x_n, y_1, \ldots, y_m]).$$

We can express the product as an intersection

$$V(I_1) \times V(I_2) = \Pi_1^{-1}(V(I_1)) \cap \Pi_2^{-1}(V(I_2))$$

Proposition 3.10 then yields

$$V(I_1) \times V(I_2) = V(J). \qquad \square$$

3.1.1 A warning about our definitions

We have defined an affine variety as the locus in $\mathbb{A}^n(k)$ where a collection of polynomials vanish. Over some base fields, it can be very difficult to determine precisely where a polynomial is zero!

Example 3.17 Fermat's Last Theorem, as proven by Andrew Wiles and Richard Taylor, asserts that for any integers x, y, z with

$$x^N + y^N = z^N, \quad N \geq 3,$$

at least one of the three integers is zero. We may as well assume $x, y, z \in \mathbb{Q}$; multiplying through by the least common multiple of the denominators would yield an integral solution. In our notation, Fermat's Last Theorem takes the following form: If $N \geq 3$ and $V = V(x^N + y^N - z^N) \subset \mathbb{A}^3(\mathbb{Q})$ then $xyz \in I(V)$.

For any ideal $I \subset k[x_1, \ldots, x_n]$ we have (cf. Exercise 3.3)

$$I(V(I)) \supset I.$$

Whether equality holds is a subtle problem, depending both on the base field and the geometry of $V(I)$. There are many general theorems asserting that $I(V(f)) \supsetneq \langle f \rangle$ for certain classes of polynomials f. When $k = \mathbb{Q}$, a number field, or a finite field, these problems have a strong number-theoretic flavor. This area is known as *Diophantine geometry* or *Arithmetic algebraic geometry*. On the other hand, when k is algebraically closed the *Nullstellensatz* (Theorem 7.3) allows precise descriptions of $I(V(I))$ in terms of I.

3.2 Closed sets and the Zariski topology

Definition 3.18 The algebro-geometric *closure* of a subset $S \subset \mathbb{A}^n(k)$ is defined

$$\overline{S} = \{a \in \mathbb{A}^n(k) : f(a) = 0 \quad \text{for each } f \in I(S)\} = V(I(S)).$$

A subset $S \subset \mathbb{A}^n(k)$ is *closed* if $S = \overline{S}$; $U \subset \mathbb{A}^n(k)$ is *open* if its complement $\mathbb{A}^n(k) \setminus U$ is closed in $\mathbb{A}^n(k)$.

Example 3.19

1. The closure of $\mathbb{N} \subset \mathbb{A}^1(\mathbb{C})$ is the complex line $\mathbb{A}^1(\mathbb{C})$.
2. The closure of $\{(x, y) : x^2 + y^2 = 1, x \neq 0\} \subset \mathbb{A}^2(\mathbb{R})$ is the circle $\{(x, y) : x^2 + y^2 = 1\}$.
3. The open subsets of $\mathbb{A}^1(\mathbb{C})$ are the empty set and $U \subset \mathbb{C}$ with finite complement, e.g., $U = \mathbb{C} \setminus \{a_1, \ldots, a_d\}$.

You may remember open and closed sets from calculus, e.g., $U \subset \mathbb{R}^n$ is open if, for each $x \in U$, a sufficiently small ball centered at x is contained in U. There is a very general definition underlying both usages:

Definition 3.20 A *topological space* consists of a set X and a collection of subsets $\mathcal{Z} = \{Z \subset X\}$ called the *closed subsets* of X, satisfying the following:

- $\emptyset, X \in \mathcal{Z}$;
- if $Z_1, Z_2 \in \mathcal{Z}$ then $Z_1 \cup Z_2 \in \mathcal{Z}$;
- if $\{Z_j\}_{j \in J} \subset \mathcal{Z}$ then $\cap_{j \in J} Z_j \in \mathcal{Z}$.

A subset $U \subset X$ is *open* if its complement $X \setminus U$ is closed.

Our axioms imply that a finite intersection of open subsets is open and an arbitrary union of open subsets is open.

Proposition 3.13 shows that closed subsets of affine space (in the sense of algebraic geometry) satisfy the axioms of a topological space. This is called the *Zariski topology* in recognition of Oscar Zariski (1899–1986). By Proposition 3.14, all Zariski open sets are of the form

$$U = \mathbb{A}^n(k) \setminus Z \text{ with } Z \text{ closed}$$
$$= \{a \in \mathbb{A}^n(k) : f_j(a) \neq 0, j = 1, \ldots, r, \ f_j \in k[x_1, \ldots, x_n]\},$$

i.e., where a finite set of polynomials do not simultaneously vanish.

The Zariski topology on affine space induces a topology on subsets $V \subset \mathbb{A}^n(k)$: $Z \subset V$ is *closed* if $Z = V \cap Y$ for some closed $Y \subset \mathbb{A}^n(k)$. If $V \subset \mathbb{A}^n(k)$ is an affine variety then closed subsets of V are precisely closed subsets of $\mathbb{A}^n(k)$ contained in V. This is called the Zariski topology on the affine variety.

Definition 3.21 A function of topological spaces $f : X \to Y$ is *continuous* if for each closed $Z \subset Y$ the preimage $f^{-1}(Z) = \{x \in X : f(x) \in Z\}$ is closed.

The concept of a 'Zariski continuous' function is really too weak to be of much use. For instance, any bijective function $\mathbb{C} \to \mathbb{C}$ is automatically Zariski continuous!

One useful class of Zariski continuous functions are the morphisms introduced in Chapter 1:

Proposition 3.22 *Let* $\phi : \mathbb{A}^n(k) \to \mathbb{A}^m(k)$ *be a morphism of affine spaces. Then ϕ is Zariski continuous.*

Proof Let $Z \subset \mathbb{A}^m(k)$ be closed, i.e.,

$$Z = \{b \in \mathbb{A}^m(k) : g_j(b) = 0, \{g_j\}_{j \in J} \subset k[y_1, \ldots, y_m]\}.$$

Thus

$$\phi^{-1}(Z) = \{a \in \mathbb{A}^n(k) : g_j(\phi(a)) = 0\}$$

is also closed, because $g_j \circ \phi$ is a polynomial. □

3.3 Coordinate rings and morphisms

We elaborate on algebraic aspects of morphisms of affine space.

Definition 3.23 Choose coordinates $x_1, \ldots, x_n$ and $y_1, \ldots, y_m$ on $\mathbb{A}^n(k)$ and $\mathbb{A}^m(k)$. Let $\phi : \mathbb{A}^n(k) \to \mathbb{A}^m(k)$ be a morphism given by the rule

$$\phi(x_1, \ldots, x_n) = (\phi_1(x_1, \ldots, x_n), \ldots, \phi_m(x_1, \ldots, x_n)), \quad \phi_j \in k[x_1, \ldots, x_n].$$

AFFINE VARIETIES

For each $f \in k[y_1, \ldots, y_m]$, the *pull-back* by ϕ is defined

$$\phi^* f = f \circ \phi = f(\phi_1(x_1, \ldots, x_n), \ldots, \phi_m(x_1, \ldots, x_n)).$$

We obtain a ring homomorphism

$$\phi^* : k[y_1, \ldots, y_m] \to k[x_1, \ldots, x_n]$$
$$y_j \mapsto \phi_j(x_1, \ldots, x_n),$$

with the property that $\phi^*(c) = c$ for each constant $c \in k$, i.e., pull-back is a *k-algebra homomorphism*.

Conversely, any k-algebra homomorphism

$$\psi : k[y_1, \ldots, y_m] \to k[x_1, \ldots, x_n]$$

is determined by its values on the generators. Writing $\psi_j(x_1, \ldots, x_n) = \psi(y_j)$, we obtain a morphism

$$\mathbb{A}^n(k) \to \mathbb{A}^m(k)$$
$$(x_1, \ldots, x_n) \mapsto (\psi_1(x_1, \ldots, x_n), \ldots, \psi_m(x_1, \ldots, x_n)).$$

To summarize:

Proposition 3.24 *There is a natural correspondence between morphisms $\phi : \mathbb{A}^n(k) \to \mathbb{A}^m(k)$ and k-algebra homomorphisms*

$$\psi : k[y_1, \ldots, y_m] \to k[x_1, \ldots, x_n]$$

identifying ϕ^* and ψ.

We have already considered the ring of polynomial functions on affine space. How does this generalize to arbitrary affine varieties? Let $V \subset \mathbb{A}^n(k)$ be affine with ideal $I(V)$. We restrict polynomial functions on $\mathbb{A}^n(k)$ to V; elements of $I(V)$ are zero along V, so these functions can be identified with the quotient $k[x_1, \ldots, x_n]/I(V)$.

$$\begin{array}{ccc} I(V) \subset & k[x_1, \ldots, x_n] & \to k[x_1, \ldots, x_n]/I(V) \\ \downarrow & \downarrow & \swarrow \\ 0 & \in \text{functions on } V & \end{array}$$

Example 3.25 Consider the circle $V = \{(x, y) : x^2 + y^2 = 1\} \subset \mathbb{A}^2(\mathbb{R})$ with $I(V) = \langle x^2 + y^2 - 1 \rangle$. The polynomials x^2 and $1 - y^2$ define the same function on the circle. We have

$$x^2 \equiv 1 - y^2 \mod I(V).$$

3.3 COORDINATE RINGS AND MORPHISMS

Definition 3.26 Let $V \subset \mathbb{A}^n(k)$ be an affine variety. The *coordinate ring* is defined as the quotient ring

$$k[V] = k[x_1, \ldots, x_n]/I(V).$$

Note that $k[\mathbb{A}^n] = k[x_1, \ldots, x_n]$ provided k is infinite (see Exercise 3.2). This is one point where finite fields create difficulties. Our definition of the coordinate ring requires modification in this case.

Definition 3.27 Fix an affine variety $V \subset \mathbb{A}^n(k)$. Two morphisms $\bar{\phi}, \hat{\phi} : \mathbb{A}^n(k) \to \mathbb{A}^m(k)$ are *equivalent on V* if the induced pull-back homomorphisms

$$\bar{\phi}^* : k[\mathbb{A}^m] \to k[V], \quad \hat{\phi}^* : k[\mathbb{A}^m] \to k[V]$$

are equal. The resulting equivalence classes are called *morphisms* $\phi : V \to \mathbb{A}^m(k)$. Each $\hat{\phi} : \mathbb{A}^n(k) \to \mathbb{A}^m(k)$ in the equivalence class is called an *extension* of ϕ to affine space.

Example 3.28 Consider the circle $V = \{(x, y) : x^2 + y^2 = 1\} \subset \mathbb{A}^2(\mathbb{R})$ with $I(V) = \langle x^2 + y^2 - 1 \rangle$. The morphisms

$$\bar{\phi} : \mathbb{A}^2(\mathbb{R}) \to \mathbb{A}^1(\mathbb{R}),$$
$$(x, y) \mapsto x^2,$$
$$\hat{\phi} : \mathbb{A}^2(\mathbb{R}) \to \mathbb{A}^1(\mathbb{R}),$$
$$(x, y) \mapsto 1 - y^2$$

are equivalent on the circle.

Equivalent morphisms are equivalent as functions:

Proposition 3.29 Let $V \subset \mathbb{A}^n(k)$ be an affine variety and

$$\bar{\phi}, \hat{\phi} : \mathbb{A}^n(k) \to \mathbb{A}^m(k)$$

two morphisms equivalent on V. Then we have $\bar{\phi}(v) = \hat{\phi}(v)$ for each $v \in V$.

Proof If $\bar{\phi}(v) \neq \hat{\phi}(v)$ then they can be differentiated by coordinate functions from $k[\mathbb{A}^m]$, i.e., we have

$$y_i(\bar{\phi}(v)) \neq y_i(\hat{\phi}(v))$$

for some i. It follows that $\bar{\phi}^* y_i(v) \neq \hat{\phi}^* y_i(v)$, which violates the equivalence assumption. □

Definition 3.30 Fix affine varieties $V \subset \mathbb{A}^n(k)$ and $W \subset \mathbb{A}^m(k)$. A *morphism* $\phi : V \to W$ is defined to be morphism $\phi : V \to \mathbb{A}^m(k)$ with $\phi(V) \subset W$.

The geometric condition $\phi(V) \subset W$ is equivalent to the algebraic condition $\phi^* I(W) \subset I(V)$. Indeed first assume that $\phi^* I(W) \subset I(V)$. Given $v \in V$ and an arbitrary $g \in I(W)$, we have $g(\phi(v)) = \phi^*(g)(v) = 0$ as $\phi^*(g) \in I(V)$, so that $\phi(v) \in W$. Conversely, if $\phi(V) \subset W$ then, given $g \in I(W)$, we have $g(\phi(v)) = 0$ for each $v \in V$, and thus $\phi^* g \in I(V)$.

We are tacitly assuming that the polynomials defining ϕ have coefficients in k. If we have to make this explicit, we say that the morphism is *defined* over k.

Proposition 3.31 Let $V \subset \mathbb{A}^n(k)$ and $W \subset \mathbb{A}^m(k)$ be affine varieties. Any morphism $\phi : V \to W$ induces a k-algebra homomorphism $\phi^* : k[W] \to k[V]$. Conversely, each k-algebra homomorphism $\psi : k[W] \to k[V]$ can be expressed as ϕ^* for some morphism ϕ.

Proof Suppose we have a morphism $\phi : V \to W$. Consider the composition

$$\begin{array}{ccccc} k[y_1, \ldots, y_m] & \xrightarrow{\phi^*} & k[x_1, \ldots, x_n] & \to & k[V] \\ \cup & & \cup & & \cup \\ I(W) & \xrightarrow{\phi^*} & I(V) & \to & 0 \end{array}$$

The ideal $I(W)$ is mapped to zero in $k[V]$, so there is an induced homomorphism

$$\phi^* : k[W] = k[y_1, \ldots, y_m]/I(W) \to k[V].$$

Conversely, suppose we have a k-algebra homomorphism

$$\begin{array}{ccc} k[y_1, \ldots, y_m] & & k[x_1, \ldots, x_n] \\ \downarrow & & \downarrow \\ k[W] & \xrightarrow{\psi} & k[V] \end{array}.$$

It suffices to find a k-algebra homomorphism

$$\psi' : k[y_1, \ldots, y_m] \to k[x_1, \ldots, x_n]$$

making the diagram above commute. Indeed, Proposition 3.24 then gives a morphism $\phi' : \mathbb{A}^n(k) \to \mathbb{A}^m(k)$ such that $\psi' = \phi'^*$. The diagram guarantees $\psi'(I(W)) \subset I(V)$, so $\phi'(V) \subset W$ and the induced homomorphism on coordinate rings $k[W] \to k[V]$ is just ψ.

3.4 RATIONAL MAPS

To construct ψ', consider the elements $\psi(y_j) \in k[V]$. Lifting these to polynomials $\phi_j \in k[x_1, \ldots, x_n]$, $j = 1, \ldots, m$, we obtain a homomorphism

$$\psi' : k[y_1, \ldots, y_m] \to k[x_1, \ldots, x_n]$$
$$y_j \mapsto \phi_j$$

making the diagram commute. $\square$

The main idea here is that polynomial rings are extraordinarily flexible; we can send the generators anywhere we like!

Corollary 3.32 *Let V and W be affine varieties. There is a one-to-one correspondence between morphisms $V \to W$ and k-algebra homomorphisms $k[W] \to k[V]$.*

Definition 3.33 An *isomorphism of affine varieties* is a morphism $\phi : V \to W$ admitting an inverse morphism $\phi^{-1} : W \to V$. An *automorphism of an affine variety* is an isomorphism $\phi : V \to V$.

One important consequence of Corollary 3.32 is that automorphisms of V correspond to k-algebra isomorphisms $k[V] \to k[V]$.

Example 3.34 Consider $V = \mathbb{A}^2(k)$ and the homomorphism

$$\psi(x_1) = x_1, \quad \psi(x_2) = x_2 + g(x_1), \quad g \in k[x_1],$$

with inverse

$$\psi^{-1}(x_1) = x_1, \quad \psi^{-1}(x_2) = x_2 - g(x_1).$$

Each $\psi = \phi^*$ for some automorphism $\phi : \mathbb{A}^2(k) \to \mathbb{A}^2(k)$. Thus each polynomial $g \in k[x_1]$ yields an automorphism of the affine plane. See [37] chapter I, section 2, esp. exercise 9, for more information.

Exercise 3.14 is the classification of automorphisms of the affine line $\mathbb{A}^1(k)$. As far as I know, there is no intelligible classification of automorphisms of $\mathbb{A}^3(k)$. There is substantial research on this question: see [2], for example.

3.4 Rational maps

Let $k(x_1, \ldots, x_n)$ denote the fraction field of $k[x_1, \ldots, x_n]$, consisting of quotients f/g where $f, g \in k[x_1, \ldots, x_n]$, $g \neq 0$.

Definition 3.35 A *rational map* $\rho : \mathbb{A}^n(k) \dashrightarrow \mathbb{A}^m(k)$ is given by a rule

$$\rho(x_1, \ldots, x_n) = (\rho_1(x_1, \ldots, x_n), \ldots, \rho_m(x_1, \ldots, x_n)), \quad \rho_j \in k(x_1, \ldots, x_n).$$

A rational map does not yield a well-defined function from $\mathbb{A}^n(k)$ to $\mathbb{A}^m(k)$ – hence the dashed arrow! Represent each component ρ_j as a fraction

$$\rho_j = f_j/g_j, \quad f_j, g_j \in k[x_1, \ldots, x_n];$$

we will generally assume that f_j and g_j have no common irreducible factors, for $j = 1, \ldots, m$. Wherever any of the g_j vanish, ρ is not well-defined; the closed set $V(\{g_1, \ldots, g_m\}) \subset \mathbb{A}^n(k)$ is called the *indeterminacy locus* of ρ. However, over the complement

$$U := \{(a_1, \ldots, a_n) \in \mathbb{A}^n(k) : g_j(a_1, \ldots, a_n) \neq 0, j = 1, \ldots, m\}$$

we obtain a well-defined function to $\mathbb{A}^m(k)$.

The argument for Proposition 3.24 also yields:

Proposition 3.36 *Each rational map $\rho : \mathbb{A}^n(k) \dashrightarrow \mathbb{A}^m(k)$ defined over k induces a k-algebra homomorphism*

$$\rho^* : k[y_1, \ldots, y_m] \to k(x_1, \ldots, x_n),$$
$$y_j \mapsto \rho_j(x_1, \ldots, x_n).$$

Conversely, each k-algebra homomorphism

$$k[y_1, \ldots, y_m] \to k(x_1, \ldots, x_n)$$

arises from a rational map.

Definition 3.37 Let $W \subset \mathbb{A}^m(k)$ be an affine variety. A *rational map* $\rho : \mathbb{A}^n(k) \dashrightarrow W$ is a rational map $\rho : \mathbb{A}^n(k) \dashrightarrow \mathbb{A}^m(k)$ with $\rho^* I(W) = 0$.

Example 3.38 If $W = \{(y_1, y_2) : y_2^2 = y_1^2 + y_1^3\} \subset \mathbb{A}^2(\mathbb{Q})$ then we have the rational map

$$\rho : \mathbb{A}^1 \dashrightarrow W,$$
$$s \mapsto \left(\frac{1-s^2}{s^2}, \frac{1-s^2}{s^3}\right).$$

How do we define a rational map from a general affine variety $\rho : V \dashrightarrow \mathbb{A}^m(k)$? As in our discussion of morphisms, we realize $V \subset \mathbb{A}^n(k)$ as a closed subset. It is natural to define ρ as an equivalence class of rational maps $\rho' : \mathbb{A}^n(k) \dashrightarrow \mathbb{A}^m(k)$, restricted to V. However, rational maps can behave badly along a variety, especially when one of the denominators of the ρ'_j vanishes along that variety.

Example 3.39 Consider the rational map

$$\rho : \mathbb{A}^2(\mathbb{R}) \dashrightarrow \mathbb{A}^3(\mathbb{R}),$$
$$(x_1, x_2) \mapsto (x_1^{-3}, x_1^{-1}x_2^{-1}, x_2^{-3}),$$

3.4 RATIONAL MAPS

which is well-defined over the open subset

$$U = \{(x_1, x_2) : x_1 x_2 \neq 0\} \subset \mathbb{A}^2(\mathbb{R}).$$

However, ρ is not defined along the affine variety

$$V = \{(x_1, x_2) : x_1 = 0\} \subset \mathbb{A}^2(\mathbb{R}).$$

We formulate a definition to address this difficulty:

Definition 3.40 Let $\rho : \mathbb{A}^n(k) \dashrightarrow \mathbb{A}^m(k)$ be a rational map with components $\rho_j = f_j/g_j$ with $f_j, g_j \in k[x_1, \ldots, x_n]$ having no common irreducible factors. Let $V \subset \mathbb{A}^n(k)$ be an affine variety with ideal $I(V)$ and coordinate ring $k[V]$. Assume that the image of each g_j in $k[V]$ does not divide zero. Then we say that ρ is *admissible on V*.

We would like an algebraic description of rational maps, generalizing Proposition 3.36 in the spirit of Corollary 3.32. We require an algebraic construction generalizing the field of fractions of a domain:

Definition 3.41 The *ring of fractions* of a ring R is defined

$$K = \{r/s : r, s \in R, s \text{ not a zero divisor}\},$$

where $r_1/s_1 = r_2/s_2$ whenever $r_1 s_2 = r_2 s_1$.

We can realize $R \subset K$ as the fractions with denominator 1 (see Exercise 3.17).

Definition 3.42 For an affine variety V, let $k(V)$ denote the ring of fractions of the coordinate ring $k[V]$.

Proposition 3.43 *Let $\rho : \mathbb{A}^n(k) \dashrightarrow \mathbb{A}^m(k)$ be a rational map admissible on an affine variety $V \subset \mathbb{A}^n(k)$. Then ρ induces a k-algebra homomorphism*

$$\rho^* : k[\mathbb{A}^m] \to k(V).$$

Conversely, each such homomorphism arises from a suitable rational map.

Proof By hypothesis, each component function ρ_j of ρ can be expressed as a fraction f_j/g_j, where $f_j, g_j \in k[x_1, \ldots, x_n]$ and g_j does not divide zero in $k[V]$. In particular, each f_j/g_j goes to an element of $k(V)$ and we obtain a homomorphism

$$\rho^* : k[y_1, \ldots, y_m] \to k(V),$$
$$y_j \mapsto f_j/g_j.$$

Conversely, given such a homomorphism ψ, we can express $\psi(y_j) = r_j/s_j$ with $r_j, s_j \in k[V]$ for $j = 1, \ldots, m$. Choose polynomials $f_1, \ldots, f_m, g_1, \ldots, g_m \in k[x_1, \ldots, x_m]$ with $f_j \equiv r_j \pmod{I(V)}$ and $g_j \equiv s_j \pmod{I(V)}$. The rational map with components $\rho_j = f_j/g_j$ induces the homomorphism ψ. $\square$

With these results, we can make the general definition of a rational map:

Definition 3.44 Let $V \subset \mathbb{A}^n(k)$ be an affine variety and

$$\bar{\rho}, \hat{\rho} : \mathbb{A}^n(k) \to \mathbb{A}^m(k)$$

rational maps admissible on V. These are *equivalent* along V if the induced homomorphisms

$$\bar{\rho}^*, \hat{\rho}^* : k[\mathbb{A}^m] \to k(V)$$

are equal.

Definition 3.45 Let V and W be affine varieties realized as closed subsets of $\mathbb{A}^n(k)$ and $\mathbb{A}^m(k)$ respectively. A *rational map* $\rho : V \dashrightarrow W$ is defined as an equivalence class of rational maps $\rho' : \mathbb{A}^n(k) \dashrightarrow W$ admissible on V. Each such ρ' is called an *extension* of ρ to affine space.

The following analog of Corollary 3.32 is left as an exercise:

Corollary 3.46 *Let V and W be affine varieties. There is a one-to-one correspondence between rational maps $V \dashrightarrow W$ over k and k-algebra homomorphisms $k[W] \to k(V)$.*

3.5 Resolving rational maps

We will introduce a systematic procedure for replacing rational maps by morphisms defined on a smaller variety. This will be used to compute the image of a rational map in Chapter 4:

Proposition 3.47 *Let V and W be affine varieties realized in $\mathbb{A}^n(k)$ and $\mathbb{A}^m(k)$ respectively. Consider a rational map $\rho : V \dashrightarrow W$ obtained from a map*

$$\mathbb{A}^n(k) \dashrightarrow \mathbb{A}^n(k)$$
$$(x_1, \ldots, x_n) \mapsto (f_1/g_1, \ldots, f_m/g_m)$$

admissible on V. Write $g = g_1 \ldots g_m$ so that ρ is well-defined over the open set $U = \{v \in V : g(v) \neq 0\}$. There is an affine variety V_g and morphisms $\pi : V_g \to V$ and $\phi : V_g \to W$, with the following properties:

3.5 RESOLVING RATIONAL MAPS

1. $\pi(V_g) = U$;
2. *there is a rational map $\psi : V \dashrightarrow V_g$, well-defined on U, such that $\pi \circ \psi = \mathrm{Id}_U$ and $\psi \circ \pi = \mathrm{Id}_{V_g}$;*

$$\begin{array}{c} V_g \xrightarrow{\phi} W \\ \psi \uparrow \downarrow \pi \quad \nearrow \rho \\ V \end{array}$$

3. $\phi = \rho \circ \pi$.

Thus π is a *birational morphism*, i.e., a morphism which admits an inverse rational map. We'll discuss these more in Chapter 6.

Proof The new affine variety is obtained by imposing invertibility by fiat: take

$$\mathbb{A}^n(k)_g = \{(x_1, \ldots, x_n, z) : zg(x_1, \ldots, x_n) = 1\} \subset \mathbb{A}^{n+1}(k)$$

so that projection

$$\pi : \mathbb{A}^{n+1}(k) \to \mathbb{A}^n(k)$$
$$(x_1, \ldots, x_n, z) \mapsto (x_1, \ldots, x_n),$$

takes $\mathbb{A}^n(k)_g$ bijectively to $\mathbb{A}^n(k) \setminus \{g = 0\}$. Similarly, define

$$V_g = \{p \in \mathbb{A}^n(k)_g : \pi(p) \in V\}.$$

Abusing notation, we use π to designate the restriction of the projection to V_g; it also maps V_g bijectively to U.

These varieties have coordinate rings

$$k[\mathbb{A}^n(k)_g] = \frac{k[x_1, \ldots, x_n, z]}{\langle zg - 1 \rangle} = k[x_1, \ldots, x_n][1/g] \subset k(x_1, \ldots, x_n)$$

and

$$k[V_g] = k[V][1/g] \subset k(V).$$

The rational map inverse to π is

$$\psi(x_1, \ldots, x_n) = (x_1, \ldots, x_n, 1/g);$$

we use the same notation for its restriction to V. The morphism

$$\mathbb{A}^{n+1}(k) \to \mathbb{A}^m(k)$$
$$(x_1, \ldots, x_n, z) \mapsto (f_1 g_2 \cdots g_m z, \ldots, g_1 \cdots g_{i-1} f_i g_{i+1} \cdots g_m z, \ldots)$$

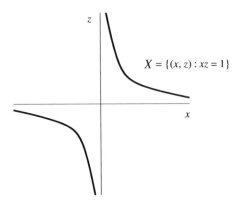

Figure 3.1 Resolving the rational map $x \mapsto \frac{1}{x}$.

restricted to $\mathbb{A}^n(k)_g$ coincides with the composition

$$\mathbb{A}^n(k)_g \xrightarrow{\pi} \mathbb{A}^n(k) \dashrightarrow^{\rho} \mathbb{A}^m(k).$$

Restricting to V_g yields $\phi : V_g \to W$ with the analogous factorization. □

The varieties $V_g \hookrightarrow V$ produced above are called *affine open subsets* of V.

Example 3.48   The rational map

$$\rho : \mathbb{A}^1(k) \dashrightarrow \mathbb{A}^1(k),$$
$$x \mapsto \frac{1}{x},$$

is defined on the open subset $U = \mathbb{A}^1 \setminus \{0\}$. We can identify U with the affine variety

$$X = \mathbb{A}^1(k)_x = \{(x, z) : xz = 1\},$$

which is a hyperbola. The morphism π is projection onto the x-axis; ϕ is projection onto the z-axis.

Example 3.49   The rational map

$$\rho : \mathbb{A}^1(\mathbb{R}) \dashrightarrow \mathbb{A}^2(\mathbb{R}),$$
$$x \mapsto \left(\frac{x^2 - 1}{x^2 + 1}, \frac{1}{x^2 + 4} \right),$$

is defined on $U = \mathbb{A}^1(\mathbb{R})$.

The rational map over the complex numbers defined by the same rule

$$\rho' : \mathbb{A}^1(\mathbb{C}) \dashrightarrow \mathbb{A}^2(\mathbb{C})$$

is defined on $U' = \mathbb{A}^1(\mathbb{C}) - \{\pm i, \pm 2i\}$.

Example 3.49 shows that the behavior of rational maps under field extensions can be subtle. Proposition 3.47 has content even when ρ is defined at each point of $\mathbb{A}^n(k)$,

3.5 RESOLVING RATIONAL MAPS

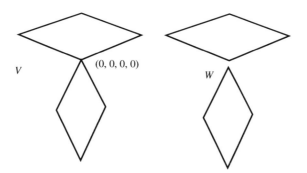

Figure 3.2 The rational map of Example 3.50.

i.e., when g has no zeros over k. The morphism $\mathbb{A}^n(k)_g \to \mathbb{A}^n(k)$ is bijective but is not an isomorphism unless g is constant. This reflects the possibility that g may acquire zeros over an extension of k. We will be able to say more about this once we have the Nullstellensatz at our disposal (see Proposition 8.40.)

Proposition 3.47 has one major drawback. The open set $U \subset V$ and the variety V_g are *not* intrinsic to the rational map $\rho: V \dashrightarrow W$. Rather they depend on the choice of extension

$$\mathbb{A}^n(k) \dashrightarrow \mathbb{A}^m(k),$$

as shown by the following example.

Example 3.50 Consider the varieties

$$V = \{x_1 = x_2 = 0\} \cup \{x_3 = x_4 = 0\} \subset \mathbb{A}^4(k)$$
$$W = \{x_1 = x_2 = x_5 = 0\} \cup \{x_3 = x_4 = x_5 - 1 = 0\} \subset \mathbb{A}^5(k)$$

with coordinate rings

$$k[V] = k[x_1, x_2, x_3, x_4]/\langle x_1 x_3, x_1 x_4, x_2 x_3, x_2 x_4 \rangle$$
$$k[W] = k[x_1, x_2, x_3, x_4, x_5]/\langle x_1(x_5 - 1), x_2(x_5 - 1), x_3 x_5, x_4 x_5, x_5(x_5 - 1) \rangle.$$

The rational maps

$$\rho': \mathbb{A}^4(k) \dashrightarrow \mathbb{A}^5(k)$$
$$(x_1, x_2, x_3, x_4, x_5) \mapsto (x_1, x_2, x_3, x_4, x_1/(x_1 + x_3))$$
$$\rho'': \mathbb{A}^4(k) \dashrightarrow \mathbb{A}^5(k)$$
$$(x_1, x_2, x_3, x_4, x_5) \mapsto (x_1, x_2, x_3, x_4, x_2/(x_2 + x_4))$$

are admissible on V and induce the same rational map $\rho: V \dashrightarrow W$. (Check this!) However, the corresponding open sets $U' = \{v \in V : x_1 + x_3 \neq 0\}$ and $U'' = \{v \in V : x_2 + x_4 \neq 0\}$ differ, as do the affine varieties $V_{x_1+x_3}$ and $V_{x_2+x_4}$.

How then should we define the indeterminacy locus of a general rational map of affine varieties $\rho : V \dashrightarrow W$? For each extension $\rho' : \mathbb{A}^n(k) \dashrightarrow \mathbb{A}^m(k)$, we have an open subset $U' \subset V$ where the denominators of the coordinate functions of ρ do not vanish. However, in some sense a map is well defined at all points $v \in V$ for which there exists *some* extension well defined at v.

Provisional Definition 3.51 (Indeterminacy locus) Let $\rho : V \dashrightarrow W$ be a rational map of affine varieties. We define the *indeterminacy locus* as

$$Z = \cap_{\rho'} (V \setminus U'),$$

where the intersection is taken over all extensions

$$\rho' : \mathbb{A}^n(k) \dashrightarrow \mathbb{A}^m(k)$$

and U' is the open set where ρ' is well defined.

Our ultimate definition will have to wait until we define the *indeterminacy ideal* in Chapter 8.

Example 3.52 In Example 3.50 the indeterminacy locus of ρ is the origin.

3.5.1 Localization

The proof of Proposition 3.47 and the definition of the 'ring of fractions' are both instances of a very common algebraic construction:

Definition 3.53 Let R be a ring and $S \subset R$ a *multiplicative subset*, i.e., for all $s_1, s_2 \in S$ the product $s_1 s_2 \in S$. The *localization* $R[S^{-1}]$ is defined as equivalence classes $\{r/s : r \in R, s \in S\}$, where we identify $r_1/s_1 \equiv r_2/s_2$ whenever there exists a $t \in S$ with $t(r_1 s_2 - r_2 s_1) = 0$. This is a ring under the operations of addition and multiplication of fractions.

We leave it to the reader to verify that the operations are compatible with the equivalence classes (see Exercise 3.19).

Example 3.54

- If R is a domain and $S = R^*$ then $R[S^{-1}]$ is the field of fractions of R.
- If R is a ring then the nonzero divisors form a multiplicative set S (Check this!) and $R[S^{-1}]$ is the ring of fractions.
- If $R = k[x_1, \ldots, x_n]$ and S is the multiplicative set generated by g then $R[S^{-1}] = k[x_1, \ldots, x_n][1/g]$ is the coordinate ring of $\mathbb{A}^n(k)_g$, the variety introduced in the proof of Proposition 3.47.

3.6 Rational and unirational varieties

We introduce a more flexible notion of parametrization, generalizing the regular parametrizations studied in Chapter 1.

Definition 3.55 Consider an affine variety $W \subset \mathbb{A}^m(k)$. A *rational parametrization* of W is a rational map

$$\rho : \mathbb{A}^n(k) \dashrightarrow W$$

such that W is the closure of the image of ρ, i.e., if $U \subset \mathbb{A}^n(k)$ is the open subset over which ρ is defined then $\overline{\rho(U)} = W$. W is *unirational* if it admits a rational parametrization.

Here ρ is defined over k; to emphasize this, we'll say that W is *unirational over k*.

Here are some examples beyond those introduced in Chapter 1.

Example 3.56

1. $W = \{(y_1, y_2) : y_1^2 + y_2^2 = 1\} \subset \mathbb{A}^2(\mathbb{Q})$ is unirational with parametrization

$$\rho : \mathbb{A}^1(\mathbb{Q}) \dashrightarrow W$$
$$s \mapsto \left(\frac{2s}{s^2+1}, \frac{s^2-1}{s^2+1} \right).$$

The image of ρ is $W \setminus \{(0, 1)\}$.

2. $W = \{(y_1, y_2, y_3) : y_1^2 + y_2^2 + y_3^2 = 1\} \subset \mathbb{A}^3(\mathbb{Q})$ is unirational with parametrization

$$\rho : \mathbb{A}^2(\mathbb{Q}) \dashrightarrow W,$$
$$(s, t) \mapsto \left(\frac{2s}{s^2+t^2+1}, \frac{2t}{s^2+t^2+1}, \frac{s^2+t^2-1}{s^2+t^2+1} \right).$$

The image of ρ is $W \setminus \{(0, 0, 1)\}$.

These formulas come from stereographic projections of the unit sphere from the north pole (the points $(0, 1)$ and $(0, 0, 1)$ respectively).

Proposition 3.57 Let W be a unirational affine variety with rational parametrization $\rho : \mathbb{A}^n(k) \dashrightarrow W$ inducing

$$\rho^* : k[W] \to k(\mathbb{A}^n) = k(x_1, \ldots, x_n).$$

Then ρ^* is injective and induces a field extension

$$j : k(W) \hookrightarrow k(x_1, \ldots, x_n).$$

Conversely, if W is an affine variety admitting an injection of k-algebras

$$\psi : k[W] \hookrightarrow k(x_1, \ldots, x_n)$$

then W is unirational.

We will put this on a systematic footing when we introduce *dominant maps*.

Proof Suppose that ρ^* were not injective, so there exists $f \neq 0 \in k[W]$ with $\rho^* f = 0$. Let $U \subset \mathbb{A}^n(k)$ denote the complement of the indeterminacy of ρ. We therefore have

$$\phi(U) \subset \{w \in W : f(w) = 0\} \subsetneq W;$$

the strict inclusion reflects the fact that f is not identically zero on W. It follows that $\overline{\phi(U)} \neq W$, a contradiction. The injection ρ^* allows us to regard $k[W] \subset k(x_1, \ldots, x_n)$; its ring of fractions $k(W) \subset k(x_1, \ldots, x_n)$ as well.

Conversely, suppose we have an injection ψ as above. Corollary 3.46 yields a rational map $\rho : \mathbb{A}^n(k) \dashrightarrow W$. If $\overline{\phi(U)} \subsetneq W$ then there must exist some $f \neq 0 \in k[W]$ vanishing on $\phi(U)$, contradicting the assumption that ψ is injective. $\square$

Example 3.58 Consider the curve

$$W = \{(y_1, y_2) : y_1^m = y_2^n\} \subset \mathbb{A}^2(k)$$

where $m, n \in \mathbb{N}$ are relatively prime. Then W is unirational via

$$k[W] \hookrightarrow k(s)$$
$$(y_1, y_2) \mapsto (s^n, s^m).$$

There is a stronger notion which is also worth mentioning:

Definition 3.59 An affine variety W is *rational* if it admits a rational parametrization $\rho : \mathbb{A}^n(k) \dashrightarrow W$ such that the induced field extension

$$j : k(W) \hookrightarrow k(x_1, \ldots, x_n)$$

is an isomorphism.

We have the following algebraic characterization:

Corollary 3.60 *An affine variety W is rational if and only if $k(W) \simeq k(x_1, \ldots, x_n)$ as k-algebras.*

Proof The isomorphism restricts to an injective homomorphism

$$k[W] \hookrightarrow k(x_1, \ldots, x_n).$$

Proposition 3.57 yields the rational parametrization $\rho : \mathbb{A}^n \dashrightarrow W$.

Example 3.61 Both instances of Example 3.56 are rational. For the first case, we must check that $j : \mathbb{Q}(W) \hookrightarrow \mathbb{Q}(s)$ is surjective. However, the fraction $y_1/(1 - y_2)$ goes to s. In the second case, $j : \mathbb{Q}(W) \hookrightarrow \mathbb{Q}(s, t)$ is surjective because $y_1/(1 - y_3)$ goes to s and $y_2/(1 - y_3)$ goes to t.

Example 3.58 is also rational: m and n are relatively prime so $k(s^m, s^n) = k(s)$.

Here are some open problems in the field:

Problem 3.62 (Unirationality of small degree hypersurfaces) Let $f \in \mathbb{C}[y_1, \ldots, y_m]$ be an irreducible polynomial of degree $d \leq m$, and $V(f) \subset \mathbb{A}^m(\mathbb{C})$ the corresponding hypersurface. Is $V(f)$ unirational?

For general such hypersurfaces, there are no techniques for disproving unirationality. However, unirationality has been established only when $d = 2, m \geq 2, d = 3, m \geq 3$ [26], or $m \gg d$ [18] [34]. Indeed, m grows very rapidly as a function of d; for fixed $d \geq 4$, there are many values of m for which unirationality is an open problem.

On the other hand, general degree d hypersurfaces in $\mathbb{A}^m(\mathbb{C})$ do not admit rational parametrizations when $d > m$ [25] §4. For instance, the hypersurface

$$\{(y_1, y_2, y_3) : y_1^4 + y_2^4 + y_3^4 = 1\}$$

lacks one.

Deciding whether hypersurfaces are rational is even more difficult: There are open problems even for cubic hypersurfaces! Given $f \in \mathbb{C}[x_1, \ldots, x_5]$ of degree 3, when is $V(f) \subset \mathbb{A}^5(\mathbb{C})$ rational? [20] It would be wonderful to have an explicit test that would decide whether $\mathbb{C}(V) \simeq \mathbb{C}(t_1, \ldots, t_d)$!

3.7 Exercises

3.1 Prove Propositions 3.6 and 3.7.

3.2 (a) Let $S \subset \mathbb{A}^n(k)$. Show that

$$I(S) = k[x_1, \ldots, x_n]$$

if and only if $S = \emptyset$.

(b) Let k be an infinite field. Show that

$$I(\mathbb{A}^n(k)) = \langle 0 \rangle.$$

Hint: Use induction on n.

3.3 For each ideal $I \subset k[x_1, \ldots, x_n]$, show that

$$I(V(I)) \supset I.$$

Give an example where equality does not hold.

3.4 Given a ring R and a collection of ideals $\{I_\beta\}_{\beta \in B}$ of R, show that $\sum_{\beta \in B} I_\beta$ is an ideal. Prove it is the smallest ideal containing $\cup_{\beta \in B} I_\beta$:

$$\sum_{\beta \in B} I_\beta = \langle f \in I_\beta \text{ for some } \beta \in B \rangle.$$

3.5 Let $\{V_\beta\}_{\beta \in B}$ be a (possibly infinite) collection of affine varieties $V_\beta \subset \mathbb{A}^n(k)$. Show there exist a finite number of $\beta_1, \ldots, \beta_r \in B$ such that

$$\cap_{\beta \in B} V_\beta = V_{\beta_1} \cap V_{\beta_2} \cap \ldots V_{\beta_r}.$$

3.6 (a) Show that every finite subset $S \subset \mathbb{A}^n(k)$ is a variety. Prove that $\dim_k k[S] = |S|$.
(b) Suppose that $V \subset \mathbb{A}^n(k)$ is an affine variety, with $|V| = \infty$. Show that $\dim_k k[V]$ is not finite.

3.7 Compute the Zariski closures $\overline{S} \subset \mathbb{A}^2(\mathbb{Q})$ of the following subsets:
(a) $S = \{(n^2, n^3) : n \in \mathbb{N}\} \subset \mathbb{A}^2(\mathbb{Q})$;
(b) $S = \{(x, y) : x^2 + y^2 < 1\} \subset \mathbb{A}^2(\mathbb{Q})$;
(c) $S = \{(x, y) : x + y \in \mathbb{Z}\} \subset \mathbb{A}^2(\mathbb{Q})$.

3.8 Let $V_1, V_2 \subset \mathbb{A}^n(k)$ be affine varieties. Show that

$$I(V_1) + I(V_2) \subset I(V_1 \cap V_2).$$

Find an example where equality does not hold.

3.9 Consider the varieties $V, W \subset \mathbb{A}^2(\mathbb{C})$

$$V = \{(x_1, x_2) : x_1^4 + x_2^4 = 1\}, \quad W = \{(y_1, y_2) : y_1^2 + y_2^2 = 1\},$$

and the morphism

$$\phi : \mathbb{A}^2(\mathbb{C}) \to \mathbb{A}^2(\mathbb{C})$$
$$(x_1, x_2) \to (x_1^2, x_2^2).$$

Show that $\phi(V) \subset W$.

3.10 (Diagonal morphism) Let V be an affine variety. The *diagonal map* is defined

$$\Delta : V \to V \times V$$
$$v \mapsto (v, v)$$

with image $\Delta_V := \Delta(V) \subset V \times V$.
(a) Show that Δ is a morphism.
(b) Let $V = \mathbb{A}^n(k)$ and fix coordinates $x_1, \ldots, x_n$ and $y_1, \ldots, y_n$ on $\mathbb{A}^n(k) \times \mathbb{A}^n(k)$. Show that

$$I(\Delta_{\mathbb{A}^n(k)}) = \langle x_1 - y_1, x_2 - y_2, \ldots, x_n - y_n \rangle.$$

(c) For general V, show that Δ_V is closed in $V \times V$ and hence an affine variety.

3.7 EXERCISES

(d) Show that $\Delta : V \to \Delta_V$ is an isomorphism. *Hint:* Use the projections $\pi_1, \pi_2 : V \times V \to V$.

3.11 Consider the following lines in affine space $\mathbb{A}^3(\mathbb{R})$:

$$\ell_1 = \{x_1 = x_2 = 0\}, \quad \ell_2 = \{x_1 = x_3 = 0\}, \quad \ell_3 = \{x_2 = x_3 = 0\}.$$

Compute generators for $I(\ell_1 \cup \ell_2 \cup \ell_3)$. Justify your answer.

3.12 Identify 2×3 matrices with entries in k

$$A = \begin{pmatrix} a_{11} & a_{12} & a_{13} \\ a_{21} & a_{22} & a_{23} \end{pmatrix}$$

with the affine space $\mathbb{A}^6(k)$ with coordinates $a_{11}, a_{12}, a_{13}, a_{21}, a_{22}, a_{23}$. Show that the matrices of rank 2 are open. Write explicit polynomials vanishing along the matrices of rank ≤ 1.

3.13 Show there is a one-to-one correspondence between morphisms $V \to \mathbb{A}^1(k)$ and functions $f \in k[V]$.

3.14 Show that every automorphism of the affine line $\mathbb{A}^1(\mathbb{Q})$ takes the form

$$x \to ax + b, \quad a, b \in \mathbb{Q}, \quad a \neq 0.$$

3.15 Let $Y \subset \mathbb{A}^2(\mathbb{C})$ be the variety

$$\{(y_1, y_2) : y_1^3 = y_2^4\}.$$

(a) Show there is a bijective morphism $\phi : \mathbb{A}^1(\mathbb{C}) \to Y$. *Hint:* Try $y_1 = x^4$.
(b) Show that ϕ is not an isomorphism, i.e., ϕ does not have an inverse morphism.

3.16 Let W_1, W_2, and V be affine varieties. Given a morphism $\phi : V \to W_1 \times W_2$, show there exist unique morphisms $\phi_1 : V \to W_1$ and $\phi_2 : V \to W_2$ such that

$$\phi_1 = \pi_1 \circ \phi, \quad \phi_2 = \pi_2 \circ \phi.$$

Conversely, given $\phi_1 : V \to W_1$ and $\phi_2 : V \to W_2$, show there exists a unique morphism $\phi : V \to W_1 \times W_2$ satisfying these identities.

3.17 Let R be a ring with ring of fractions K. Show that the rule

$$R \to K$$
$$r \mapsto r/1$$

defines an injective homomorphism $R \hookrightarrow K$.

3.18 Describe the rings of fractions of the following:
(a) $R = \mathbb{Z}/12\mathbb{Z}$;
(b) $R = k[x_1, x_2]/\langle x_1 x_2 \rangle$.

3.19 Let R be a ring and $S \subset R$ a multiplicative subset. Show that the operations of addition and multiplication for the localization $R[S^{-1}]$ are well-defined on equivalence classes of fractions.

3.20 Prove Corollary 3.46.

3.21 Let R be a ring, $g_1, \ldots, g_m \in R$, and $g = g_1 g_2 \ldots g_m$. Let S and T be the multiplicative sets generated by g and $\{g_1, \ldots, g_m\}$ respectively. For example, $T = \{g_1^{e_1} \ldots g_m^{e_m} : e_1, \ldots, e_m \geq 0\}$. Show that the localizations $R[S^{-1}]$ and $R[T^{-1}]$ are isomorphic.

3.22 (General linear group) Identify $\mathbb{A}^{n^2}(k)$ with the space of $n \times n$ matrices $A = (a_{ij})$, with coordinate ring

$$k[a_{11}, \ldots, a_{1n}, a_{21}, \ldots, a_{2n}, \ldots, a_{n1}, \ldots, a_{nn}].$$

(a) Show that matrix multiplication induces a morphism

$$\mu : \mathbb{A}^{n^2}(k) \times \mathbb{A}^{n^2}(k) \to \mathbb{A}^{n^2}(k)$$
$$(A, B) \mapsto AB.$$

(b) Show that there is a rational map

$$\iota : \mathbb{A}^{n^2}(k) \dashrightarrow \mathbb{A}^{n^2}(k)$$
$$A \mapsto A^{-1}.$$

Hint: Use Cramer's Rule.

(c) The affine variety $\mathbb{A}^{n^2}(k)_{\det(A)}$ constructed in Proposition 3.47 is called the *general linear group* and denoted $\mathrm{GL}_n(k)$. Show there are multiplication and inversion morphisms

$$\mu : \mathrm{GL}_n(k) \times \mathrm{GL}_n(k) \to \mathrm{GL}_n(k), \quad \iota : \mathrm{GL}_n(k) \to \mathrm{GL}_n(k).$$

A variety G with multiplication and inversion operations

$$\mu : G \times G \to G, \quad \iota : G \to G$$

satisfying the axioms of a group is called an *algebraic group*.

3.23 Verify that ρ' and ρ'' in Example 3.50 are admissible along V and define the same rational map $V \dashrightarrow W$.

3.24 (a) Verify the images of the maps in Example 3.56.

(b) Given nonzero numbers r_1, r_2, r_3, show that

$$\left(\frac{y_1}{r_1}\right)^2 + \left(\frac{y_2}{r_2}\right)^2 + \left(\frac{y_3}{r_3}\right)^2 = 1$$

is unirational.

3.25 Show that

$$W = \{(y_1, y_2, y_3) : y_1 y_2 y_3 = 1\} \subset \mathbb{A}^3(\mathbb{Q})$$

is rational.

4 Elimination

> Eliminate, eliminate, eliminate
> Eliminate the eliminators of elimination theory
> From Shreeram S. Abhyankar, *Polynomials and Power Series* [7, pp. 783]

Elimination theory is the systematic reduction of systems of polynomial equations in many variables to systems in a subset of these variables. For example, when a system of polynomials admits a finite number of solutions, we would like to express these as the roots of a single polynomial in one of the original variables.

The language of affine varieties, morphisms, and rational maps allows us to understand elimination theory in more conceptual terms. Recall that one of our original guiding problems concerned implicitization: describe equations for the image of a morphism $\phi : \mathbb{A}^n(k) \to \mathbb{A}^m(k)$. In light of the theory we have developed, it makes sense to recast this in a more general context:

Problem 4.1 (Generalized Implicitization Problem) Consider an affine variety $V \subset \mathbb{A}^n(k)$, with ideal $I(V)$, and a morphism $\phi : V \to \mathbb{A}^m(k)$. Describe generators for the ideal $I(\phi(V))$ in terms of generators for $I(V)$.

A warning is in order: the image of a polynomial morphism is not necessarily closed, so the best we can do is to find equations for the *closure* of the image. We will come back to this point when we discuss projective elimination theory in Chapter 10.

In this chapter, we continue to assume that the base field is infinite.

4.1 Projections and graphs

We start with an example illustrating how images of morphisms can fail to be closed:

Example 4.2 Consider the variety $V = \{(x_1, x_2) : x_1 x_2 = 1\}$ and the morphism

$$\phi : V \to \mathbb{A}^1(k)$$
$$(x_1, x_2) \mapsto x_1.$$

The image $\phi(V) = \{x_1 : x_1 \neq 0\}$, which is not closed.

Here ϕ is induced by a *projection morphism* $\mathbb{A}^2(k) \to \mathbb{A}^1(k)$; projections play an important role in elmination. Initially, we will focus on finding images of varieties under projection:

Theorem 4.3 *Let $V \subset \mathbb{A}^{m+n}(k)$ be an affine variety with ideal $J = I(V)$. Consider the projection morphism*

$$\pi : \mathbb{A}^{m+n}(k) \to \mathbb{A}^m(k)$$
$$(x_1, \ldots, x_n, y_1, \ldots, y_m) \mapsto (y_1, \ldots, y_m).$$

Then we have

$$\overline{\pi(V)} = V(J \cap k[y_1, \ldots, y_m]).$$

Proof Given a polynomial $f \in k[y_1, \ldots, y_m]$, $\pi^* f$ is the polynomial regarded as an element in $k[x_1, \ldots, x_n, y_1, \ldots, y_m]$.

To establish the forward inclusion, it suffices to check that $\pi(V) \subset V(J \cap k[y_1, \ldots, y_m])$. This is the case if each $f \in J \cap k[y_1, \ldots, y_m]$ vanishes on $\pi(V)$. For each $p = (a_1, \ldots, a_m) \in \pi(V)$, choose $q = (b_1, \ldots, b_n, a_1, \ldots, a_m) \in V$ with $\pi(q) = p$. We have

$$f(p) = f(a_1, \ldots, a_m)$$
$$= \pi^* f(b_1, \ldots, b_n, a_1, \ldots, a_m)$$
$$= \pi^* f(q) = 0$$

as f vanishes on V.

We prove the reverse inclusion

$$V(J \cap k[y_1, \ldots, y_m]) \subset \overline{\pi(V)}.$$

Pick $p = (a_1, \ldots, a_m) \in V(J \cap k[y_1, \ldots, y_m])$ and $f \in I(\pi(V))$. Polynomials vanishing on $\pi(V)$ pull back to polynomials vanishing on V, i.e.,

$$\pi^* I(\pi(V)) \subset I(V) = J.$$

In particular, $\pi^* f \in J \cap k[y_1, \ldots, y_m]$ so that $f(a_1, \ldots, a_m) = 0$. $\square$

The key in passing to general morphisms is the graph construction:

Definition 4.4 The *graph* Γ_ϕ of a morphism $\phi : V \to W$ is the locus of pairs

$$\{(v, \phi(v)) : v \in V\} \subset V \times W.$$

4.1 PROJECTIONS AND GRAPHS

There are projections

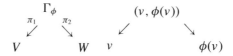

where π_1 is invertible and $\pi_2(\Gamma_\phi) = \phi(V)$.

The graph of a morphism is itself an affine variety. We will prove a more precise statement:

Proposition 4.5 *Consider affine varieties $V \subset \mathbb{A}^n(k)$ and $W \subset \mathbb{A}^m(k)$ and a morphism $\phi : V \to W$. Then Γ_ϕ is naturally an affine variety.*

Precisely, choose a morphism $\hat{\phi} : \mathbb{A}^n(k) \to \mathbb{A}^m(k)$ extending ϕ with coordinate functions $\phi_1, \ldots, \phi_m \in k[x_1, \ldots, x_n]$. Then we have

$$I(\Gamma_\phi) = \pi_1^* I(V) + \langle y_1 - \phi_1, \ldots, y_m - \phi_m \rangle,$$
$$V(I(\Gamma_\phi)) = \Gamma_\phi,$$

where $\pi_1 : \mathbb{A}^{m+n}(k) \to \mathbb{A}^n(k)$ is the projection morphism.

Proof Here the projections are induced by:

$$\begin{array}{ccc} & (x_1, \ldots, x_n, y_1, \ldots, y_m) & \\ \swarrow & & \searrow \\ (x_1, \ldots, x_n) & & (y_1, \ldots, y_m) \end{array}$$

The inclusions

$$\Gamma_\phi \subset V \times W \subset \mathbb{A}^n(k) \times \mathbb{A}^m(k) = \mathbb{A}^{m+n}(k)$$

yield $\pi_1^* I(V) \subset I(\Gamma_\phi)$. If $v = (v_1, \ldots, v_n) \in V$ then

$$\phi(v) = (\phi_1(v_1, \ldots, v_n), \ldots, \phi_m(v_1, \ldots, v_n)),$$

so $y_j - \phi_j$ vanishes at $(v, \phi(v))$. This proves that

$$I(\Gamma_\phi) \supset \pi_1^* I(V) + \langle y_1 - \phi_1, \ldots, y_m - \phi_m \rangle.$$

It remains to check that

$$I(\Gamma_\phi) \subset \pi_1^* I(V) + \langle y_1 - \phi_1, \ldots, y_m - \phi_m \rangle.$$

For each $f \in k[x_1, \ldots, x_n, y_1, \ldots y_m]$, we have (see Exercise 2.12)

$$f \equiv f(x_1, \ldots, x_n, \phi_1(x_1, \ldots, x_n), \ldots, \phi_m(x_1, \ldots, x_n))$$
$$(\mathrm{mod}\ \langle y_1 - \phi_1, \ldots, y_m - \phi_m \rangle),$$

i.e., each element is congruent modulo $\langle y_1 - \phi_1, \ldots, y_m - \phi_m \rangle$ to a polynomial in $x_1, \ldots, x_n$. However, if $f \in k[x_1, \ldots, x_n]$ vanishes along Γ_ϕ then $f \in I(V)$.

Finally, suppose we are given $(v, w) \in V(I(\Gamma_\phi))$. We have $\pi_1^* I(V) \subset I(\Gamma_\phi)$ so that $V(I(\Gamma_\phi)) \subset \pi_1^{-1}(V)$, i.e., $v \in V$. The remaining equations

$$y_j = \phi_j(x_1, \ldots, x_n), \quad j = 1, \ldots, m$$

imply that $w = \phi(v)$, hence $(v, w) \in \Gamma_\phi$. $\square$

Definition 4.6 A monomial order on $k[x_1, \ldots, x_n, y_1, \ldots, y_m]$ is an *elimination order* for $x_1, \ldots, x_n$ if each polynomial with leading monomial in $k[y_1, \ldots, y_m]$ is actually contained in $k[y_1, \ldots, y_m]$, i.e.,

$$\mathrm{LM}(g) \in k[y_1, \ldots, y_m] \Rightarrow g \in k[y_1, \ldots, y_m].$$

Example 4.7

1. Lexicographic order with $x_i > y_j$ for each i, j is an elimination order for $x_1, \ldots, x_n$. However, this is usually relatively inefficient computationally (see Exercise 2.10).
2. Fix monomial orders $>_x$ and $>_y$ on $x_1, \ldots, x_n$ and $y_1, \ldots, y_m$ respectively. Then the *product order* is defined as follows: We have $x^\alpha y^\beta > x^\gamma y^\delta$ if

$$\begin{cases} x^\alpha >_x x^\gamma \text{ or} \\ x^\alpha =_x x^\gamma \text{ and } y^\beta >_y y^\delta. \end{cases}$$

3. One fairly efficient elimination order is the product of graded reverse lexicographic orders $>_x$ and $>_y$.

Theorem 4.8 (Elimination Theorem) *Let $J \subset k[x_1, \ldots, x_n, y_1, \ldots, y_m]$ be an ideal and $>$ an elimination order for $x_1, \ldots, x_n$. Let $\{f_1, \ldots, f_r\}$ be a Gröbner basis for J with respect to $>$. Then $J \cap k[y_1, \ldots, y_m]$ is generated by the elements of the Gröbner basis contained in $k[y_1, \ldots, y_m]$, i.e.,*

$$J \cap k[y_1, \ldots, y_m] = \langle f_j : f_j \in k[y_1, \ldots, y_m] \rangle \subset k[y_1, \ldots, y_m].$$

Proof It suffices to show that each element $g \in J \cap k[y_1, \ldots, y_m]$ is generated by the $f_j \in k[y_1, \ldots, y_m]$. Choose

$$g \in (J \cap k[y_1, \ldots, y_m]) \setminus \langle f_j : f_j \in k[y_1, \ldots, y_m] \rangle$$

with $\mathrm{LM}(g)$ minimal. Apply the division algorithm to g and let f_j be a Gröbner basis element with $\mathrm{LM}(f_j) | \mathrm{LM}(g)$. Hence $\mathrm{LM}(f_j) \in k[y_1, \ldots, y_m]$ so the definition of the elimination order implies $f_j \in k[y_1, \ldots, y_m]$. Thus

$$\tilde{g} := g - f_j \mathrm{LT}(g)/\mathrm{LT}(f_j)$$

is an element of $(J \cap k[y_1, \ldots, y_m]) \setminus \langle f_j : f_j \in k[y_1, \ldots, y_m]\rangle$ with $\mathrm{LM}(\tilde{g}) < \mathrm{LM}(g)$, a contradiction. $\square$

Example 4.9 (Solvability of varying equations) For which values of $a \in \mathbb{C}$ is the system

$$x + y = a,$$
$$x^2 + y^2 = a^3,$$
$$x^3 + y^3 = a^5$$

solvable? Let $Z \subset \mathbb{A}^3(\mathbb{C})$ be the solution set and

$$\pi : \mathbb{A}^3(\mathbb{C}) \to \mathbb{A}^1(\mathbb{C})$$
$$(x, y, a) \mapsto a,$$

projection onto the coordinate a. We want to compute the image $\pi(Z)$.

We compute a Gröbner basis for the ideal

$$I = \langle x + y - a, x^2 + y^2 - a^3, x^3 + y^3 - a^5 \rangle$$

with respect to lexicographic order:

$$x + y - a, \ 2y^2 - a^3 - 2ya + a^2, \ 2a^5 + a^3 - 3a^4.$$

Thus $\pi(Z)$ is given by the solutions to the last polynomial

$$a = 0, 1, \tfrac{1}{2}.$$

Here are some corresponding solutions:

$$(x, y, a) = (0, 0, 0), \quad (0, 1, 1), \quad \left(\tfrac{1}{4}, \tfrac{1}{4}, \tfrac{1}{2}\right).$$

4.2 Images of rational maps

Consider a rational map

$$\rho : \mathbb{A}^n(k) \dashrightarrow \mathbb{A}^m(k)$$
$$(x_1, \ldots, x_n) \mapsto (f_1/g_1, \ldots, f_m/g_m),$$

well-defined over the open set $U = \{g = g_1 \ldots g_m \neq 0\}$. Proposition 3.47 yields an affine variety $\mathbb{A}^n(k)_g$ and a morphism $\phi : \mathbb{A}^n(k)_g \to W$ such that $\phi(\mathbb{A}^n(k)_g) = \rho(U)$. Recall that $\mathbb{A}^n(k)_g \subset \mathbb{A}^{n+1}(k)$ is given by

$$\{(x_1, \ldots, x_n, z) : zg(x_1, \ldots, x_n) = 1\}.$$

We write down equations for the graph of ϕ in $\mathbb{A}^{n+1}(k) \times \mathbb{A}^m(k)$. Since
$$f_j/g_j = g_1 \cdots g_{j-1} f_j g_{j+1} \cdots g_m/g$$
we have
$$I(\Gamma_\phi) = \langle zg - 1, y_j - g_1 \cdots g_{j-1} f_j g_{j+1} \cdots g_m z, j = 1, \ldots, m \rangle.$$

The equations for the image $\rho(U)$ are obtained by eliminating $x_1, \ldots, x_n, z$, i.e., we find generators for $I(\Gamma_\phi) \cap k[y_1, \ldots, y_m]$.

Example 4.10 Compute the image of the rational map
$$\mathbb{A}^1(k) \dashrightarrow \mathbb{A}^2(k)$$
$$x \mapsto \left(\frac{x^2+1}{x^2-1}, \frac{1}{x} \right).$$

The associated affine variety is
$$Y := \mathbb{A}^1(k)_{x(x^2-1)} = \{(x, z) : z(x^2 - 1)x - 1 = 0\}$$
and the graph of $\phi : Y \to \mathbb{A}^2(k)$ has equations
$$I(\Gamma_\phi) := \langle z(x^2-1)x - 1, y_1 - zx(x^2+1), y_2 - z(x^2 - 1) \rangle.$$

Using lexicographic order, we get a Gröbner basis
$$\{y_2^2 - y_1 + 1 + y_2^2 y_1, -y_2 y_1 + y_2 + 2z, -1 + xy_2, xy_1 - x - y_2 y_1 - y_2\}$$
and the first entry is the equation for the image $\rho(\mathbb{A}^1(k))$.

Computing the image of a rational map from a general affine variety looks trickier – it is hard to describe the locus where the map is well-defined (cf. Example 3.50). Luckily, a complete description of the indeterminacy is not necessary.

Proposition 4.11 Let $\rho : V \dashrightarrow W$ be a rational map of affine varieties $V \subset \mathbb{A}^n(k)$ and $W \subset \mathbb{A}^m(k)$. Let $Z \subset V$ be the indeterminacy locus for ρ. Choose an extension
$$\rho' : \mathbb{A}^n(k) \dashrightarrow \mathbb{A}^m(k),$$
write $g' = g'_1 \cdots g'_m$, and let $\phi : V_{g'} \to W$ be the morphism given by Proposition 3.47. Then we have
$$\overline{\rho(V \setminus Z)} = \overline{\phi(V_{g'})}.$$

Proof Recall the indeterminacy locus is the set of points in V where each extension ρ' fails to be well-defined
$$V \setminus Z = \cup_{\text{extensions } \rho'} \{v \in V : g'(v) \neq 0\}$$
where g' is the denominator of ρ'.

4.2 IMAGES OF RATIONAL MAPS

It suffices to show that the closure of the image is independent of the choice of extension. Given

$$\rho', \rho'' : \mathbb{A}^n(k) \dashrightarrow \mathbb{A}^m(k)$$
$$(x_1, \ldots, x_n) \overset{\rho'}{\mapsto} (f'_1/g'_1, \ldots, f'_m/g'_m),$$
$$(x_1, \ldots, x_n) \overset{\rho''}{\mapsto} (f''_1/g''_1, \ldots, f''_m/g''_m)$$

with $g' = g'_1 \ldots g'_m$, $g'' = g''_1, \ldots, g''_m$, and $\phi' : V_{g'} \to W$ and $\phi'' : V_{g''} \to W$ the morphisms coming from Proposition 3.47, we will show that

$$\overline{\phi'(V_{g'})} = \overline{\phi''(V_{g''})}.$$

We can also consider $\phi : V_{g'g''} \to W$, by inverting the product $g'g''$. Since

$$\phi(V_{g'g''}) \subset \phi'(V_{g'}), \phi''(V_{g''})$$

is suffices to show that

$$\overline{\phi(V_{g'g''})} = \overline{\phi'(V_{g'})}.$$

This is a special case of the following:

Lemma 4.12 Let $\psi : Y \to W$ be a morphism of varieties, h an element of $k[Y]$ which does not divide zero, and $U = \{x \in Y : h(x) \neq 0\}$. Then

$$\overline{\psi(U)} = \overline{\psi(Y)}.$$

Proof of lemma If $\overline{\psi(U)} \subsetneq \overline{\psi(Y)}$ then there would exist an $f \in k[W]$ such that $\psi^* f \neq 0$ but $f(\psi(u)) = 0$ for each $u \in U$. But then $h\psi^* f = 0 \in k[Y]$, contradicting the assumption that h does not divide zero. □

Thus to compute the image of a rational map

$$\rho : V \dashrightarrow W,$$

it suffices to compute the image of *any* morphism

$$\phi : V_g \to W$$

given by Proposition 3.47. The graph of ϕ has equations

$$I(\Gamma_\phi) = \langle y_j - g_1 \ldots g_{j-1} f_j g_{j+1} \ldots g_m z, zg - 1 \rangle + I(V)$$

and equations of the image are generators for $I(\Gamma_\phi) \cap k[y_1, \ldots, y_m]$.

Definition 4.13 Let $\rho : V \dashrightarrow W$ be a rational map well-defined outside the closed subset $Z \subset V$. The *graph* of ρ is the locus

$$\{(v, \rho(v))\} \subset (V \setminus Z) \times W.$$

The graph of a rational map may not be an affine variety – this will be a crucial point when we discuss abstract varieties.

Example 4.14 The graph of

$$\rho : \mathbb{A}^2(k) \dashrightarrow \mathbb{A}^2(k)$$
$$(x_1, x_2) \mapsto (x_1, x_2/x_1)$$

satisfies the equations

$$y_1 - x_1 = y_2 x_1 - x_2 = 0.$$

The corresponding variety contains the line

$$x_1 = x_2 = y_1 = 0,$$

which lies over the indeterminacy of ρ.

Definition 4.15 Let $\rho : V \dashrightarrow W$ be a rational map with indeterminacy locus Z. The *closed graph* of ρ is the closure

$$\Gamma_\rho = \overline{\{(v, \rho(v)) : v \in V \setminus Z\}} \subset V \times W,$$

which is an affine variety.

Again, we have projections $\pi_1 : \Gamma_\rho \to V$ and $\pi_2 : \Gamma_\rho \to W$.

Equations of Γ_ρ can be obtained by computing the image of the rational map

$$(\mathrm{Id}, \rho) : V \dashrightarrow V \times W$$
$$v \mapsto (v, \rho(v)).$$

Example 4.16 Consider the rational map

$$\rho : \mathbb{A}^1(k) \dashrightarrow \mathbb{A}^2(k)$$
$$t \mapsto (t/(t+1), t^2/(t-1))$$

and the induced map

$$(\mathrm{Id}, \rho) : \mathbb{A}^1(k) \dashrightarrow \mathbb{A}^1(k) \times \mathbb{A}^2(k).$$

The corresponding morphisms are

$$\phi : \mathbb{A}^1(k)_{t^2-1} \to \mathbb{A}^2(k)$$
$$(t, z) \mapsto (t(t-1)z, t^2(t+1)z)$$
$$(\pi, \phi) : \mathbb{A}^1(k)_{t^2-1} \to \mathbb{A}^1(k) \times \mathbb{A}^2(k)$$
$$(t, z) \mapsto (t, t(t-1)z, t^2(t+1)z)$$

where $z(t^2 - 1) = 1$. The equations of the graph are given by generators of

$$k[t, y_1, y_2] \cap \langle y_1 - t(t-1)z, y_2 - t^2(t+1)z, z(t^2-1) - 1 \rangle$$

which equals

$$\langle -2y_1y_2 + t - y_1 + y_2, 2y_1^2 y_2 + y_1^2 - 3y_1 y_2 + y_2 \rangle.$$

4.3 Secant varieties, joins, and scrolls

In this section we describe some classical geometric constructions and how elimination techniques can be applied to write down their equations.

Consider the variety

$$\Delta_N = \{(t_1, \ldots, t_N) : t_1 + t_2 + \cdots + t_N = 1\} \subset \mathbb{A}^N(k).$$

For each finite set of points $S = \{p_1, \ldots, p_N\} \subset \mathbb{A}^n(k)$, we have a morphism

$$\sigma_S : \Delta_N \to \mathbb{A}^n$$
$$(t_1, \ldots, t_N) \mapsto t_1 p_1 + \cdots + t_N p_N,$$

where we add the p_j as vectors in k^n. The image is called the *affine span of S* in $\mathbb{A}^n(k)$ and denoted affspan(S). We leave it to the reader to verify this is closed (cf. Exercise 4.9.)

Example 4.17 Given distinct points $p_1, p_2 \in \mathbb{A}^2(\mathbb{R})$, affspan($p_1, p_2$) is the unique line joining them. Given distinct noncollinear points $p_1, p_2, p_3 \in \mathbb{A}^3(\mathbb{R})$, affspan($p_1, p_2, p_3$) is the unique plane containing them.

Proposition 4.18 *The set $S = \{p_1, \ldots, p_N\}$ imposes independent conditions on polynomials of degree ≤ 1 if and only if σ_S is injective. We say that S is in* linear general position.

Proof σ_S is not injective if there are distinct $(t_1, \ldots, t_N), (t'_1, \ldots, t'_N) \in \Delta_N$ with

$$t_1 p_1 + \cdots + t_N p_N = t'_1 p_1 + \cdots + t'_N p_N.$$

ELIMINATION

Reordering indices, if necessary, we can assume $t_1 \neq t_1'$; we can write

$$p_1 = \frac{t_2' - t_2}{t_1 - t_1'} p_2 + \cdots + \frac{t_N' - t_N}{t_1 - t_1'} p_N,$$

i.e., $p_1 \in \text{affspan}(\{p_2, \ldots, p_N\})$. It follows that every linear polynomial vanishing at $p_2, \ldots, p_N$ also vanishes at p_1 (see Exercise 4.9) so S fails to impose independent conditions on polynomials of degree ≤ 1.

Conversely, suppose S fails to impose independent conditions on polynomials of degree ≤ 1. After reordering, we find

$$I_1(p_2, \ldots, p_N) = I_1(p_1, p_2, \ldots, p_N),$$

which implies (see Exercise 4.9)

$$p_1 \in \text{affspan}(p_2, \ldots, p_N).$$

We can therefore write $p_1 = t_2 p_2 + \cdots + t_N p_N$ with $t_2 + \cdots + t_N = 1$. In particular,

$$\sigma_S(1, 0, \ldots, 0) = \sigma_S(0, t_2, \ldots, t_N)$$

so σ_S is not injective. $\square$

Definition 4.19 Given a variety $V \subset \mathbb{A}^n(k)$ and points $p_1, \ldots, p_N \in V$ in linear general position, $\text{affspan}(p_1, \ldots, p_N)$ is called an *N-secant subspace* to V.

Examples include 2-secant lines and 3-secant planes.

All the N-secants are contained in the image of the morphism

$$\sigma_N : \underbrace{V \times \ldots \times V}_{N \text{ times}} \times \Delta_N \to \mathbb{A}^n$$

$$(v_1, \ldots, v_N, (t_1, \ldots, t_N)) \mapsto t_1 v_1 + \cdots + t_N v_N.$$

The closure of the image is the *N-secant variety* of V

$$\text{Sec}_N(V) = \overline{\sigma_N(V \times \ldots \times V \times \Delta_N)}.$$

Example 4.20 (Secants of twisted cubic curves) Consider the curve $V \subset \mathbb{A}^3(k)$ satisfying the equations

$$\langle x_3 - x_1 x_2, x_2 - x_1^2 \rangle.$$

Show that $\text{Sec}_2(V) = \mathbb{A}^3(k)$.

We compute the image of $V \times V \times \Delta_2$ under σ_2. Let $\{z_1, z_2, z_3\}$ designate the coordinates on the first V and $\{w_1, w_2, w_3\}$ the coordinates on the second V. The homomorphism σ_N^* is induced by

$$x_i \mapsto t_1 z_i + t_2 w_i, \quad i = 1, 2, 3.$$

4.3 SECANT VARIETIES, JOINS, AND SCROLLS

The defining ideal of the graph is

$$J = \langle z_3 - z_1 z_2, z_2 - z_1^2, w_3 - w_1 w_2, w_2 - w_1^2, t_1 + t_2 - 1,$$
$$x_i - (t_1 z_i + t_2 w_i), i = 1, 2, 3 \rangle.$$

Take a Gröbner basis with respect to the product of graded reverse lexicographic orders on $\{t_1, t_2, w_1, w_2, w_3, z_1, z_2, z_3\}$ and $\{x_1, x_2, x_3\}$. This has more terms than can be produced here, but none of them involve just x_1, x_2, x_3. It follows that

$$J \cap k[x_1, x_2, x_3] = 0$$

and $\text{Sec}_2(V) = \mathbb{A}^3(k)$.

The computation is a bit easier if we use the parametrization $\mathbb{A}^1(k) \to V$ given in Example 1.5: $x_1 = t, x_2 = t^2, x_3 = t^3$. Then we get a morphism

$$\mathbb{A}^1 \times \mathbb{A}^1 \times \Delta_2 \to \mathbb{A}^3$$
$$(s, u, (t_1, t_2)) \mapsto (t_1 s + t_2 u, t_1 s^2 + t_2 u^2, t_1 s^3 + t_2 u^3)$$

with graph defined by

$$I = \langle x_1 - (t_1 s + t_2 u), x_2 - (t_1 s^2 + t_2 u^2), x_3 - (t_1 s^3 + t_2 u^3), t_1 + t_2 - 1 \rangle.$$

The Gröbner basis with respect to the product of graded reverse lexicographic orders on $\{t_1, t_2, s, u\}$ and $\{x_1, x_2, x_3\}$ has no terms involving x_1, x_2, x_3.

Our analysis of the twisted cubic curve suggests the following variation:

Definition 4.21 Let V be an affine variety and

$$\phi(1), \ldots, \phi(N) : V \to \mathbb{A}^n$$

morphisms to affine space. The *scroll*

$$\text{Scroll}(V; \phi(1), \ldots, \phi(N)) \subset \mathbb{A}^n$$

is defined as the closure of the image of

$$V \times \Delta_N \to \mathbb{A}^n$$
$$(v, (t_1, \ldots, t_N)) \mapsto t_1 \phi(1)(v) + \cdots + t_N \phi(N)(v).$$

Example 4.22 Let $V = \mathbb{A}^1$ with coordinate s and consider morphisms

$$\phi(1) : \mathbb{A}^1 \to \mathbb{A}^3 \qquad \phi(2) : \mathbb{A}^1 \to \mathbb{A}^3$$
$$s \to (s, 0, 0), \qquad\qquad s \to (1, 1, s).$$

Equations for $\text{Scroll}(V; \phi(1), \phi(2))$ are given by computing the intersection

$$k[x_1, x_2, x_3] \cap \langle x_1 - (t_1 s + t_2), x_2 - t_2, x_3 - t_2 s, t_1 + t_2 - 1 \rangle.$$

ELIMINATION

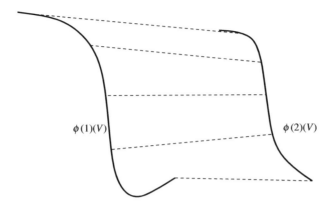

Figure 4.1 Scroll over two curves.

Compute a Gröbner basis using the product of graded reverse lexicographic orders on $\{t_1, t_2, s\}$ and $\{x_1, x_2, x_3\}$:

$$\left(-x_3 + x_1x_2 - x_2^2 + x_2x_3,\, s - x_1 + x_2 - x_3,\, -x_2 + t_2,\, x_2 + t_1 - 1\right).$$

The equation is

$$-x_3 + x_1x_2 - x_2^2 + x_2x_3 = 0.$$

Definition 4.23 Let $V(1), \ldots, V(N) \subset \mathbb{A}^n$ be affine varieties. The *join* $\mathrm{Join}(V(1), \ldots, V(N)) \subset \mathbb{A}^n$ is defined as the closure of the image of

$$V(1) \times V(2) \ldots \times V(N) \times \Delta_N \to \mathbb{A}^n$$
$$(v(1), \ldots, v(N), (t_1, \ldots, t_N)) \mapsto t_1 v(1) + \cdots + t_N v(N).$$

Let $V \subset \mathbb{A}^n$ be affine and $p \in \mathbb{A}^n$. The *cone over V with vertex p* is defined

$$\mathrm{Cone}(V, p) = \mathrm{Join}(V, p).$$

4.4 Exercises

4.1 The *cardioid* is defined as the curve $C \subset \mathbb{R}^2$ with parametric representation

$$x(\theta) = \cos\theta + \tfrac{1}{2}\cos 2\theta, \quad y(\theta) = \sin\theta + \tfrac{1}{2}\sin 2\theta, \quad 0 \le \theta < 2\pi.$$

Show that C can be defined by a polynomial equation $p(x, y) = 0$. *Hint*: Introduce auxiliary variables u and v satisfying $u^2 + v^2 = 1$. Express x and y as polynomials in u and v; eliminate u and v to get the desired equation in x and y.

4.2 Consider the image of the circle

$$V = \left\{(x_1, x_2) : x_1^2 + x_2^2 = 1\right\} \subset \mathbb{R}^2$$

4.4 EXERCISES

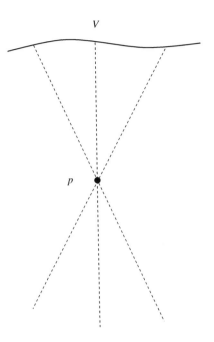

Figure 4.2 Cone over a curve V with vertex p.

under the map $\phi : V \to \mathbb{A}^2(\mathbb{R})$ given by

$$(x_1, x_2) \to \left(\frac{x_1}{1 + x_2^2}, \frac{x_1 x_2}{1 + x_2^2} \right).$$

Compute the equation(s) of the image. *Bonus:* Produce a nice graph of the real points of the lemniscate.

4.3 Consider the morphism

$$\phi : \mathbb{A}^1(\mathbb{C}) \to \mathbb{A}^2(\mathbb{C})$$
$$x \to (x^2, x^3).$$

Write equations for the graph of ϕ and compute the image of ϕ using elimination theory. *Advice:* Throw out superfluous generators from the ideal as you go along.

4.4 Suppose that $x, y \in \mathbb{C}$ and satisfy the relation

$$\sqrt[3]{x} + \sqrt{y} = 1,$$

where we allow each of the possible roots of x and y. Show that x and y satisfy the polynomial relation

$$x^2 - 2x - 6xy - y^3 + 3y^2 - 3y + 1 = 0.$$

ELIMINATION

Using a computer algebra system, extract the polynomial relation correponding to

$$\sqrt[3]{x} + \sqrt{y} = \sqrt{x+y}.$$

Hint: For the second problem, introduce auxilliary variables s, t and u with $s^3 = x$, $t^2 = y$, and $u^2 = x + y$.

4.5 Prove the *Descartes Circle Theorem*: Given four mutually tangent circles $C_1, \ldots, C_4 \subset \mathbb{R}^2$ with radii r_1, r_2, r_3, r_4. Take r_i to be negative if the other three circles are in the interior of C_i and positive otherwise. Show that

$$2\left(r_1^{-2} + r_2^{-2} + r_3^{-2} + r_4^{-2}\right) = \left(r_1^{-1} + r_2^{-1} + r_3^{-1} + r_4^{-1}\right)^2.$$

Hint: If (x_i, y_i) is the center of the ith circle, the relevant equations are

$$(x_i - x_j)^2 + (y_i - y_j)^2 = (r_i + r_j)^2.$$

4.6 Consider cubic polynomials

$$p(x) = x^3 + ax^2 + bx + c$$

over $\mathbb{C}$; we regard these as an affine space with coordinates (a, b, c). We say that p has a *double* (resp. *triple*) root if there exists an $\alpha \in \mathbb{C}$ such that

$$(x - \alpha)^2 | p(x) \text{ (resp. } (x - \alpha)^3 | p(x)).$$

(a) Find equations in a, b, c for the locus of cubic polynomials with a triple root.
Hint: We must have $a = -3\alpha, b = 3\alpha^2, c = -\alpha^3$.
(b) Find equations in a, b, c for the locus of cubic polynomials with a double root.
(c) Show that $x^3 + x^2 + x + 1$ has no multiple roots.

4.7 Consider the ideal

$$I = \langle x_1 + x_2 + x_3 - a, x_1 + 2x_2 + 4x_3 - b, x_1 - x_2 + x_3 - c \rangle$$

and the corresponding variety

$$V(I) \subset \mathbb{A}^6(\mathbb{C}).$$

Consider the projection morphism

$$\pi : V(I) \to \mathbb{A}^3$$
$$(x_1, x_2, x_3, a, b, c) \mapsto (a, b, c).$$

Determine the image of π.

4.8 Write down equations for the image and the graph of the rational maps

$$\mathbb{A}^4(\mathbb{Q}) \dashrightarrow \mathbb{A}^6(\mathbb{Q})$$

(a)
$$(x_1, x_2, x_3, x_4) \mapsto \left(\frac{1}{x_1 x_2}, \frac{1}{x_1 x_3}, \frac{1}{x_1 x_4}, \frac{1}{x_2 x_3}, \frac{1}{x_2 x_4}, \frac{1}{x_3 x_4}\right).$$

$$\rho : \mathbb{A}^2(k) \dashrightarrow \mathbb{A}^3(k)$$

(b)
$$(x_1, x_2) \mapsto \left(\frac{1}{x_1 + 1}, \frac{1}{x_1 + x_2}, \frac{x_2}{x_2^2 + 1}\right).$$

4.4 EXERCISES

4.9 For $S = \{p_1, \ldots, p_N\} \subset \mathbb{A}^n(k)$, show that
(a) $I_1(\text{affspan}(S)) = I_1(S)$, i.e., the polynomials of degree ≤ 1 vanishing on affspan(S) equal the polynomials of degree ≤ 1 vanishing on S;
(b) affspan(S) = $V(I_1(S))$.
Conclude affspan(S) is an affine-linear subspace (and hence a closed subset) of $\mathbb{A}^n(k)$.

4.10 Let $N \leq n + 1$ and consider the subset

$$U := \{(p_1, \ldots, p_N) : p_1, \ldots, p_N \text{ in linear general position}\}$$
$$\subset \underbrace{\mathbb{A}^n(k) \times \ldots \times \mathbb{A}^n(k)}_{N \text{ times}} \simeq \mathbb{A}^{nN}(k).$$

Show that U is open and nonempty. *Hint:* Write each $p_i = (a_{i1}, \ldots, a_{in})$ consider the $N \times (n+1)$ matrix

$$A = \begin{pmatrix} a_{11} & a_{12} & \cdots & a_{1n} & 1 \\ a_{21} & a_{22} & \cdots & a_{2n} & 1 \\ \vdots & \vdots & \vdots & \vdots & \vdots \\ a_{N1} & a_{N2} & \cdots & a_{Nn} & 1 \end{pmatrix}.$$

Argue that $p_1, \ldots, p_N$ are in linear general position if and only if some $N \times N$ minor of A has nonvanishing determinant.

4.11 Let $V \subset \mathbb{A}^4(k)$ denote the image of

$$\mathbb{A}^1(k) \to \mathbb{A}^4(k)$$
$$t \mapsto (t, t^2, t^3, t^4).$$

(a) Extract equations for $\overline{V}$ by computing

$$\langle x_1 - t, x_2 - t^2, x_3 - t^3, x_4 - t^4 \rangle \cap k[x_1, x_2, x_3, x_4].$$

(b) Show that $V = \overline{V}$.
(c) Show that $\text{Sec}_2(V)$ satisfies the equation

$$-x_2 x_4 + x_2^3 - 2x_1 x_2 x_3 + x_1^2 x_4 + x_3^2 = 0.$$

4.12 (a) Consider morphisms $\phi(i) : \mathbb{A}^2 \to \mathbb{A}^6$, $i = 1, 2, 3$

$$\phi_1(s_1, s_2) = (s_1, 0, 0, s_2, 0, 0),$$
$$\phi_2(s_1, s_2) = (0, s_1, 0, 0, s_2, 0),$$
$$\phi_3(s_1, s_2) = (0, 0, s_1, 0, 0, s_2).$$

Show that $\text{Scroll}(\mathbb{A}^2; \phi(1), \phi(2), \phi(3))$ is given by the 2×2 minors of the matrix

$$\begin{pmatrix} x_1 & x_2 & x_3 \\ x_4 & x_5 & x_6 \end{pmatrix}.$$

(b) Consider morphisms $\Phi(i) : \mathbb{A}^3 \to \mathbb{A}^6, i = 1, 2$ given by

$$\Phi(1)(r_1, r_2, r_3) = (r_1, 0, r_2, 0, r_3, 0),$$
$$\Phi(2)(r_1, r_2, r_3) = (0, r_1, 0, r_2, 0, r_3).$$

Compute equations for $\text{Scroll}(\mathbb{A}^3; \Phi(1), \Phi(2))$.

4.13 Consider morphisms

$$\phi(1) : \mathbb{A}^1 \to \mathbb{A}^3 \qquad \phi(2) : \mathbb{A}^1 \to \mathbb{A}^3$$
$$s \to (s, 0, 0), \qquad\qquad s \to (1, s, s^2).$$

Compute an equation for $\text{Scroll}(\mathbb{A}^1; \phi(1), \phi(2))$.

4.14 Let $\ell(1), \ell(2) \subset \mathbb{A}^3(\mathbb{R})$ be disjoint non-parallel lines. Show that $\text{Join}(\ell(1), \ell(2)) = \mathbb{A}^3(\mathbb{R})$.

4.15 Let $p = (0, 0, 0)$ and

$$V = V(x_3 - 1, x_1^2 + x_2^2 + x_3^2 - 2).$$

Write down an equation for $\text{Cone}(V, p)$.

5 Resultants

In this chapter, we develop criteria for deciding whether systems of equations have solutions. These take the form of polynomials in the coefficients of the equations that vanish whenever they can be solved. The prototype is the determinant of a system of linear equations: let $A = (a_{ij})$ be an $N \times N$ matrix with entries in a field k. The system $Ax = b$ can be solved for each $b \in k^N$ if and only if $\det(A) \neq 0$, in which case we can put $x = A^{-1}b$. However, in general our methods will not give an explicit formula for the solution.

For higher-degree equations, we allow solutions in some extension of k. Finding solutions in a given field like $k = \mathbb{Q}$ or $\mathbb{F}_p$ is really a problem of number theory rather than algebraic geometry.

Most of the algebraic techniques in this chapter apply over an arbitrary field. However, the geometric interpretations via elimination theory are valid only when k is infinite.

5.1 Common roots of univariate polynomials

We translate the search for common solutions to a problem in linear algebra, albeit over an infinite-dimensional space:

Proposition 5.1 *Consider polynomials*

$$f = a_m x^m + a_{m-1} x^{m-1} + \cdots + a_0, \quad g = b_n x^n + b_{n-1} x^{n-1} + \cdots + b_0 \in k[x]$$

of degrees m and n, i.e., with $a_m, b_n \neq 0$. The following conditions are equivalent:

- *f and g have a common solution over some extension of k;*
- *f and g share a common nonconstant factor $h \in k[x]$;*
- *there are no polynomials $A, B \in k[x]$ with $Af + Bg = 1$;*
- *$\langle f, g \rangle \subsetneq k[x]$.*

Proof We prove the first two are equivalent. Suppose f and g have a common solution $\alpha \in L$, where L/k is a field extension. Let $k[\alpha] \subset L$ denote the k-algebra generated by α. It is a quotient

$$q : k[x] \twoheadrightarrow k[\alpha]$$
$$x \mapsto \alpha$$

with kernel generated by a polynomial h (see Theorem A.9). Since $f(\alpha) = g(\alpha) = 0$, $f, g \in \langle h(x) \rangle$ and h is nonzero and divides f and g. But if h were constant then q would be zero, which is impossible.

Conversely, suppose that f and g share a common factor $h \in k[x]$. We may assume h is irreducible, so that $k[x]/\langle h \rangle$ is a field. Since f and g are in the kernel of the quotient homomorphism

$$k[x] \to k[x]/\langle h \rangle,$$

f and g both have roots over that field.

We prove the equivalence of the second and third conditions. If $h \mid f, g$ then $h \mid (Af + Bg)$ for any A, B, whence $Af + Bg \neq 1$. Conversely, assume $Af + Bg \neq 1$ for each $A, B \in k[x]$. Since $k[x]$ is a principal ideal domain (PID; see §A.5) we can write

$$\langle f, g \rangle := \langle h \rangle \subsetneq k[x]$$

and h is a divisor of f and g.

The last two conditions are equivalent because $\langle f, g \rangle = k[x]$ if and only $1 \in \langle f, g \rangle$. $\square$

Example 5.2 Consider the case $m = n = 1$, i.e.,

$$f = a_1 x + a_0, \quad g = b_1 x + b_0,$$

where $a_1, b_1 \neq 0$. These have common roots if and only if $a_1 b_0 - a_0 b_1 = 0$.

Example 5.3 (Geometric approach) Consider the variety $V \subset \mathbb{A}^5(k)$ given as

$$V = \{(x, a_0, a_1, b_0, b_1) : a_0 + a_1 x = b_0 + b_1 x = 0\}$$

and the projection

$$\pi : V \to \mathbb{A}^4(k)$$
$$(x, a_0, a_1, b_0, b_1) \mapsto (a_0, a_1, b_0, b_1).$$

The image $\pi(V)$ corresponds to the f and g that have common roots, as in Example 5.2. Note that the closure

$$\overline{\pi(V)} = \left\{ (a_0, a_1, b_0, b_1) : \det \begin{pmatrix} a_1 & a_0 \\ b_1 & b_0 \end{pmatrix} = 0 \right\}.$$

We shall pursue this further in Section 5.3.

5.1 COMMON ROOTS OF UNIVARIATE POLYNOMIALS

Definition 5.4 The *resultant* of polynomials of positive degrees

$$f = a_m x^m + a_{m-1} x^{m-1} + \cdots + a_0,$$
$$g = b_n x^n + b_{n-1} x^{n-1} + \cdots + b_0 \in k[x], a_m, b_n \neq 0,$$

is defined as the $(m+n) \times (m+n)$ determinant

$$\mathrm{Res}(f, g) = \det \begin{pmatrix} a_m & a_{m-1} & \cdots & a_1 & a_0 & 0 & \cdots & 0 \\ 0 & a_m & a_{m-1} & \cdots & a_1 & a_0 & \ddots & \vdots \\ \vdots & \ddots & \ddots & \ddots & \ddots & \ddots & \ddots & 0 \\ 0 & \cdots & 0 & a_m & a_{m-1} & \cdots & a_1 & a_0 \\ b_n & b_{n-1} & \cdots & b_1 & b_0 & 0 & \cdots & 0 \\ 0 & b_n & b_{n-1} & \cdots & b_1 & b_0 & \ddots & \vdots \\ \vdots & \ddots & \ddots & \ddots & \ddots & \ddots & \ddots & 0 \\ 0 & \cdots & 0 & b_n & b_{n-1} & \cdots & b_1 & b_0 \end{pmatrix}.$$

The first n rows involve the a_is and the last m rows involve the b_j's.

This is sometimes called the *Sylvester resultant*, in honor of James Joseph Sylvester (1814–1897).

Theorem 5.5 *Two nonconstant polynomials $f, g \in k[x]$ have a common factor if and only if $\mathrm{Res}(f, g) = 0$.*

Proof Our first task is to explain the determinantal form of the resultant. For each $d \geq 0$, let $P_{1,d}$ denote the polynomials in $k[x]$ of degree $\leq d$. Recall (see Exercise 1.4) that $\dim P_{1,d} = d + 1$ with distinguished basis $\{1, x, \ldots, x^d\}$. When $d < 0$ we define $P_{1,d} = 0$. Consider the linear transformation

$$\delta_0(d) : P_{1,d-m} \oplus P_{1,d-n} \to P_{1,d}$$
$$(A, B) \mapsto Af + Bg.$$

We have

$$\mathrm{image}(\delta_0(d)) = P_{1,d} \cap \langle f, g \rangle$$

so the following are equivalent:

- $\delta_0(d)$ is surjective for some $d \geq 0$;
- $1 \in \mathrm{image}(\delta_0(d))$ for some d;
- $\langle f, g \rangle = k[x]$.

We compute a matrix for $\delta_0(d)$. Since

$$\dim P_{1,d-m} \oplus P_{1,d-n} = 2d - m - n + 2$$

this is a $(d+1) \times (2d - m - n + 2)$ matrix. Write

$$A = r_{d-m}x^{d-m} + \cdots + r_0, \quad B = s_{d-n}x^{d-n} + \cdots + s_0$$

so that

$$Af + Bg = \sum_{j=0}^{d} x^{d-j} \left[\sum_{i_1+i_2=d-j} r_{i_1} a_{i_2} + s_{i_1} b_{i_2} \right].$$

We can represent

$$\delta_0(d)(A, B) = (r_{d-m}, \ldots, r_0, s_{d-n}, \ldots, s_0) \cdot$$

$$\begin{pmatrix}
\overbrace{a_m \quad a_{m-1} \quad \cdots \quad a_1 \quad a_0}^{m+1 \text{ columns}} \quad \overbrace{0 \quad \cdots \quad 0}^{d-m \text{ columns}} \\
0 \quad a_m \quad a_{m-1} \quad \cdots \quad a_1 \quad a_0 \quad \ddots \quad \vdots \\
\vdots \quad \ddots \quad \ddots \quad \ddots \quad \ddots \quad \ddots \quad \ddots \quad 0 \\
0 \quad \cdots \quad 0 \quad a_m \quad a_{m-1} \quad \cdots \quad a_1 \quad a_0 \\
b_n \quad b_{n-1} \quad \cdots \quad b_1 \quad b_0 \quad 0 \quad \cdots \quad 0 \\
0 \quad b_n \quad b_{n-1} \quad \cdots \quad b_1 \quad b_0 \quad \ddots \quad \vdots \\
\vdots \quad \ddots \quad \ddots \quad \ddots \quad \ddots \quad \ddots \quad \ddots \quad 0 \\
0 \quad \cdots \quad 0 \quad b_n \quad b_{n-1} \quad \cdots \quad b_1 \quad b_0 \\
\underbrace{}_{d-n \text{ columns}} \quad \underbrace{}_{n+1 \text{ columns}}
\end{pmatrix}
\begin{pmatrix} x^d \\ x^{d-1} \\ \vdots \\ \vdots \\ \vdots \\ \vdots \\ x \\ 1 \end{pmatrix}.$$

The linear transformation $\delta_0(d)$ is represented by a square matrix precisely when $d = m + n - 1$, in which case

$$\det(\delta_0(m + n - 1)) = \text{Res}(f, g). \tag{5.1}$$

The next ingredient is a precise form of Proposition 5.1:

Lemma 5.6 *Consider nonconstant polynomials*

$$f = a_m x^m + a_{m-1} x^{m-1} + \cdots + a_0, g = b_n x^n + b_{n-1} x^{n-1} + \cdots + b_0 \in k[x]$$

with $a_m, b_n \neq 0$. The following conditions are equivalent:

- *there exist polynomials $A, B \in k[x]$ with $Af + Bg = 1$;*
- *there exist polynomials $A, B \in k[x]$, with $\deg(A) \leq n - 1$ and $\deg(B) \leq m - 1$, such that $Af + Bg = 1$;*
- *for each polynomial $p \in k[x]$ of degree d, there exist $C, D \in k[x]$ with $\deg(C) \leq \max(d - m, n - 1)$ and $\deg(D) \leq \max(d - n, m - 1)$, such that $Cf + Dg = p$.*

The argument is quite similar to the proof of Theorem 2.34; we study possible cancellations in the expressions $Cf + Dg$.

5.1 COMMON ROOTS OF UNIVARIATE POLYNOMIALS

Proof of lemma It is clear that the third condition implies the second, and the second implies the first. We prove the first implies the third.

Suppose there exist $A, B \in k[x]$ with $Af + Bg = 1$. Since $(pA)f + (pB)g = p$ there exist polynomials $C, D \in k[x]$ with $Cf + Dg = p$. Choose these such that

$$M := \max(\deg(C) + m, \deg(D) + n)$$

is minimal. We claim that $M \leq \max(d, m+n-1)$. Assume on the contrary that $M > d$ and $M \geq m + n$. The leading terms of Cf and Dg are therefore of degree $> d$ and necessarily cancel

$$\text{LT}(C)\text{LT}(f) + \text{LT}(D)\text{LT}(g) = 0.$$

This cancellation implies that $\deg(C) + m = \deg(D) + n = M$; since $M \geq m + n$ it follows that $\deg(C) \geq n$ and $\deg(D) \geq m$.

We define polynomials

$$C' = C - (\text{LT}(C)/\text{LT}(g))g, \ D' = D - (\text{LT}(D)/\text{LT}(f))f \in k[x]$$

so that $\deg(C') < \deg(C)$, $\deg(D') < \deg(D)$, and

$$C'f + D'g = Cf + Dg - (fg)(\text{LT}(C)/\text{LT}(g) + \text{LT}(D)/\text{LT}(f)) = Cf + Dg = p.$$

Since

$$M' := \max(\deg(C') + m, \deg(D') + n) < M$$

we have a contradiction. □

We rephrase this in terms of the linear transformation $\delta_0(d)$. The following are equivalent:

- $1 \in \text{image}(\delta_0(d))$ for some d;
- $1 \in \text{image}(\delta_0(m + n - 1))$;
- $\delta_0(d)$ is surjective for each $d \geq m + n - 1$.

We complete the proof of Theorem 5.5: by Proposition 5.1, f and g have a common factor if and only if there exist no $A, B \in k[x]$ with $Af + Bg = 1$. It follows that $1 \notin \text{image}(\delta_0(m + n - 1))$, $\delta_0(m + n - 1)$ is not surjective, and $\det(\delta_0(m + n - 1)) = 0$; Equation 5.1 implies $\text{Res}(f, g) = 0$. Conversely, if $\text{Res}(f, g) = 0$ then $\delta_0(m + n - 1)$ is not surjective, and the previous lemma implies that

$$1 \notin \text{image}(\delta_0(d)) = \langle f, g \rangle \cap P_{1,d}$$

for any d. In particular, there exist no $A, B \in k[x]$ with $Af + Bg = 1$. □

5.1.1 Application to discriminants

Recall that α is a root of $f = a_m x^m + a_{m-1} x_{m-1} + \cdots + a_0$ if and only if $(x - \alpha) | f$; it is a *multiple root* if $(x - \alpha)^2 | f$. A polynomial f has a multiple root if and only if f and its derivative

$$f' = m a_m x^{m-1} + (m-1) a_{m-1} x^{m-2} + \cdots + 2 a_2 x + a_1$$

have a common root. Indeed, if $f = (x - \alpha)^e g$ where $g(\alpha) \neq 0$ then

$$f' = e(x - \alpha)^{e-1} g + (x - \alpha)^e g'.$$

If $e \geq 2$ then $f'(\alpha) = 0$; if $e = 1$ then $f'(\alpha) \neq 0$.

Therefore, a polynomial f has multiple roots only if $\mathrm{Res}(f, f') = 0$. For example, in the case $m = 2$ we have

$$\mathrm{Res}(f, f') = \det \begin{pmatrix} a_2 & a_1 & a_0 \\ 2a_2 & a_1 & 0 \\ 0 & 2a_2 & a_1 \end{pmatrix}$$
$$= a_2(-a_1^2 + 4 a_0 a_2).$$

In general, the leading term a_m occurs in each nonzero entry of the first column of the matrix computing $\mathrm{Res}(f, f')$, so we have

$$\mathrm{Res}(f, f') = (-1)^{m(m-1)/2} a_m \mathrm{disc}(f),$$

where $\mathrm{disc}(f)$ is a polynomial in $a_0, \cdots, a_m$ called the *discriminant* of f.

Remark 5.7 There is disagreement as to the sign of the discriminant: some books omit the power of (-1).

5.1.2 Resultants for homogeneous polynomials

In our definition of resultants, we required that the polynomials have non-vanishing leading terms. There is an alternate approach which does not require any assumptions on the coefficients.

Consider homogeneous forms in two variables

$$F = a_m x_0^m + \cdots + a_0 x_1^m, \quad G = b_n x_0^n + \cdots + b_0 x_1^n \in k[x_0, x_1]$$

of degrees m and n. We define $\mathrm{Res}(F, G)$ using the formula of Definition 5.4. We can reformulate Theorem 5.5 as follows:

Theorem 5.8 *Let $F, G \in k[x_0, x_1]$ be nonconstant homogeneous forms. Then $\mathrm{Res}(F, G) = 0$ if and only if F and G have a common nonconstant homogeneous factor.*

In particular, F and G have a nontrivial common zero over some extension of k. Here the trivial zero is $(0, 0)$.

Proof Write $f(x) = F(x, 1)$ and $g(x) = G(x, 1)$; these are called the *dehomogenizations* of F and G with respect to x_1.

Suppose that F and G have a common factor H. If H is divisible by x_1 then $a_m = b_n = 0$ and the first column of the matrix defining $\text{Res}(F, G)$ is zero, so the resultant vanishes. If x_1 does not divide H, then $h(x) = H(x, 1)$ is nonconstant and divides $f(x)$ and $g(x)$, so $\text{Res}(f, g) = 0$ by Theorem 5.5. Of course, $\text{Res}(F, G)$ vanishes as well.

Now suppose $\text{Res}(F, G) = 0$. If $a_m = b_n = 0$ then x_1 divides both F and G and we are done. If $a_m \neq 0$ and $b_n \neq 0$ then

$$\text{Res}(f, g) = \text{Res}(F, G),$$

so f and g have a common factor $h = c_d x^d + \cdots + c_0, c_d \neq 0$. The homogeneous polynomial

$$H = c_d x_0^d + c_{d-1} x_0^{d-1} x_1 + \cdots + c_0 x_1^d$$

divides F and G. (This is called the *homogenization* of h.)

We therefore focus on the case where just one of the leading coefficients is zero, e.g., $a_m = 0$ but $b_n \neq 0$. Then we can express

$$F(x_0, x_1) = x_1 E(x_0, x_1), \quad E(x_0, x_1) = a_{m-1} x_0^{m-1} + a_{m-2} x_0^{m-2} x_1 + \cdots + a_0 x_1^{m-1}.$$

Compute $\text{Res}(F, G)$ using expansion-by-minors along the first column. The only nonzero entry in that column is b_n (in the $(n+1)$th row) and the corresponding minor is

$$\begin{pmatrix} a_{m-1} & \cdots & a_1 & a_0 & 0 & \cdots & 0 \\ 0 & a_{m-1} & \cdots & a_1 & a_0 & \ddots & \vdots \\ \vdots & \ddots & \ddots & \ddots & \ddots & \ddots & 0 \\ 0 & \cdots & 0 & a_{m-1} & \cdots & a_1 & a_0 \\ b_n & b_{n-1} & \cdots & b_1 & b_0 & \ddots & \vdots \\ \ddots & \ddots & \ddots & \ddots & \ddots & \ddots & 0 \\ \cdots & 0 & b_n & b_{n-1} & \cdots & b_1 & b_0 \end{pmatrix}$$

which is the matrix for $\text{Res}(E, G)$. We therefore have

$$\text{Res}(F, G) = (-1)^n b_n \text{Res}(E, G),$$

which implies $\text{Res}(E, G) = 0$. If $a_{m-1} \neq 0$ then we stop; otherwise, we iterate until we obtain

$$F(x_0, x_1) = x_1^e E_e(x_0, x_1), \quad E_e(x_0, x_1) = a_{m-e} x_0^{m-e} + \cdots + a_0 x_1^{m-e}$$

with $a_{m-e} \neq 0$ and $\text{Res}(E_e, G) = 0$. Then the previous argument applies, and we find a common factor H dividing E_e and G. □

5.2 The resultant as a function of the roots

We continue our discussion of the resultant of two polynomials

$$f = a_m x^m + \cdots + a_0, \quad g = b_n x^n + \cdots + b_0$$

by finding an expression for $\mathrm{Res}(f, g)$ in terms of the roots of f and g. Pass to a finite extension of k over which we have factorizations (see Theorem A.17 and Exercise A.14)

$$f = a_m \prod_{i=1}^m (x - \alpha_i), \quad g = b_n \prod_{j=1}^n (x - \beta_j).$$

The coefficients can be expressed in terms of the roots

$$a_{m-k} = (-1)^k a_m \sum_{i_1 < \ldots < i_k} \alpha_{i_1} \ldots \alpha_{i_k},$$

$$b_{n-k} = (-1)^k b_n \sum_{j_1 < \ldots < j_k} \beta_{j_1} \ldots \beta_{j_k}.$$

These yield a k-algebra homomorphism

$$\psi : k[a_0, \ldots, a_m, b_0, \ldots, b_n] \to k[a_m, b_n, \alpha_1, \ldots, \alpha_m, \beta_1, \ldots, \beta_n]$$

and a morphism

$$\phi : \mathbb{A}^{m+n+2} \to \mathbb{A}^{m+n+2}$$

with $\phi^* = \psi$. Write $R = \psi(\mathrm{Res}(f, g))$, i.e., the resultant written as a function of the roots.

Example 5.9 When $m = 1$ and $n = 2$ we have

$$f = a_1(x - \alpha_1), \quad g = b_2(x - \beta_1)(x - \beta_2)$$

so that

$$R = \det \begin{pmatrix} a_1 & -a_1\alpha_1 & 0 \\ 0 & a_1 & -a_1\alpha_1 \\ b_2 & -b_2(\beta_1 + \beta_2) & b_2\beta_1\beta_2 \end{pmatrix}$$

$$= a_1^2 b_2 (\beta_1 - \alpha_1)(\beta_2 - \alpha_1).$$

Proposition 5.10 R is divisible by

$$S := a_m^n b_n^m \prod_{\substack{i=1,\ldots,m \\ j=1,\ldots,n}} (\alpha_i - \beta_j).$$

5.2 THE RESULTANT AS A FUNCTION OF THE ROOTS

Proof   The determinant defining R has n rows divisible by a_m and m rows divisible by b_n, so R is divisible by $a_m^n b_n^m$. If some root α_i of f equals some root β_j of g then R is zero. This implies that $\alpha_i - \beta_j$ divides R as well.

The polynomial ring $k[a_m, b_n, \alpha_1, \ldots, \alpha_m, \beta_1, \ldots, \beta_n]$ is a unique factorization domain (UFD; see Theorem A.14), so each element has a unique expression as a product of irreducible elements. Since a_m, b_n, and the $\alpha_i - \beta_j$ are irreducible and distinct, the product S is part of the factorization for R.   □

Proposition 5.11   *The polynomials S and R coincide up to a constant factor.*

Proof   The homomorphism ψ admits factorizations

$$
\begin{array}{ccc}
k[a_0, \ldots, a_m, b_0, \ldots, b_n] & \xrightarrow{\sigma(\alpha)} & k[a_m, \alpha_1, \ldots, \alpha_m, b_0, \ldots, b_n] \\
\sigma(\beta) \downarrow & & \downarrow \tau(\beta) \\
k[a_0, \ldots, a_m, b_n, \beta_1, \ldots, \beta_n] & \xrightarrow{\tau(\alpha)} & k[a_m, b_n, \alpha_1, \ldots, \beta_n].
\end{array}
$$

All these homomorphisms are injective (see Exercise 5.5). We write $R_a = \sigma(\beta)(\text{Res}(f, g))$ and $R_b = \sigma(\alpha)(\text{Res}(f, g))$.

We express the polynomial S in two different ways:

$$S = [a_m \prod_{i=1}^m (\beta_1 - \alpha_i)] \ldots [a_m \prod_{i=1}^m (\beta_n - \alpha_i)](-1)^{mn} b_n^m$$

$$= \tau(\alpha)\left(b_n^m (-1)^{mn} f(\beta_1) \ldots f(\beta_n)\right) := \tau(\alpha)(S_a)$$

$$S = a_m^n \left[b_n \prod_{j=1}^n (\alpha_1 - \beta_j)\right] \ldots \left[b_n \prod_{j=1}^n (\alpha_m - \beta_j)\right]$$

$$= \tau(\beta)\left(a_m^n g(\alpha_1) \ldots g(\alpha_m)\right) := \tau(\beta)(S_b)$$

The first representation shows that S_a is homogeneous of degree n in the $a_0, \ldots, a_m$. However, R_a is also homogeneous of degree n in $a_0, \ldots, a_m$; we find

$$R = p(b_n, \beta_1, \ldots, \beta_n)S.$$

The second representation shows that S_b is homogeneous of degree m in the b_j. Since R_b is also of degree m in the b_j, we conclude p is constant.   □

Proposition 5.12   *$\text{Res}(f, g)$ is irreducible.*

Proof   Suppose we can write the resultant as a product of nonconstant polynomials

$$\text{Res}(f, g) = P_1 P_2, \quad P_1, P_2 \in k[a_0, \ldots, a_m, b_0, \ldots b_n].$$

Up to multiplying by a constant, we have

$$\psi(P_1)\psi(P_2) = a_m^n b_n^m \prod_{i,j}(\alpha_i - \beta_j).$$

The coefficients a_i and b_j are symmetric in the roots $\alpha_1, \ldots, \alpha_m$ and $\beta_1, \ldots, \beta_n$ respectively. Thus both $\psi(P_1)$ and $\psi(P_2)$ are symmetric in both the α_i and the β_j. In particular, if just *one* $\alpha_i - \beta_j$ divides $\psi(P_1)$, the entire product

$$\prod_{i=1,\cdots,m,\, j=1,\cdots,n}(\alpha_i - \beta_j)$$

also does. (We are using the fact that polynomial rings are UFDs.) It follows that

$$\mathrm{Res}(f,g) \mid a_m^p b_n^q P_1$$

for suitable p, q.

However, a_m and b_n do not divide $\mathrm{Res}(f,g)$. Indeed, if $a_m | \mathrm{Res}(f,g)$ then the resultant would vanish whenever $a_m = 0$, but when

$$a_m = \ldots = a_1 = b_{n-1} = \ldots = b_0 = 0, \quad a_0 = b_n = 1$$

it can be evaluated directly via row operations

$$\mathrm{Res}(f,g) = \pm 1.$$

We conclude then that $\mathrm{Res}(f,g) | P_1$. $\square$

5.3 Resultants and elimination theory

Consider the variety $V \subset \mathbb{A}^{m+n+3}(k)$ given as

$$V = \{(x, a_0, \ldots, a_m, b_0, \ldots, b_n) : a_0 + a_1 x + \cdots + a_m x^m = b_0 + b_1 x + \cdots + b_n x^n = 0\}$$

and the projection

$$\pi : V \to \mathbb{A}^{m+n+2}(k)$$
$$(x, a_0, \ldots, a_m, b_0, \ldots, b_n) \mapsto (a_0, \ldots, a_m, b_0, \ldots, b_n).$$

Our goal is to describe the closure $\overline{\pi(V)}$ using elimination techniques (e.g. Theorem 4.3):

Theorem 5.13 *If $I = \langle f, g \rangle \subset k[x, a_0, \ldots, a_m, b_0, \ldots, b_n]$ is the ideal generated by*

$$f = a_m x^m + a_{m-1} x^{m-1} + \cdots + a_0, \quad g = b_n x^n + b_{n-1} x^{n-1} + \cdots + b_0$$

then

$$I \cap k[a_0, \ldots, a_m, b_0, \ldots, b_n] = \langle \mathrm{Res}(f,g) \rangle.$$

5.3 RESULTANTS AND ELIMINATION THEORY

Remark 5.14   In principle, one could prove the theorem directly with Gröbner bases by eliminating the variable x. For instance, in the case $m = 2, n = 3$ the resultant is the last term of the Gröbner basis below.

$\langle b_3 x^3 + b_2 x^2 + b_1 x + b_0, a_2 x^2 + a_1 x + a_0,$
$\quad x^2 b_3 a_1 + x b_3 a_0 - a_2 b_1 x - a_2 b_0 + b_2 a_1 x + b_2 a_0,$
$\quad - a_1 b_0 + x^2 b_3 a_0 - x a_2 b_0 + x b_2 a_0 + b_1 a_0,$
$\quad b_3 a_1^2 x + b_3 a_1 a_0 - a_2 x b_3 a_0 + a_2^2 b_1 x + a_2^2 b_0 - a_2 b_2 a_1 x - a_2 b_2 a_0,$
$\quad - a_2 b_1 a_0 + x b_3 a_1 a_0 + x a_2^2 b_0 - x a_2 b_2 a_0 + b_3 a_0^2 + a_1 a_2 b_0,$
$\quad - a_2 b_0 a_0 - x a_2 b_1 a_0 + x b_3 a_0^2 + b_2 a_0^2 + a_2^2 b_0 + a_2 b_0 a_1 x - a_1 b_1 a_0,$
$\quad b_3 a_0^2 a_2 b_0 + b_3 a_0^2 x a_2 b_1 - b_3^2 a_0^3 x - b_3 a_0^3 b_2 - 2 b_3 a_0 a_1^2 b_0$
$\quad + 2 b_3 a_0^2 a_1 b_1 - b_0 b_3 a_1^3 x - a_1 a_2^2 b_0^2 + 2 a_1 b_0 a_2 b_2 a_0 - a_1 a_2 b_1^2 a_0$
$\quad + b_1 x b_3 a_1^2 a_0 + b_1 a_1^2 a_2 b_0 - a_1 b_2 x b_3 a_0^2 - a_1 b_2^2 a_0^2 - b_2 a_1^3 b_0 + b_2 a_1^2 b_1 a_0,$
$\quad - b_0 b_3 a_1^2 x - b_0 b_3 a_1 a_0 + b_0 a_2 x b_3 a_0 - a_2^2 b_0^2$
$\quad + 2 b_0 a_2 b_2 a_0 - a_2 b_1^2 a_0 + b_1 x b_3 a_1 a_0 + b_1 b_3 a_0^2$
$\quad + b_1 a_1 a_2 b_0 - b_2 x b_3 a_0^2 - b_2^2 a_0^2 - b_2 a_1^2 b_0 + b_2 a_1 b_1 a_0,$
$\quad - 2 a_1 b_0^2 a_2 b_2 a_0 - b_1 a_0 a_2^2 b_0^2 + b_1^2 a_0^2 b_2 a_1 - b_3 a_0^2 a_2 b_0^2 + 2 b_3 a_0 a_1^2 b_0^2 + b_0^2 b_3 a_1^3 x$
$\quad - b_1 a_1^2 a_2 b_0^2 + b_0 b_3^2 a_0^3 x + b_0 b_3 a_0^3 b_2 + b_0 a_1 b_2^2 a_0^2 + b_0 a_1 b_2 x b_3 a_0^2$
$\quad + 2 b_1 a_0^2 b_0 a_2 b_2 + b_1^2 a_0^2 x b_3 a_1 + 2 b_1^2 a_0 a_1 a_2 b_0 - b_1 a_0^3 b_2 x b_3 - 2 b_1 a_0 b_2 a_1^2 b_0$
$\quad - 2 b_1 a_0 b_0 b_3 a_1^2 x - b_1^3 a_0^2 a_2 + b_1^2 a_0^3 b_3 - b_1 a_0^3 b_2^2 + a_1 a_2^2 b_0^3 + b_2 a_1^3 b_0^2 - 3 b_3 a_0^2 b_1 a_1 b_0,$
$\quad - 2 b_3 a_0^2 a_2 b_1 + b_3^2 a_0^3 + 3 a_2 b_0 a_1 b_3 a_0 + a_2^2 b_0^2 - 2 b_0 a_2^2 b_2 a_0 + a_2^2 b_1^2 a_0$
$\quad - b_1 a_1 a_2^2 b_0 + a_2 b_2^2 a_0^2 + a_2 b_2 a_1^2 b_0 - a_2 b_2 a_1 b_1 a_0 - a_1 b_2 b_3 a_0^2 - b_0 b_3 a_1^3 + b_1 b_3 a_1^2 a_0 \rangle$

Proof of Theorem 5.13   We first establish

$$\mathrm{Res}(f, g) \in I \cap k[a_0, \ldots, a_m, b_0, \ldots, b_n].$$

This is an application of *Cramer's Rule*: Let M be an $N \times N$ matrix with entries taken from a ring R. The *classical adjoint* $\mathrm{ad}(M)$ is the matrix with entries

$$\mathrm{ad}(M)_{i,j} = (-1)^{i+j} \det(M_{j,i}),$$

where $M_{j,i}$ is the $(N - 1) \times (N - 1)$ minor obtained by removing the jth row and ith column. For example, when $N = 2$

$$M = \begin{pmatrix} m_{11} & m_{12} \\ m_{21} & m_{22} \end{pmatrix}, \quad \mathrm{ad}(M) = \begin{pmatrix} m_{22} & -m_{12} \\ -m_{21} & m_{11} \end{pmatrix}.$$

There is a universal identity

$$M \, \mathrm{ad}(M) = \mathrm{ad}(M) M = \det(M) \mathrm{Id}.$$

In particular, for *any* $v \in R^N$ we can write

$$\det(M) v = M \cdot \mathrm{ad}(M) v \in \mathrm{image}(M).$$

When det(M) is invertible, we obtain the formula

$$M^{-1} = \text{ad}(M)/\det(M),$$

which is Cramer's Rule from linear algebra.

Apply Cramer's Rule to the matrix for $\delta_0(m+n-1)$ to obtain

$$\text{Res}(f,g)1 = \delta_0(m+n-1) \cdot \text{ad}(\delta_0(m+n-1))1 \in \langle f, g \rangle.$$

Since $\text{Res}(f,g)$ is a polynomial in the a_i and b_j, the proof of the first part is complete.

We prove the reverse inclusion of Thereom 5.13:

$$I \cap k[a_0, \ldots, a_m, b_0, \ldots, b_n] \subset \langle \text{Res}(f,g) \rangle.$$

Again given $P \in I \cap k[a_0, \ldots, b_n]$, after substituting the roots α_i and β_j we obtain a polynomial $\psi(P)$ vanishing whenever some $\alpha_i = \beta_j$. Thus

$$\prod_{i=1,\ldots,m,\, j=1,\ldots,n} (\alpha_i - \beta_j) | \psi(P)$$

so $\text{Res}(f,g) | a_m^p b_n^q P$ for some p, q. Since a_m and b_n do not divide the resultant, we conclude that $\text{Res}(f,g) | P$. $\square$

Remark 5.15 The trick here – replacing a polynomial by its factorization – is an example of *faithfully flat base change* or passing to a *faithfully flat neighborhood*. Intuitively, the roots of a polynomial can be expressed locally as a function of the coefficients. For example, when $m = 2$ and $k = \mathbb{C}$ we can write

$$\alpha_1 = \frac{-a_1 + \sqrt{a_1^2 - 4a_0a_2}}{2a_2}, \quad \alpha_2 = \frac{-a_1 - \sqrt{a_1^2 - 4a_0a_2}}{2a_2};$$

this is only valid in a neighborhood of $[f] \in \mathbb{A}^3(\mathbb{C}) \setminus \{a_2 \neq 0\}$, as there is no consistent way to choose the sign of the square root over the entire complex plane.

Writing the roots as explicit functions of the coefficients is tricky in general, even with the help of complex analysis. Algebraic geometers formally introduce new variables to represent the roots. This approach works well over more general base fields.

5.4 Remarks on higher-dimensional resultants

Suppose that $F_1, F_2, F_3 \in k[x_0, x_1, x_2]$ are homogeneous polynomials of degrees m_1, m_2, m_3. When do the equations

$$F_1 = F_2 = F_3 = 0$$

admit nontrivial common solutions over some extension of k?

5.4 REMARKS ON HIGHER-DIMENSIONAL RESULTANTS

In light of our analysis in the univariate case, it is natural to consider the linear transformation

$$\delta_0(d) : S_{d-m_1} \oplus S_{d-m_2} \oplus S_{d-m_3} \to S_d$$
$$(A_1, A_2, A_3) \mapsto A_1 F_1 + A_2 F_2 + A_3 F_3$$

where S_d is the vector space of homogeneous polynomials of degree d in x_0, x_1, x_2 and (cf. Exercise 1.4)

$$\dim S_d = \binom{d+2}{d} = \frac{(d+2)(d+1)}{2}.$$

Naively, one might expect image($\delta_0(d)$) to have dimension

$$\binom{d-m_1+2}{2} + \binom{d-m_2+2}{2} + \binom{d-m_3+2}{2}.$$

However, this does not take into account syzygies among F_1, F_2, F_3, which correspond to elements in ker $\delta_0(d)$ for various values of d. For example, we have the following obvious syzygies

$$(F_2, -F_1, 0), \quad (0, F_3, -F_2), \quad (F_3, 0, -F_1).$$

Therefore, defining

$$\delta_1(d) : S_{d-m_1-m_2} \oplus S_{d-m_2-m_3} \oplus S_{d-m_3-m_1} \to S_{d-m_1} \oplus S_{d-m_2} \oplus S_{d-m_3}$$
$$(A_{12}, A_{23}, A_{31}) \mapsto (A_{12} F_2 + A_{31} F_3, -A_{12} F_1 + A_{23} F_3, -A_{23} F_2 - A_{31} F_1)$$

we have

$$\text{image}(\delta_1(d)) \subset \ker(\delta_0(d)).$$

Less naively, one might expect image($\delta_0(d)$) to have dimension

$$\binom{d-m_1+2}{2} + \binom{d-m_2+2}{2} + \binom{d-m_3+2}{2}$$
$$- \binom{d-m_1-m_2+2}{2} - \binom{d-m_2-m_3+2}{2} - \binom{d-m_3-m_1+2}{2}.$$

There are still syzygies among the syzygies! We have

$$\delta_2(d) : S_{d-m_1-m_2-m_3} \to S_{d-m_1-m_2} \oplus S_{d-m_2-m_3} \oplus S_{d-m_3-m_1}$$
$$(A_{123}) \mapsto (A_{123} F_1, A_{123} F_2, A_{123} F_3)$$

so that

$$\text{image}(\delta_2(d)) \subset \ker(\delta_1(d)).$$

These force $\ker(\delta_1(d)) \neq 0$ when $d \geq m_1 + m_2 + m_3$.

Remark 5.16 In the discussion above we are essentially defining the *Koszul complex* associated to F_1, F_2, F_3:

$$0 \longrightarrow S_{d-m_1-m_2-m_3} \xrightarrow{\delta_2(d)} S_{d-m_1-m_2} \oplus S_{d-m_2-m_3} \oplus S_{d-m_3-m_1}$$
$$\xrightarrow{\delta_1(d)} S_{d-m_1} \oplus S_{d-m_2} \oplus S_{d-m_3} \xrightarrow{\delta_0(d)} S_d.$$

This is one place where cohomological methods have profoundly influenced algebraic geometry. For more information, see [9, ch. 17] (cf. [12, p. 52]).

We expect the image of $\delta_0(d)$ should have dimension

$$\Xi(d, m_1, m_2, m_3)$$
$$:= \binom{d-m_1+2}{2} + \binom{d-m_2+2}{2} + \binom{d-m_3+2}{2}$$
$$- \binom{d-m_1-m_2+2}{2} - \binom{d-m_2-m_3+2}{2} - \binom{d-m_3-m_1+2}{2}$$
$$+ \binom{d-m_1-m_2-m_3+2}{2}.$$

We shall see more expressions like this in Chapter 12, when we discuss Hilbert polynomials and the Bezout Theorem. Observe that the identity

$$\binom{m_1+m_2+m_3}{2} = \binom{m_2+m_3}{2} + \binom{m_1+m_3}{2} + \binom{m_1+m_2}{2}$$
$$- \binom{m_3}{2} - \binom{m_1}{2} - \binom{m_2}{2} \quad (5.2)$$

implies

$$\Xi(m_1+m_2+m_3-2, m_1, m_2, m_3) = \dim S_{m_1+m_2+m_3-2}.$$

We therefore focus on the case $d = m_1 + m_2 + m_3 - 2$, where $\delta_2(d) = 0$.

Our generalized resultant should define the locus where $\delta_0(m_1 + m_2 + m_3 - 2)$ fails to have 'expected' rank. There is a generalization of the determinant, the *determinant of a complex*, which can be applied in this context. This approach, pioneered by Arthur Cayley (1821–1895) [6] (see also [28] or [12, ch. 14]), yields a resultant which can be expressed as a quotient of determinants of large matrices.

There has been quite a bit of recent work on finding simple determinantal formulas for resultants of multivariate polynomials (see [23], for example.)

Bibliographic note: Our discussion is inspired by early editions of van der Waerden's *Moderne Algebra* [39]. However, the fourth German edition of Volume II expunges elimination theory, which gives context for the quote by Abhyankar in the previous chapter.

5.5 Exercises

5.1 Consider the polynomials $f = x^2 + 3x + 1$ and $g = x^2 - 4x + 1$. Compute the resultant Res(f, g). Do the polynomials have a common zero over $\mathbb{C}$? Also compute Res(fg^2, f^2g).

5.2 Compute the discriminant of the polynomial
$$f(x) = x^4 + px + q, \quad p, q \in k.$$

5.3 Consider the polynomial
$$x^2 + 2xy^2 + y + 1$$
as a polynomial in x with coefficients in $\mathbb{C}[y]$. Compute the discriminant of x. How do you interpret its roots?

5.4 Consider polynomials
$$f = a_m x^m + a_{m-1} x^{m-1} + \cdots + a_0, \quad g = b_n x^n + b_{n-1} x^{n-1} + \cdots + b_0$$
with $a_m \neq 0$ and $b_n \neq 0$. Suppose that the $(m+n) \times (m+n)$ matrix defining Res(f, g) has rank $m + n - 1$. Show that f and g share a common linear factor, but not a common quadratic factor.

5.5 Let L be a field and consider the L-algebra homomorphism
$$\sigma(\alpha) : L[a_0, \ldots, a_m] \to L[a_m, \alpha_1, \ldots, \alpha_m]$$
$$a_{m-k} \mapsto (-1)^k a_m \sum_{i_1 < \cdots < i_k} \alpha_{i_1} \cdots \alpha_{i_k}$$
for $k = 1, \cdots, m$.
(a) Show this is injective. *Hint:* Apply the method of Exercise 2.12.
(b) Deduce the same conclusion when L is a domain.

5.6 Show that the polynomial
$$a_m x^m + a_{m-1} x^{m-1} + \cdots + a_0 \in k[x, a_0, \cdots, a_m]$$
is irreducible. *Hint:* Introduce its roots as formal variables.

5.7 Consider a matrix
$$T = \begin{pmatrix} t_{00} & t_{01} \\ t_{10} & t_{11} \end{pmatrix}$$
acting on homogeneous forms of degree d in x_0, x_1:
$$T^* : S_d \to S_d$$
$$F(x_0, x_1) \mapsto F(x_0 t_{00} + x_1 t_{10}, x_0 t_{01} + x_1 t_{11}).$$

Show that T^* is linear (cf. Exercise 1.11) and compute its determinant in terms of det(T).

5.8 Consider homogeneous polynomials

$$F(x_0, x_1) = a_m x_0^m + \cdots + a_0 x_1^m, \quad G(x_0, x_1) = b_n x_0^n + \cdots + b_0 x_1^n \in k[x_0, x_1].$$

(a) Set

$$\widehat{F}(x_0, x_1) = F(x_1, x_0), \quad \widehat{G}(x_0, x_1) = G(x_1, x_0)$$

and show that

$$\operatorname{Res}(\widehat{F}, \widehat{G}) = (-1)^{\epsilon(m,n)} \operatorname{Res}(F, G),$$
$$\epsilon(m, n) = \tfrac{m(m-1) + n(n-1) + (m+n)(m+n-1)}{2}.$$

(b) Given $t \in k$, set

$$F_t(x_0, x_1) = F(x_0 + tx_1, x_1), \quad G_t(x_0, x_1) = G(x_0 + tx_1, x_1)$$

and show that

$$\operatorname{Res}(F_t, G_t) = \operatorname{Res}(F, G).$$

Hint: Exercise 5.7 might be useful. If the general case is tricky, work out $(m, n) = (1, 1), (1, 2),$ and $(2, 2)$.

5.9 Verify the combinatorial identity (5.2).

5.10 Consider the polynomials

$$f = x^2 + a_{10} x + a_{01} y + a_{00},$$
$$g = xy + b_{10} x + b_{01} y + b_{00},$$
$$h = y^2 + c_{10} x + c_{01} y + c_{00}$$

over a field k. Find a nonzero polynomial

$$R \in k[a_{10}, a_{01}, a_{00}, b_{10}, b_{01}, b_{00}, c_{10}, c_{01}, c_{00}]$$

such that $R = 0$ whenever $f = g = h = 0$ has a common solution over some extension of k. *Hint:* Use a computer algebra system to analyze

$$\langle f, g, h \rangle \cap k[a_{10}, a_{01}, a_{00}, b_{10}, b_{01}, b_{00}, c_{10}, c_{01}, c_{00}].$$

6 Irreducible varieties

Factorization is ubiquitous in algebra and number theory. We decompose integers as products of prime numbers and polynomials into irreducible factors. Special techniques are available for analyzing the resulting irreducible objects. Here we shall develop geometric notions of irreducibility applicable in algebraic geometry. We can decompose arbitrary varieties into *irreducible components*, which are generally much easier to understand. Surprisingly, these notions are more robust than traditional algebraic factorization techniques, which are most effective for special classes of rings. In Chapter 8 we will return to the topic of algebraic factorization, with a view to translating the techniques developed here into algebraic terms.

Recall that the polynomial ring $k[x_1, \ldots, x_n]$ is a unique factorization domain (see Theorem A.14): each nonconstant $f \in k[x_1, \ldots, x_n]$ can be written as a product of irreducible polynomials

$$f = f_1 f_2 \cdots f_R.$$

This representation is unique, up to permuting the factors or rescaling them by constants. The factors need not be distinct, i.e., two of them might be proportional. However, we can choose distinct factors $f_{j_1}, \ldots, f_{j_r}$ so that

$$f = c f_{j_1}^{e_1} \cdots f_{j_r}^{e_r},$$

where $c \in k^*$ and e_i is the number of times f_{j_i} appears in the factorization.

We recast this in geometric terms. The hypersurface

$$V(f) = \{(a_1, \ldots, a_n) : f(a_1, \ldots, a_n) = 0\}$$

can be expressed as a union

$$V(f) = V(f_{j_1}) \cup \cdots \cup V(f_{j_r}).$$

Indeed, f vanishes if and only if one of its factors f_j vanishes. Of course, $V(cf_j) = V(f_j)$ and $V(c) = \emptyset$ when $c \in k^*$.

IRREDUCIBLE VARIETIES

In this chapter, we put such decompositions into a general framework. Again, we assume the base field k is infinite (cf. Remark 6.3).

6.1 Existence of the decomposition

Definition 6.1 A variety V is *reducible* if we can write it as a union of closed proper subsets

$$V = V_1 \cup V_2, \quad V_1, V_2 \subsetneq V.$$

It is *irreducible* if there is no such representation.

If V is irreducible and $V = \cup_{i=1}^r V_i$, a union of closed subsets, then $V = V_i$ for some i.

Example 6.2

1. The affine variety $V = \{(x, y) : xy(x - y)(x + y) = 0\} \subset \mathbb{A}^2(\mathbb{Q})$ is reducible

$$V = \{x = 0\} \cup \{y = 0\} \cup \{x = y\} \cup \{x = -y\}.$$

2. Any finite set $\{p_1, \ldots, p_n\}$, $n > 1$, is reducible; it is a union of singleton sets.

Remark 6.3 When k is finite our definition is problematic: any subset of $\mathbb{A}^n(k)$ is the union of its points!

Theorem 6.4 *Let $V \subset \mathbb{A}^n(k)$ be a variety. Then we can write*

$$V = V_1 \cup V_2 \cup \cdots \cup V_r$$

as a finite union of irreducible closed subsets. This representation is unique up to permutation provided it is irredundant, *i.e., $V_i \not\subset V_j$ for any $i \neq j$.*

The V_j are called the *irreducible components* of V.

Proof (Existence) We consider the process of decomposing a variety into a union of closed proper subsets:

$$V = W_1 \cup W_1', \quad W_1, W_1' \subsetneq V.$$

Since V is not irreducible, such a decomposition is possible. Either W_1 and W_1' are irreducible or, after reordering, we can write

$$W_1 = W_2 \cup W_2', \quad W_2, W_2' \subsetneq W_1.$$

There are two possibilities. Either this process terminates with an expression of V as a union of irreducibles, or there is an infinite descending sequence of closed subsets

$$V \supsetneq W_1 \supsetneq W_2 \cdots$$

and an infinite ascending sequence of ideals

$$I(V) \subsetneq I(W_1) \subsetneq I(W_2) \cdots .$$

However, such a sequence violates the 'ascending chain condition' for ideals, derived from the Hilbert Basis Theorem (see Proposition 2.24).

(Uniqueness) Suppose we have two representations

$$V = V_1 \cup \cdots \cup V_r \quad V = V_1' \cup \cdots \cup V_s'$$

with $V_i \not\subset V_j$ and $V_i' \not\subset V_j'$ for any distinct i, j. We have

$$V_j = V_j \cap V = \cup_{i=1}^{s}(V_j \cap V_i')$$

so that $V_j = V_j \cap V_i'$ for some i, i.e., $V_j \subset V_i'$. Similarly, $V_i' \subset V_m$ for some m. The irredundancy assumption implies that $j = m$, so we have $V_j \subset V_i' \subset V_j$, and the two sets are equal. □

6.2 Irreducibility and domains

We translate the geometric condition of irreducibility into algebraic terms. Unfortunately, the algebraic notion of irreducible elements is not adequate for this purpose; see Exercise 6.2 for instances where irreducible elements give rise to reducible varieties.

Theorem 6.5 Let $V \subset \mathbb{A}^n(k)$ be a variety. The following are equivalent:

1. V is irreducible;
2. the coordinate ring $k[V]$ is a domain;
3. $I(V)$ is prime.

Proof The equivalence of the last two conditions follows from Proposition A.7.

We prove 1 ⇔ 2: Let $\overline{f}_1, \overline{f}_2 \neq 0 \in k[V]$ with $\overline{f}_1 \overline{f}_2 = 0$. Consider the closed subsets

$$V_i = \{v \in V : \overline{f}_i(v) = 0\}.$$

We have $V_i \subsetneq V$ because $\overline{f}_i \neq 0$, and

$$V_1 \cup V_2 = \{v \in V : \overline{f}_1 \overline{f}_2(v) = 0\} = V.$$

Thus V is reducible.

Conversely, if V is reducible we can write

$$V = V_1 \cup V_2, \quad V_1, V_2 \subsetneq V.$$

Thus $I(V_i) \supsetneq I(V)$, and we can take $f_i \in I(V_i) \setminus I(V)$. The product $f_1 f_2$ is zero on $V_1 \cup V_2 = V$ and is thus in $I(V)$. The resulting elements $\overline{f}_i \in k[V]$ are divisors of zero. □

Definition 6.6 Let V be an irreducible variety, so that $k[V]$ is a domain. The field $k(V)$ is called the *function field* of V.

Our geometric notion of irreducibility is nevertheless related to irreducible polynomials:

Proposition 6.7 *A nonzero principal ideal $\langle f \rangle \subsetneq k[x_1, \cdots, x_n]$ is prime if and only if f is irreducible.*

Corollary 6.8 *Consider a nontrivial hypersurface $V \subset \mathbb{A}^n(k)$, i.e., a variety with $I(V) = \langle f \rangle$ with $f \in k[x_1, \cdots, x_n]$ nonconstant. Then V is irreducible if and only if f is irreducible.*

Proof Theorem A.14 implies $k[x_1, \cdots, x_n]$ is a unique factorization domain. Thus every irreducible element f generates a prime ideal by Proposition A.11.

Suppose that $f \neq 0$ is reducible and is not a unit. We can write $f = f_1 f_2$, where f_1 and f_2 are neither zero nor units. We claim f_1 and f_2 are not contained in $\langle f \rangle$, hence $\langle f \rangle$ is not prime. Indeed, suppose the contrary, e.g., $f_1 = gf$ for some $g \in k[x_1, \cdots, x_n]$. Then $f = f_1 f_2 = gff_2$, which implies $1 = gf_2$, contradicting the assumption that f_2 is not a unit. $\square$

Example 6.9 Let $f \in k[x_1, \cdots, x_n]$ be an irreducible polynomial. We have *not* shown that $V(f)$ is irreducible. Indeed, consider the irreducible polynomial

$$f = x^2(x-1)^2 + y^2 + z^2 \in \mathbb{Q}[x, y, z].$$

Then $V(f) = \{(0, 0, 0), (1, 0, 0)\}$, which is reducible as a variety!

Over an algebraically closed field, $V(f)$ is irreducible if and only if f is irreducible. We will deduce this from the *Nullstellensatz* in Theorem 7.23.

6.3 Dominant morphisms

We develop tools for proving certain varieties are irreducible.

Definition 6.10 A morphism of affine varieties $\phi : V \to W$ is *dominant* if $\overline{\phi(V)} = W$, i.e., W is the smallest affine variety containing the image of V.

Proposition 6.11 *A morphism of affine varieties $\phi : V \to W$ is dominant if and only if $\phi^* : k[W] \to k[V]$ is injective.*

Proof Recall that $\phi(V)$ is contained in a subvariety $Z \subset W$ if and only if $\phi^*(I_Z) \subset I_V$. If $\phi(V)$ is contained in a proper subvariety $Z \subset W$ then there exists a

6.3 DOMINANT MORPHISMS

nonzero $f \in I_Z \subset k[W]$ vanishing on the image $\phi(V)$. In particular, $\phi^* f = 0 \in k[V]$. Conversely, suppose that $\phi^* f = 0$ for some $f \neq 0 \in k[W]$. Then we have

$$\phi(V) \subset W \cap \{w \in W : f(w) = 0\} \subsetneq W.$$

$\square$

Example 6.12

1. Any surjective morphism is dominant.
2. Consider the projection morphism

$$\pi_i : V := \{(x_1, x_2) : x_1 x_2 = 1\} \to \mathbb{A}^1(k)$$
$$(x_1, x_2) \mapsto x_i.$$

The image consists of the subset $\mathbb{A}^1(k) \setminus \{0\}$, which is not contained in a proper closed subset of $\mathbb{A}^1(k)$.

Keeping Corollary 3.46 in mind, Proposition 6.11 suggests the following extension:

Definition 6.13 A rational map $\rho : V \dashrightarrow W$ is *dominant* if the induced homomorphism $\rho^* : k[W] \to k(V)$ is injective.

This is compatible with our original definition: If ρ is a morphism then ρ^* factors

$$k[W] \to k[V] \hookrightarrow k(V).$$

Proposition 6.14 *Let V be an irreducible variety and $\rho : V \dashrightarrow W$ a dominant map. Then W is also irreducible.*

Proof Since V is irreducible, $k[V]$ has no divisors of zero. However, the dominance assumption implies that

$$\rho^* : k[W] \to k(V)$$

has trivial kernel. In particular, any zero divisors in $k[W]$ would yield zero divisors of function field $k(V)$, a contradiction. $\square$

Finally, we have the following extension of Proposition 3.57:

Proposition 6.15 *Let $\rho : V \dashrightarrow W$ be a dominant map of irreducible varieties. Then $\rho^* : k[W] \to k(V)$ induces a k-algebra homomorphism $k(W) \hookrightarrow k(V)$. Conversely, given a homomorphism of k-algebras*

$$\psi : k(W) \to k(V)$$

there exists a dominant rational map defined over k, $\rho : V \dashrightarrow W$, with ρ^* inducing ψ.

Proof By definition, $\rho^* : k[W] \to k(V)$ is an injection. Since $k(V)$ is a field, ρ^* extends to a field extension $k(W) \hookrightarrow k(V)$.

The homomorphism ψ restricts to an injective homomorphism $k[W] \to k(V)$. Corollary 3.46 yields a rational map $V \dashrightarrow W$, which is dominant by definition. □

Definition 6.16 Let V and W be affine varieties defined over k. We say that V and W are *birational over k* if $k(V)$ and $k(W)$ are isomorphic as k-algebras.

Example 6.17 An affine variety W is rational if and only if it is birational to $\mathbb{A}^n(k)$ (see Corollary 3.60).

Proposition 6.15 implies that two irreducible varieties V and W are birational if and only if there are rational maps

$$\rho : V \dashrightarrow W, \quad \xi : W \dashrightarrow V$$

which are inverse to each other (see Exercise 6.9 for discussion of compositions of rational maps). A rational map that induces an isomorphism of function fields is called a *birational map*.

6.4 Algorithms for intersections of ideals

Consider the decomposition of a variety into irreducible components

$$V = V_1 \cup V_2 \cup \cdots \cup V_r.$$

Given equations for V, it is fairly difficult to compute equations for each of the irreducible components. We will discuss methods for doing this when we discuss primary decomposition. For the moment, we will focus on the reverse problem:

Problem 6.18 Let V_1 and V_2 be affine varieties in $\mathbb{A}^n(k)$, with ideals $I(V_1)$ and $I(V_2)$. How do we compute $I(V_1 \cup V_2)$?

This boils down to computing the *intersection* of the ideals $I(V_1)$ and $I(V_2)$. Indeed, Proposition 3.7 implies

$$I(V_1 \cup V_2) = I(V_1) \cap I(V_2).$$

We shall require the following general result:

Proposition 6.19 Let $I, J \subset R$ be ideals in a ring, and consider the ideal

$$tI + (1-t)J := \langle t \rangle I + \langle 1-t \rangle J \subset R[t],$$

generated by elements tf and $(1-t)g$ where $f \in I$ and $g \in J$. Then we have

$$(tI + (1-t)J) \cap R = I \cap J.$$

6.4 ALGORITHMS FOR INTERSECTIONS OF IDEALS

Proof The inclusion ($\supset$) is straightforward: for any $h \in I \cap J$, we can write
$$h = th + (1-t)h \in tI + (1-t)J.$$
For the reverse inclusion ($\subset$), suppose we are given an element $h \in tI + (1-t)J \cap R$, which may be expressed
$$h = \sum_i tf_i r_i + \sum_j (1-t)g_j s_j, \quad f_i \in I, g_j \in J, r_i, s_j \in R[t].$$
The left-hand side is constant as a polynomial in t. Setting $t = 0$ we obtain
$$h = 0 + \sum_j s_j|_{t=0} g_j \in J$$
and setting $t = 1$ we obtain
$$h = \sum_i r_i|_{t=1} f_i + 0 \in I.$$
This proves $h \in I \cap J$. $\square$

This result reduces the computation of intersections to an elimination problem:

Algorithm 6.20 (Computing intersections) *Given ideals*
$$I = \langle f_1, \ldots, f_r \rangle, J = \langle g_1, \ldots, g_s \rangle \subset k[x_1, \ldots, x_n]$$
the intersection $I \cap J$ is obtained as follows: Compute a Gröbner basis $h_1, \ldots, h_q$ for
$$tI + (1-t)J = \langle tf_1, \ldots, tf_r, (1-t)g_1, \ldots, (1-t)g_s \rangle \subset k[x_1, \ldots, x_n, t]$$
using an elimination order for t. The basis elements $h_j \in k[x_1, \ldots, x_n]$ generate $I \cap J$.

Example 6.21 Compute equations for $\{(0, 1), (1, 0)\}$.
We set $I = \langle x_1, x_2 - 1 \rangle$ and $J = \langle x_1 - 1, x_2 \rangle$, so that
$$tI + (1-t)J = \langle tx_1, t(x_2 - 1), (1-t)(x_1 - 1), (1-t)x_2 \rangle.$$
With respect to lexicographic order, we have Gröbner basis
$$\langle t - x_2, x_1 + x_2 - 1, x_2^2 - x_2 \rangle$$
so we recover the equations $x_1 + x_2 = 1$ and $x_2^2 = x_2$.

The algorithm computing the intersection of two ideals has a geometric interpretation. Given varieties $V, W \subset \mathbb{A}^n(k)$ with ideals I and J, the ideal tI corresponds to
$$(V \times \mathbb{A}^1(k)) \cup (\mathbb{A}^n(k) \times \{0\}) \subset \mathbb{A}^n(k) \times \mathbb{A}^1(k) = \mathbb{A}^{n+1}(k),$$

while $(1-t)J$ corresponds to

$$(W \times \mathbb{A}^1(k)) \cup (\mathbb{A}^n(k) \times \{1\}) \subset \mathbb{A}^n(k) \times \mathbb{A}^1(k) = \mathbb{A}^{n+1}(k).$$

Adding ideals corresponds to intersecting varieties, so $tI + (1-t)J$ defines the variety

$$[(V \cap W) \times \mathbb{A}^1(k)] \cup [V \times \{1\}] \cup [W \times \{0\}].$$

The image under projection to $\mathbb{A}^n(k)$ is the union $V \cup W$.

6.5 Domains and field extensions

Extending the base field can affect whether a quotient ring is a domain. Let $f \in k[t]$ be irreducible; recall that

$$k \hookrightarrow k[t]/\langle f(t) \rangle$$

is a field extension (see Section A.6). However, if L/k is a field extension over which we have a factorization

$$f = g_1 g_2, \quad g_1, g_2 \in L[t],$$

then $L[t]/\langle f(t) \rangle$ has zero divisors.

On the other hand, extending the base field will not cause a reducible variety to become irreducible:

Proposition 6.22 *Let $f_1, \ldots, f_r \in k[x_1, \ldots, x_n]$ and L/k a field extension. If $L[x_1, \ldots, x_n]/\langle f_1, \ldots, f_r \rangle$ is a domain then $k[x_1, \ldots, x_n]/\langle f_1, \ldots, f_r \rangle$ is also a domain.*

Proof The key ingredient is the following fact:

Lemma 6.23 *The induced ring homomorphism*

$$\eta : k[x_1, \ldots, x_n]/\langle f_1, \ldots, f_r \rangle \to L[x_1, \ldots, x_n]/\langle f_1, \ldots, f_r \rangle$$

is injective.

Assuming this, the proposition is straightforward: suppose we have $g, h \in k[x_1, \ldots, x_n]$ with

$$gh \equiv 0 \mod \langle f_1, \ldots, f_r \rangle.$$

The fact that $L[x_1, \ldots, x_n]/\langle f_1, \ldots, f_r \rangle$ is a domain implies $\eta(g)$ or $\eta(h)$ is zero. Since η has trivial kernel, either g or h is zero.

6.5 DOMAINS AND FIELD EXTENSIONS

We prove the lemma: choose a monomial order on $k[x_1, \ldots, x_n]$, which automatically induces an order on $L[x_1, \ldots, x_n]$. We claim that

$$\operatorname{LT} \langle f_1, \ldots, f_r \rangle_k = \operatorname{LT} \langle f_1, \ldots, f_r \rangle_L;$$

the subscripts designate the coefficient fields in each case. The lemma follows by analyzing normal forms: let $\Xi := \{x^\alpha\}$ denote the monomials *not* appearing in $\operatorname{LT} \langle f_1, \ldots, f_r \rangle$. Every element of $k[x_1, \ldots, x_n]/\langle f_1, \ldots, f_r \rangle_k$ has a unique expression as a k-linear combination of monomials from Ξ; elements of $L[x_1, \ldots, x_n]/\langle f_1, \ldots, f_r \rangle_L$ have unique expressions as L-linear combinations of the same monomials. An element of $k[x_1, \ldots, x_n]$ is zero in the first quotient if and only if it is zero in the second quotient.

We establish the claim: the inclusion $\subset$ is evident, so we focus on the reverse inclusion $\supset$. Suppose x^α is the leading term of

$$h_1 f_1 + \cdots + h_r f_r, \quad h_1, \ldots, h_r \in L[x_1, \ldots, x_n];$$

choose d such that $h_1, \ldots, h_r$ all have degree $\leq d$.

We assert that there exist $g_1, \ldots, g_r \in P_{n,d}$ (polynomials in $k[x_1, \ldots, x_n]$ of degree $\leq d$) with

$$\operatorname{LT}(g_1 f_1 + \cdots + g_r f_r) = x^\alpha.$$

This is just linear algebra. Let $m = \max\{\deg(f_j)\}_{j=1,\ldots,r}$ and consider the linear transformation

$$\Phi : P_{n,d}^r \to P_{n,d+m}$$
$$(g_1, \ldots, g_r) \mapsto g_1 f_1 + \cdots + g_r f_r.$$

We want an element

$$p = \sum_\beta c_\beta x^\beta \in \operatorname{image}(\Phi)$$

with $c_\alpha = 1$ and $c_\beta = 0$ for each $\beta > \alpha$. Note that each c_β is linear as a function of the coefficients of $g_1, \ldots, g_r$, which we interpret as free variables. Finding p boils down to solving over k a system of linear equations with coefficients in k. However, we already know that this system of equations has a solution over L, i.e., $(h_1, \ldots, h_r)$.

Our claim then is a consequence of the following result, left as an exercise:

Lemma 6.24 *Let L/k be a field extension, A an $M \times N$ matrix with entries in k, and $b \in k^M$ a column vector. Consider the system of linear equations*

$$At = b, \quad t = \begin{pmatrix} t_1 \\ \vdots \\ t_N \end{pmatrix},$$

and assume there is a solution with $t_1, \ldots, t_N \in L$. Then there is a solution with $t_1, \ldots, t_N \in k$.

Remark 6.25 Lemma 6.24 reflects the fact that L is *flat* as a module over k. We will not discuss flatness systematically here, but [29, pp. 49] describes how flatness can be expressed in terms of systems of linear equations.

6.6 Exercises

6.1 Is the intersection of two irreducible varieties irreducible? Give a proof or counterexample.

6.2 Irreducible elements can give rise to reducible varieties:
(a) Show that
$$V = \{(x, y, z) : xy - z^2 = 0 \subset \mathbb{A}^3(\mathbb{R})\}$$
is irreducible. *Hint:* It's the image of a morphism
$$\mathbb{A}^2(\mathbb{R}) \to \mathbb{A}^3(\mathbb{R})$$
$$(s, t) \mapsto (s^2, t^2, st).$$
(b) Show that $z \in \mathbb{R}[V]$ is irreducible as an element in $\mathbb{R}[V]$.
(c) Show that $\{z = 0\} \subset V$ is reducible.

6.3 Let $S \subset \mathbb{A}^n(k)$ be a finite set. Show that the affine span affspan(S) is irreducible.

6.4 Suppose that $V \subset \mathbb{A}^n(k)$ is irreducible. Show that $V \times \mathbb{A}^m(k)$ is irreducible.

6.5 Consider a matrix with entries in k
$$A = \begin{pmatrix} a_{11} & a_{12} \\ a_{21} & a_{22} \end{pmatrix}$$
with $\det(A) = a_{11}a_{22} - a_{12}a_{21} \neq 0$. Show that
$$\rho : \mathbb{A}^1(k) \dashrightarrow \mathbb{A}^1(k)$$
$$t \mapsto \frac{a_{11}t + a_{12}}{a_{21}t + a_{22}}$$
is dominant.

6.6 Consider the ideal $I \subset \mathbb{C}[x_1, \ldots, x_6]$ generated by two-by-two minors of
$$\begin{pmatrix} x_1 & x_2 & x_3 \\ x_4 & x_5 & x_6 \end{pmatrix}.$$
Show that I is prime. *Hint:* Use Proposition 6.14 and Exercise 4.12.

6.7 Let $V \subset \mathbb{A}^n(k)$ be irreducible and choose nonzero $p_1, \ldots, p_r \in k[V]$. Consider
$$W = \{(v, y_1, \ldots, y_r) : p_1 y_1 = p_2 y_2 = \cdots = p_r y_r = 1\} \subset V \times \mathbb{A}^r(k),$$

6.6 EXERCISES

where $y_1, \ldots, y_r$ are coordinates of $\mathbb{A}^r(k)$. Prove that W is irreducible and birational to V.

6.8 Show that the composition of two dominant morphisms is itself a dominant morphism.

6.9 Let V, W, X denote irreducible varieties and

$$V \xdashrightarrow{\rho} W \xdashrightarrow{\xi} X$$

dominant rational maps. Show that there is a rational map $V \dashrightarrow X$ which equals the composition $\xi \circ \rho$ over a suitable nonempty open subset $U \subset V$.

6.10 Consider the affine varieties

$$V_1 = \{(x, y) : y^2 = x^3\} \subset \mathbb{A}^2(\mathbb{Q})$$

and

$$V_2 = \{(u, v) : u^3 = v^4\} \subset \mathbb{A}^2(\mathbb{Q}).$$

with function fields $\mathbb{Q}(V_1)$ and $\mathbb{Q}(V_2)$.
(a) Show that these function fields are isomorphic as $\mathbb{Q}$-algebras.
(b) Construct an explicit birational map $V_1 \dashrightarrow V_2$.

6.11 (Intersections versus products)
(a) We generally have $I_1 \cap I_2 \neq I_1 I_2$, even when I_1 and I_2 define distinct varieties. Consider the lines

$$V_1 = \{x_1 = x_2 = 0\}, \quad V_2 = \{x_2 = x_3 = 0\}, \quad V_3 = \{x_3 = x_1 = 0\}.$$

Show that the product ideal $I(V_1)I(V_2)I(V_3)$ is smaller than the intersection $I(V_1) \cap I(V_2) \cap I(V_3)$, even though they define the same variety.
(b) Suppose that $I, J \subset R$ are ideals such that $I + J = R$. Show that $IJ = I \cap J$.

6.12 Let $V, W \subset \mathbb{A}^n(k)$ denote affine varieties. Define the *disjoint union*

$$V \sqcup W \subset \mathbb{A}^n \times \mathbb{A}^1 \simeq \mathbb{A}^{n+1}$$

as the union of affine varieties

$$(V \times \{0\}) \cup (W \times \{1\}).$$

Writing $k[\mathbb{A}^n] = k[x_1, \ldots, x_n]$, $k[\mathbb{A}^1] = k[t]$, and $R = k[x_1, \ldots, x_n, t]$, show that

$$I(V \sqcup W) = (I(V)R + \langle t \rangle) \cap (I(W)R + \langle t - 1 \rangle)$$

and

$$k[V \sqcup W] = k[V] \oplus k[W].$$

The *direct sum* of rings R_1 and R_2 is defined

$$R_1 \oplus R_2 = \{(r_1, r_2) : r_1 \in R_1, r_2 \in R_2\}$$

with addition and multiplication taken componentwise.

6.13 Let $X = \mathbb{A}^1(k) \sqcup \mathbb{A}^1(k)$ be the disjoint union of two affine lines. Express X explicitly as an affine variety, compute the coordinate ring $k[X]$, and find examples of zero divisors.

6.14 Consider monomial ideals

$$I = \langle x^{\alpha(1)}, \ldots, x^{\alpha(r)} \rangle, \quad J = \langle x^{\beta(1)}, \ldots, x^{\beta(s)} \rangle \subset k[x_1, \ldots, x_n].$$

Show that

$$I \cap J = \langle \mathrm{LCM}(x^{\alpha(i)}, x^{\beta(j)}) \rangle_{i=1,\ldots,r,\, j=1,\ldots,s};$$

in particular, any intersection of monomial ideals is monomial. *Hint:* You may not have to use Proposition 6.19; try a direct approach via Exercise 2.3.

6.15 Prove Lemma 6.24. *Hint:* Any system of linear equations over k with a solution in L is *consistent*.

7 Nullstellensatz

'Nullstellensatz' is a German term translated literally as 'Zero places theorem'. It is associated with a problem first identified in Chapter 3: given an ideal $I \subset k[x_1, \ldots, x_n]$ defining a variety $V(I)$, what are the polynomials vanishing on $V(I)$? Generally, we have the inclusion (cf. Exercise 3.3)

$$I(V(I)) \supset I.$$

When does equality hold? Where there is a strict inclusion, can we obtain $I(V(I))$ directly from I?

Raising a polynomial to a power does not change where it vanishes, i.e.,

$$V(f) = V(f^N)$$

for each $N \geq 1$. A general definition will help us utilize this fact:

Definition 7.1 The *radical* of an ideal I in a ring R is defined

$$\sqrt{I} = \{g \in R : g^N \in I \text{ for some } N \in \mathbb{N}\}.$$

An ideal J is said to be radical if $\sqrt{J} = J$.

The reader should verify that $\sqrt{I}$ is automatically an ideal (see Exercise 7.3). Our observation then translates into the following result:

Proposition 7.2 *If $I \subset k[x_1, \ldots, x_n]$ is an ideal then*

$$\sqrt{I} \subset I(V(I)).$$

Proof For each $f \in \sqrt{I}$, there exists an $N \gg 0$ such that $f^N \in I$. We have

$$V(f) = V(f^N) \supset V(I),$$

hence f vanishes over $V(I)$. □

7.1 Statement of the Nullstellensatz

Theorem 7.3 (Hilbert Nullstellensatz) If k is algebraically closed and $I \subset k[x_1, \ldots, x_n]$ is an ideal then $I(V(I)) = \sqrt{I}$.

In other words, given a function vanishing at each point of a variety, some power of that function can be written in terms of the defining equations for the variety.

Example 7.4 The relationship between $\sqrt{I}$ and $I(V(I))$ is still quite subtle over nonclosed fields. Consider

$$I = \langle x^{2n} + y^{2n} + 1 \rangle \subset \mathbb{R}[x, y]$$

so that

$$\emptyset = V(I) \subset \mathbb{A}^2(\mathbb{R})$$

and $I(V(I)) = \mathbb{R}[x, y]$. On the other hand, $\sqrt{I} \subsetneq \mathbb{R}[x, y]$. Indeed, if $\sqrt{I} = \mathbb{R}[x, y]$ then $1 \in \sqrt{I}$, which would imply that $1 \in I$, a contradiction. Thus we have

$$\sqrt{I} \subsetneq I(V(I)).$$

Here is another very useful statement also known as the Nullstellensatz:

Theorem 7.5 (Nullstellensatz II) Let k be algebraically closed and $I = \langle f_1, \ldots, f_r \rangle \subset k[x_1, \ldots, x_n]$. Then either

1. $I = k[x_1, \ldots, x_n]$; or
2. there exists a common solution $(a_1, \ldots, a_n)$ for the system

$$f_1 = f_2 = \ldots = f_r = 0.$$

In other words, over an algebraically closed field every consistent system of polynomials has a solution. Of course, an inconsistent system has no solutions over any field: if $f_1, \ldots, f_r$ have a common zero then $\langle f_1, \ldots, f_r \rangle$ does not contain 1 and

$$\langle f_1, \ldots, f_r \rangle \subsetneq k[x_1, \ldots, x_n].$$

7.1.1 Effective results

It is very desirable to have constructive approaches to these questions. We refer the reader to [24, 4] for more discussion and results.

Problem 7.6 (Effective Nullstellensatz) Consider polynomials $f_1, \ldots, f_r \in k[x_1, \ldots, x_n]$. Find explicit constants N and N' depending only on the degrees of the f_i and the number of variables, such that

7.2 CLASSIFICATION OF MAXIMAL IDEALS

- if $g \in \sqrt{\langle f_1, \ldots, f_r \rangle}$ then $g^N \in \langle f_1, \ldots, f_r \rangle$;
- if $f_1, \ldots, f_r$ have no common zeros over an algebraically closed field then there exist polynomials $g_1, \ldots, g_r$ of degree $\leq N'$ such that $f_1 g_1 + \cdots + f_r g_r = 1$.

The *a priori* bound on the degrees allows us to solve for the coefficients of $g_1, \ldots, g_r$ using linear algebra.

The Hilbert Basis Theorem does yield a non-effective result for the first part of the problem:

Proposition 7.7 *Let R be a Noetherian ring and $I \subset R$ an ideal. There exists an integer N such that*

$$g \in \sqrt{I} \Rightarrow g^N \in I.$$

Proof Let $h_1, \ldots, h_s$ be generators for $\sqrt{I}$ and choose M such that $h_1^M, \ldots, h_s^M \in I$. We take $N = Ms$. For each

$$g = g_1 h_1 + \cdots + g_s h_s \in \sqrt{I}$$

we have

$$g^N = g^{Ms} = \sum_{e_1 + \cdots + e_s = Ms} n_{e_1 \ldots e_s} (g_1 h_1)^{e_1} \ldots (g_s h_s)^{e_s},$$

where the $n_{e_1 \ldots e_s}$ are suitable positive integers. Each term has some $e_j \geq M$ and thus is contained in I. We conclude that $g^N \in I$ as well. □

7.2 Classification of maximal ideals

We start with a general result, valid over any field:

Proposition 7.8 *Let $a_1, \ldots, a_n \in k$ and consider the ideal*

$$\mathfrak{m} = \langle x_1 - a_1, x_2 - a_2, \ldots, x_n - a_n \rangle \subset k[x_1, \ldots, x_n].$$

Then $\mathfrak{m}$ is maximal and coincides with the kernel of the evaluation homomorphism

$$k[x_1, \ldots, x_n] \stackrel{ev(a)}{\to} k$$
$$x_i \mapsto a_i.$$

Proof (cf. Exercise 2.12) For *any* monomial order on $k[x_1, \ldots, x_n]$ we have

$$\mathrm{LT}(\mathfrak{m}) = \langle x_1, \ldots, x_n \rangle,$$

so the normal form of $f \in k[x_1, \ldots, x_n]$ modulo $\mathfrak{m}$ is constant and equal to $f(a_1, \ldots, a_n) = ev(a)f$. □

Theorem 7.9 (Nullstellensatz I) Let k be an algebraically closed field. Then any maximal ideal $\mathfrak{m} \subset k[x_1, \ldots, x_n]$ takes the form $\langle x_1 - a_1, \ldots, x_n - a_n \rangle$.

Theorems 7.5 and 7.3 follow easily from this statement.

Proof of Theorem 7.5 assuming Theorem 7.9 Suppose that $1 \notin I$ so that $I \subsetneq k[x_1, \ldots, x_n]$. Then there exists a maximal ideal $\mathfrak{m} \supset I$, and Theorem 7.9 implies $I \subset \langle x_1 - a_1, \ldots, x_n - a_n \rangle$, for some $(a_1, \ldots, a_n) \in \mathbb{A}^n(k)$. Hence $(a_1, \ldots, a_n) \in V(I)$. □

Proof of Theorem 7.3 assuming Theorem 7.9 Consider the affine variety

$$\mathbb{A}^n(k)_g = \{(x_1, \ldots, x_n, z) : zg(x_1, \ldots, x_n) = 1\} \subset \mathbb{A}^{n+1}(k)$$

introduced in the proof of Proposition 3.47. Projection onto the variables $x_1, \ldots, x_n$ takes this variety to the open subset

$$U = \{(x_1, \ldots, x_n) : g(x_1, \ldots, x_n) \neq 0\} \subset \mathbb{A}^n(k).$$

Our hypothesis says that $U \cap V = \emptyset$, so consider the ideal

$$J = \langle f_1, \ldots, f_r, gz - 1 \rangle,$$

which is the full polynomial ring $k[x_1, \ldots, x_n, z]$ by Nullstellensatz II. Therefore, we may write

$$h_1 f_1 + \cdots + h_r f_r + h(gz - 1) = 1, \quad h_1, \ldots, h_r, h \in k[x_1, \ldots, x_n, z].$$

Substitute $z = 1/g$ to get

$$1 = h_1(x_1, \ldots, x_n, 1/g) f_1 + \cdots + h_r(x_1, \ldots, x_n, 1/g) f_r$$

so that, on clearing denominators, we have for some N

$$g^N = \tilde{h}_1 f_1 + \cdots + \tilde{h}_r f_r, \quad \tilde{h}_j \in k[x_1, \ldots, x_n].$$ □

7.3 Transcendence bases

Let F/k denote a field extension.

Definition 7.10 F/k is *finitely generated* if $F = k(z_1, \ldots, z_N)$ for some choice of $z_1, \ldots, z_N \in F$.

Example 7.11 Let $I \subset k[x_1, \ldots, x_n]$ be prime so that $k[x_1, \ldots, x_n]/I$ is an integral domain with fraction field F. Then F/k is finitely generated, e.g., by $x_1, \ldots, x_n$. In particular, the function field of an affine variety is finitely generated.

7.3 TRANSCENDENCE BASES

Definition 7.12   An element $z \in F$ is *transcendental over k* if it is not algebraic. A collection of elements $z_1, \ldots, z_d \in F$ is *algebraically independent* over k if there is no nonzero polynomial $f(x_1, \ldots, x_d) \in k[x_1, \ldots, x_d]$ with $f(z_1, \ldots, z_d) = 0$. It is *algebraically dependent* otherwise.

The following result is left as an exercise:

Proposition 7.13   *Let F/k be a field extension. Then the elements $z_1, \ldots, z_d$ are algebraically independent over k if and only if*

- z_1 *is transcendental over k;*
- z_2 *is transcendental over $k(z_1)$;*
- $\vdots$
- z_d *is transcendental over $k(z_1, \ldots, z_{d-1})$.*

Definition 7.14   A *transcendence basis* for F/k is a collection of algebraically independent elements

$$z_1, \ldots, z_d \in F$$

such that F is algebraic over $k(z_1, \ldots, z_d)$. By convention, the empty set is a transcendence base for an algebraic field extension.

Proposition 7.15   *Every finitely generated field extension admits a transcendence basis; we may take a suitable subset of the generators. Any two transcendence bases have the same number of elements.*

The resulting invariant of the extension is called its *transcendence degree*.

Proof   Express $F = k(z_1, \ldots, z_N)$ for some $z_j \in F$. If the z_j are all algebraic over k then F is algebraic over k (by Proposition A.16), and the transcendence basis is empty. We therefore assume some of the z_j are transcendental.

After reordering $z_1, \ldots, z_N$, we have

$$\{z_1, \ldots, z_d\} \subset \{z_1, \ldots, z_N\}$$

as a maximal algebraically independent subset, i.e., for each $j > d$ the set $\{z_1, \ldots, z_d, z_j\}$ is algebraically dependent. We therefore have a nonzero polynomial

$$f \in k[x_1, \ldots, x_d, x_{d+1}]$$

such that $f(z_1, \ldots, z_d, z_j) = 0$; expand

$$f = \sum_i f_i(x_1, \ldots, x_d) x_{d+1}^i$$

where $f_i \neq 0$ for some $i > 0$. (Otherwise, $z_1, \ldots, z_d$ would be algebraically dependent.) Thus each z_j, $j > d$, is algebraic over $k(z_1, \ldots, z_d)$, and F is algebraic over $k(z_1, \ldots, z_d)$.

Remark 7.16 The remainder of the argument is analogous to a *linear* independence result: given linearly independent sets

$$\{z_1, \ldots, z_d\}, \{w_1, \ldots, w_e\}$$

generating the same subspace, we necessarily have $d = e$. The proof entails exchanging elements of the second set for elements of the first, one at a time. After reordering the z_is and w_js, we obtain a sequence of linearly independent sets

$$\{z_1, z_2, \ldots, z_d\}, \{w_1, z_2, \ldots, z_d\}, \{w_1, w_2, z_3, \ldots, z_d\}, \ldots, \{w_1, w_2, w_3, \ldots, w_d\}$$

all generating the same subspace. If there were a w_j with $j > d$ it would have to be contained in the subspace generated by $\{w_1, \ldots, w_d\}$.

Suppose we have algebraically independent sets

$$\{z_1, \ldots, z_d\}, \{w_1, \ldots, w_e\} \subset F$$

such that F is algebraic over $k(z_1, \ldots, z_d)$ and $k(w_1, \ldots, w_e)$. We want to show that $d = e$. For simplicity, we assume $d \leq e$.

We know that w_1 is algebraic over $k(z_1, \ldots, z_d)$, i.e., there is a nonzero $f \in k[x_1, \ldots, x_d, x_{d+1}]$ such that $f(z_1, \ldots, z_d, w_1) = 0$. Since w_1 is not algebraic over k, f must involve at least one of the variables $x_1, \ldots, x_d$, say x_1. Then z_1 is algebraic over $k(w_1, z_2, \ldots, z_d)$, and hence F is algebraic over $k(w_1, z_2, \ldots, z_d)$ by Proposition A.16.

We know then that w_2 is algebraic over $k(w_1, z_2, \ldots, z_d)$. Proposition 7.13 implies w_2 is not algebraic over $k(w_1)$. We then have $g \neq 0 \in k[x_1, \ldots, x_d, x_{d+1}]$ such that $g(w_1, z_2, \ldots, z_d, w_2) = 0$; g must involve at least one of the variables $x_2, \ldots, x_d$, say x_2. But then z_2 is algebraic over $k(w_1, w_2, z_3, \ldots, z_d)$ and F is algebraic over $k(w_1, w_2, z_3, \ldots, z_d)$.

We continue in this way, deducing that F is algebraic over $k(w_1, \ldots, w_d)$. In particular, if there were a w_j with $j > d$ then this would be algebraic over $k(w_1, \ldots, w_d)$. This contradicts our assumption on algebraic independence. □

7.4 Integral elements

Let R and F be Noetherian domains with $R \subset F$. An element $\alpha \in F$ is *integral* over R if either of the following equivalent conditions are satisfied:

1. α is a root of a monic polynomial $t^D + r_{D-1}t^{D-1} + \cdots r_0 \in R[t]$;
2. the R-submodule $R[\alpha] \subset F$ generated by α is finitely generated.

7.4 INTEGRAL ELEMENTS

The equivalence is easy: $R[\alpha]$ is finitely generated if and only if we can write

$$-\alpha^D = r_{D-1}\alpha^{D-1} + r_{D-2}\alpha^{D-2} + \cdots + r_0, \quad r_j \in R,$$

for some D.

Proposition 7.17 *Assume that the fraction field L of R is contained in F. If $w \in F$ is algebraic over L then there exists a nonzero $p \in R$ such that pw is integral over R.*

Proof Let g denote the irreducible polynomial for w over L. Clearing denominators if necessary, we may assume $g \in R[t]$. Write

$$g = a_N t^N + \cdots + a_0, \quad a_i \in R, a_N \neq 0,$$

and observe that

$$a_N^{N-1} g = \sum_{i=0}^{N} a_i a_N^{N-i-1}(ta_N)^i.$$

The polynomial

$$\sum_{i=0}^{N} a_i a_N^{N-i-1} u^i = u^N + a_{N-1} a_N u^{N-1} + \cdots + a_0 a_N^{N-1} \in R[u]$$

is monic and has $a_N w$ as a root. In particular, $a_N w$ is integral over R. □

Proposition 7.18 *Let R be a unique factorization domain with fraction field L. Then every element of L integral over R is contained in R.*

Proof Suppose α is algebraic over the fraction field L of R. Consider the evaluation homomorphism

$$L[t] \to L(\alpha)$$
$$t \mapsto \alpha$$

with kernel generated by an irreducible $g \in L[t]$. Clearing denominators, we may assume $g \in R[t]$ with coefficients having no common irreducible factor.

If α is integral over R then there exists a monic irreducible $h \in R[t]$ such that $h(\alpha) = 0$. We know that $g|h$ in $L[t]$, so Gauss' Lemma (Proposition A.12) implies $h = fg$ for $f \in R[t]$. The leading coefficients of f and g are units multiplying to 1; g becomes monic after multiplying by the leading coefficient of f.

Thus we may assume the irreducible polynomial for α over L is a monic polynomial $g \in R[t]$. If $\alpha \in L$ then g has degree 1, i.e., $g(t) = t - \alpha$. In particular, $\alpha \in R$. □

Proposition 7.19 *Let $S \subset F$ denote the elements in F that are integral over R. Then S is a subring of F.*

Proof We recall Theorem 2.36: A submodule of a finitely generated module over a Noetherian ring is itself finitely generated. Suppose that $\alpha, \beta \in F$ are integral over R. Then $R[\alpha]$ is finitely generated over R and β remains integral over $R[\alpha]$, so $R[\alpha][\beta]$ is finitely generated over $R[\alpha]$ and R. It follows that the subrings $R[\alpha + \beta]$ and $R[\alpha\beta]$ are also finitely generated over R and thus $\alpha + \beta$ and $\alpha\beta$ are integral. $\square$

7.5 Proof of Nullstellensatz I

Suppose $\mathfrak{m} \subset k[x_1, \ldots, x_n]$ is maximal. Consider the field $F = k[x_1, \ldots, x_n]/\mathfrak{m}$, which is finitely generated *as an algebra* over k. By definition, a k-algebra R is *finitely generated* if it can be expressed as a quotient of a polynomial ring over k.

Lemma 7.20 *Let k be an algebraically closed field. Then any field extension F/k, finitely generated as an algebra over k, is trivial.*

Assuming Lemma 7.20, we obtain the theorem: consider the induced map $\pi_1 : k[x_1] \to F$. Since F is trivial over k, $\pi_1(x_1) = \alpha_1$ for some $\alpha_1 \in k$, thus $x_1 - \alpha_1 \in \mathfrak{m}$.

We prove Lemma 7.20: first, observe that, if $\alpha \in F$ is algebraic over k, then it must have an irreducible polynomial $f(t) \in k[t]$. Since k is algebraically closed this is necessarily of the form $t - \alpha$ (see Exercise A.14), and $\alpha \in k$.

If F/k is nontrivial then Proposition 7.15 gives a nonempty subset

$$\{x_{j_1}, \ldots, x_{j_d}\} \subset \{x_1, \ldots, x_n\}$$

such that the images

$$z_1 = \overline{x}_{j_1}, \ldots, z_d = \overline{x}_{j_d} \in F$$

are a transcendence basis for F over k. For notational simplicity, we reorder the x_j so that $z_j = \overline{x}_j$ for $j = 1, \ldots, d$. Write $R = k[z_1, \ldots, z_d] \subset F$, which is a polynomial ring because $z_1, \ldots, z_d$ are algebraically independent. Take the remaining generators for F over k

$$w_1 = x_{d+1}, \ldots, w_{n-d} = x_{n-d},$$

which are algebraic over $L := k(z_1, \ldots, z_d)$.

Let $S \subset F$ denote the ring of elements of F integral over $k[z_1, \ldots, z_d]$ (see Proposition 7.19). Proposition 7.17 yields nonzero $p_1, \ldots, p_{n-d} \in k[z_1, \ldots, z_d]$ such that each $t_j := p_j w_j$ is integral over $k[z_1, \ldots, z_d]$, i.e., $t_j \in S$.

Pick an element $f/g \in k(z_1, \ldots, z_d)$, with $f, g \in k[z_1, \ldots, z_d]$, such that g is relatively prime to $p_1, \ldots, p_{n-d}$. It follows that $p_1^{e_1} \ldots p_{n-d}^{e_{n-d}}(f/g)$ is not in $k[z_1, \ldots, z_d]$ for any $e_1, \ldots, e_{n-d} \in \mathbb{N}$. However, we can represent $f/g \equiv q(x_1, \ldots, x_n)$ for some

7.6 APPLICATIONS

polynomial q with coefficients in k, i.e.,

$$f/g = q(z_1, \ldots, z_d, t_1/p_1, \ldots, t_{n-d}/p_{n-d}).$$

Let e_j, $j = 1, \ldots, n-d$, denote the highest power of x_{d+j} appearing in q, so that multiplying through by $p_1^{e_1} \ldots p_{n-d}^{e_{n-d}}$ clears the denominators in

$$q(z_1, \ldots, z_d, t_1/p_1, \ldots, t_{n-d}/p_{n-d}).$$

It follows that there exists a polynomial q' over k such that

$$p_1^{e_1} \ldots p_{n-d}^{e_{n-d}} f/g = q'(z_1, \ldots, z_d, t_1, \ldots, t_{n-d}) \in S.$$

We also have

$$p_1^{e_1} \ldots p_{n-d}^{e_{n-d}}(f/g) \in k(z_1, \ldots, z_d),$$

thus Proposition 7.18 implies

$$p_1^{e_1} \ldots p_{n-d}^{e_{n-d}}(f/g) \in k[z_1, \ldots, z_d],$$

a contradiction. □

We sketch an alternate (and much easier!) proof of Lemma 7.20 over $k = \mathbb{C}$. Any finitely generated algebra over $\mathbb{C}$ has a countable basis, e.g., a subset of the monomials $x_1^{i_1} \ldots x_n^{i_n}$. On the other hand, if $x_1 \neq \alpha_1$ for any $\alpha_1 \in \mathbb{C}$, the uncountable collection of rational functions $\frac{1}{x_1 - \alpha} \in F$, $\alpha \in \mathbb{C}$ are linearly independent over $\mathbb{C}$.

7.6 Applications

7.6.1 When is a polynomial in the radical of an ideal?

Unfortunately, computing over algebraically closed fields presents significant technical difficulties. How can a computer represent a general complex number? The Hilbert Nullstellensatz gives us a procedure for deciding whether a polynomial vanishes over the complex points of variety *without ever computing over the complex numbers!* We just need to check whether the polynomial is contained in the radical of the ideal generated by some defining set of equations. This can be checked over any field containing the coefficients of the polynomials at hand.

We no longer assume that k is algebraically closed. To test whether a polynomial $g \in \sqrt{I}$, where $I = \langle f_1, \ldots, f_r \rangle$, we use the following criterion.

Proposition 7.21 *Given an ideal $I = \langle f_1, \ldots, f_r \rangle \subset k[x_1, \ldots, x_n]$, $g \in \sqrt{I}$ if and only if $\langle f_1, \ldots, f_r, zg - 1 \rangle = k[x_1, \ldots, x_n, z]$.*

Proof The proof of the Hilbert Nullstellensatz gives that

$$1 \in \langle f_1, \ldots, f_r, zg - 1 \rangle \Rightarrow g^N \in I \text{ for some } N.$$

Conversely, if $g^N = f_1 h_1 + \cdots + f_r h_r$ then $z^N g^N = f_1(h_1 z^N) + \cdots + f_r(h_r z^N)$ and
$$1 = f_1(h_1 z^N) + \cdots + f_r(h_r z^N) + (1 - z^N g^N).$$
Since $(1 - z^N g^N) = (1 - zg)(1 + zg + \cdots + z^{N-1} g^{N-1})$, we conclude $1 \in \langle f_1, \ldots, f_r, zg - 1 \rangle$. □

Algorithm 7.22 To decide whether $g \in \sqrt{I}$, compute a Gröber basis for $\langle f_1, \ldots, f_r, zg - 1 \rangle \subset k[x_1, \ldots, x_n, z]$. If it contains 1 then $g \in \sqrt{I}$.

Note, however, that we have *not* given an algorithm for computing generators for $\sqrt{I}$ for an arbitrary ideal I.

7.6.2 Irreducibility of hypersurfaces

Let k be algebraically closed. Given $f \in k[x_1, \ldots, x_n]$ with associated hypersurface $V(f)$, we can factor
$$f = f_1^{a_1} \ldots f_r^{a_r}$$
as a product of irreducibles. This yields a decomposition
$$V(f) = V(f_1^{a_1}) \cup V(f_2^{a_2}) \ldots V(f_r^{a_r})$$
$$= V(f_1) \cup V(f_2) \ldots V(f_r).$$

Are these the irreducible components of $V(f)$? In other words, does the algebraic formulation of irreducibility for polynomials coincide with the geometric notion of irreducibility for varieties?

Theorem 7.23 Let k be algebraically closed. If $f \in k[x_1, \ldots, x_n]$ is irreducible as a polynomial then $V(f)$ is irreducible as a variety.

Proof Since $k[x_1, \ldots, x_n]$ is a unique factorization domain, the ideal $\langle f \rangle$ is prime by Proposition A.11. By the Nullstellensatz, $I(V(f)) = \sqrt{\langle f \rangle}$, so we just need to check that $\sqrt{\langle f \rangle} = \langle f \rangle$. Given $g \in \sqrt{\langle f \rangle}$ then $g^N \in \langle f \rangle$ and $f | g^N$. Since f is irreducible and $k[x_1, \ldots, x_n]$ is a unique factorization domain, we conclude that $f | g$ and $g \in \langle f \rangle$. □

Corollary 7.24 If $f_1, \ldots, f_r$ are irreducible and distinct in $k[x_1, \ldots, x_n]$, with k algebraically closed, then
$$\sqrt{\langle f_1^{a_1} \ldots f_r^{a_r} \rangle} = \langle f_1 \ldots f_r \rangle.$$

Remark 7.25 The assumption that k is algebraically closed is necessary. We have seen (Example 6.9) the irreducibility of f does not guarantee the irreducibility of V!

7.7 Dimension

Definition 7.26   Let $V \subset \mathbb{A}^n(k)$ be an irreducible affine variety. The *dimension* $\dim V$ is defined as the transcendence degree of $k(V)$ over k.

We outline an effective procedure to compute the dimension of a variety. Let $I \subset k[x_1, \ldots, x_n]$ be a prime ideal, F the quotient field of $k[x_1, \ldots, x_n]/I$, and d the transcendence degree of F over k.

By Proposition 7.15, there exist indices

$$1 \leq i_1 < i_2 < \ldots < i_d \leq n$$

such that $x_{i_1}, \ldots, x_{i_d}$ form a trascendence basis of F over k. Indeed, any maximal algebraically independent subset will do. We therefore focus on determining whether a subset of the variables is algebraically independent. For notational simplicity, we take the first few variables.

Proposition 7.27   *The elements $x_1, \ldots, x_e \in F$ are algebraically independent over k if and only if $I \cap k[x_1, \ldots, x_e] = 0$.*

The intersection can be effectively computed using the Elimination Theorem (Theorem 4.8)

Proof   If $x_1, \ldots, x_e$ are algebraically dependent then there exists a nonzero polynomial $f \in k[t_1, \ldots, t_e]$ such that $f(x_1, \ldots, x_e) \equiv 0 \pmod{I}$. This gives a nontrivial element of $I \cap k[x_1, \ldots, x_e]$. Conversely, each such element gives an algebraic dependence relation among $x_1, \ldots, x_e$. $\square$

Corollary 7.28 $x_1, \ldots, x_d \in F$ are a transcendence basis for F/k if and only if $I \cap k[x_1, \ldots, x_d] = 0$ and $I \cap k[x_1, \ldots, x_d, x_j] \neq 0$ for each $j > d$.

Nonzero elements $g(x_1, \ldots, x_d, x_j) \in I \cap k[x_1, \ldots, x_d, x_j]$ show that x_j is algebraically dependent on $x_1, \ldots, x_d$.

This suggests that to check whether an algebraically independent set of variables $x_1, \ldots, x_d$ is a transcendence basis, we should carry out $n - d$ distinct eliminations. We can do a bit better:

Proposition 7.29   *For each $j = 1, \ldots, n$ let*

$$I_j = (I \cap k[x_1, \ldots, x_j])k[x_1, \ldots, x_n].$$

Then $x_1, \ldots, x_d$ form a transcendence basis for F/k if and only if

$$0 = I_1 = I_2 = \ldots = I_d \subsetneq I_{d+1} \ldots \subsetneq I_n.$$

These ideals can be extracted by computing a Gröbner basis with respect to a monomial order which is simultaneously an elimination order for the sets $\{x_1, \ldots, x_j\}$, $j = d, \ldots, n$. Pure lexicographic order has this property.

Proof Suppose that $h \in k[x_1, \ldots, x_n]$ is contained in I_j. By Exercise 7.13, each coefficient of h, as a polynomial in $x_{j+1}, \ldots, x_n$, is also contained in I_j.

The vanishing of $I_1, \ldots, I_d$ is equivalent to the independence of $x_1, \ldots, x_d$ by Proposition 7.27.

Suppose that $x_1 \ldots, x_d$ is a transcendence basis for F/k. For each $j > d$ there exists a nonzero $g_j(t_1, \ldots, t_d, s) \in k[t_1, \ldots, t_d, s]$ such that

$$g_j(x_1, \ldots, x_d, x_j) \equiv 0 \pmod{I}.$$

If $g_j \in I_{j-1}$ then each coefficient of g_j (regarded as a polynomial in s) is contained in I_d, which is zero. Thus $g_j \in I_j$ and $g_j \notin I_{j-1}$.

Conversely, suppose we have the tower of ideals as described in the proposition. For $j > d$ choose $h \in k[x_1, \ldots, x_n] \cap k[x_1, \ldots, x_j]$ with $h \notin I_{j-1}$; we may expand

$$h = \sum_{e=0}^{N} h_e(x_1, \ldots, x_{j-1}) x_j^e$$

with some $h_e \not\equiv 0 \pmod{I}$. This implies x_j is algebraic over $k(x_1, \ldots, x_{j-1})$. Iterating Proposition A.16, we deduce that x_j is algebraic over $k(x_1, \ldots, x_d)$. □

7.8 Exercises

7.1 Let $F \in \mathbb{C}[x, y]$ be an irreducible polynomial. Consider the set

$$V = \{(x, y) \in \mathbb{C}^2 : F(x, y) = 0\}.$$

Suppose that $G \in \mathbb{C}[x, y]$ is a polynomial such that $G(u, v) = 0$ for each $(u, v) \in V$. Show that $F | G$. When $\mathbb{C}$ is replaced by $\mathbb{R}$, can we still conclude F divides G?

7.2 Let $V \subset \mathbb{A}^n(k)$ be an affine variety with coordinate ring $k[V]$. For each ideal $I \subset k[V]$, let

$$V(I) = \{v \in V : g(v) = 0 \text{ for each } g \in I\}.$$

Assume that k is algebraically closed. Prove that $V(I) = \emptyset$ if and only if $I = k[V]$.

7.3 Let $I \subset R$ be an ideal. Show that the radical

$$\sqrt{I} = \{g \in R : g^N \in I \text{ for some } N \in \mathbb{N}\}$$

is automatically an ideal.

7.4 Let $S \subset \mathbb{A}^n(k)$ be a subset. Show that $I(S)$ is radical.

7.5 Let F/k be a finitely generated extension which is algebraic. Show this extension is finite.

7.8 EXERCISES

7.6 Determine the dimensions of the following varieties:
(a) affine space $\mathbb{A}^n(k)$;
(b) a point;
(c) an irreducible hypersurface $V(f) \subset \mathbb{A}^n(\mathbb{C})$.

7.7 Let $V \subset \mathbb{A}^6(\mathbb{Q})$ be defined by the two-by-two minors of the matrix

$$\begin{pmatrix} x_1 & x_2 & x_3 \\ x_4 & x_5 & x_6 \end{pmatrix}.$$

We write $F = \mathbb{Q}(V)$. Exhibit a transcendence base $z_1, \ldots, z_d$ for F over $\mathbb{Q}$, and express F explicitly as an algebraic extension over $\mathbb{Q}(z_1, \ldots, z_d)$.

7.8 Let F be the quotient field of

$$\mathbb{C}[x_1, x_2, x_3, x_4, x_5]/\langle x_1^2 + x_2^2 + x_3^2 - 1 \rangle.$$

Exhibit a transcendence base $z_1, \ldots, z_d$ for F over $\mathbb{C}$, and express F explicitly as an algebraic extension over $\mathbb{C}(z_1, \ldots, z_d)$.

7.9 Prove Proposition 7.13.

7.10 Let R be a domain with fraction field L, and assume that α is algebraic over L. Show that

$$\{r \in R : r\alpha \text{ integral over } R\}$$

is a nonzero ideal in R.

7.11 Let $V \to W$ be a dominant morphism of affine varieties. Show that $\dim V \geq \dim W$.

7.12 Let R and S be Noetherian integral domains with $R \subset S$. Suppose that $\alpha, \beta \in S$ are roots of the monic polynomials

$$x^2 + a_1 x + a_0, \, x^2 + b_1 x + b_0 \in R[x]$$

respectively. Using Gröbner basis techniques, exhibit a monic polynomial that has $\alpha + \beta$ as a root. Do the same for $\alpha\beta$.

7.13 Let $I \subset k[x_1, \ldots, x_n]$ be an ideal and set

$$I_j = (I \cap k[x_1, \ldots, x_j])k[x_1, \ldots, x_n]$$

so that

$$I_1 \subset I_2 \subset \ldots \subset I_n.$$

Given $h \in k[x_1, \ldots, x_n]$ write

$$h = \sum_{\alpha_{j+1} \ldots \alpha_n} c_{\alpha_{j+1} \ldots \alpha_n} x_{j+1}^{\alpha_{j+1}} \ldots x_n^{\alpha_n}$$

such that each $c_{\alpha_{j+1} \ldots \alpha_n} \in k[x_1, \ldots, x_j]$. Show that if $h \in I_j$ then each $c_{\alpha_{j+1} \ldots \alpha_n} \in I_j$.

7.14 Given a monomial $x^\alpha = x_1^{\alpha_1} \ldots x_n^{\alpha_n}$, write

$$\mathrm{rad}(x^\alpha) = \prod_{i \text{ with } \alpha_i \neq 0} x_i.$$

For each monomial ideal

$$I = \langle x^{\alpha(i)} = x_1^{\alpha(i,1)} \ldots x_n^{\alpha(i,n)} \rangle_{i=1,\ldots,r}$$

show that

$$\sqrt{I} = \langle \mathrm{rad}(x^{\alpha(i)}) \rangle_{i=1,\ldots,r}.$$

Assume that the generating set of monomials for I is minimal, i.e., given distinct $i, j = 1, \cdots r$, $x^{\alpha(i)}$ does not divide $x^{\alpha(j)}$. Show that I is radical if and only if each $\alpha(i, \ell) = 0$ or 1.

Write down all the radical monomial ideals in $k[x_1, x_2, x_3]$ and describe the corresponding varieties in $\mathbb{A}^3(k)$.

7.15 Show that every maximal ideal $\mathfrak{m} \subset \mathbb{R}[x_1, x_2]$ is one of the following:
- $\mathfrak{m} = \langle x_1 - \alpha_1, x_2 - \alpha_2 \rangle$ for some $\alpha_1, \alpha_2 \in \mathbb{R}$;
- $\mathfrak{m} = \langle x_2 - rx_1 - s, x_1^2 + bx_1 + c \rangle$ for some $r, s, b, c \in \mathbb{R}$ with $b^2 - 4c < 0$;
- $\mathfrak{m} = \langle x_1 - t, x_2^2 + bx_2 + c \rangle$ for some $t, b, c \in \mathbb{R}$ with $b^2 - 4c < 0$.

7.16 Let $I = \langle f_1, f_2 \rangle \subset \mathbb{C}[x, y]$ be an ideal generated by a linear and an irreducible quadratic polynomial. Suppose that $g \in \sqrt{I}$. Show that $g^2 \in I$.

7.17 (Geometric version of Proposition 7.27) Let $V \subset \mathbb{A}^n(k)$ be an irreducible affine variety over an infinite field. Show that $\dim(V) \geq d$ if and only if there exists a subset

$$\{x_{j_1}, \ldots, x_{j_d}\} \subset \{x_1, \ldots, x_n\}$$

so the projection morphism

$$\pi : V \to \mathbb{A}^d(k)$$
$$(x_1, \ldots, x_n) \mapsto (x_{j_1}, \ldots, x_{j_d})$$

is dominant.

7.18 Let I and J be ideals in a ring R.
(a) Show that

$$\sqrt{I \cap J} = \sqrt{I} \cap \sqrt{J}.$$

(b) On the other hand, give an example in $R = k[x_1, \ldots, x_n]$ where

$$\sqrt{I + J} \neq \sqrt{I} + \sqrt{J}.$$

7.19 (Codimension-1 varieties are hypersurfaces) Show that any irreducible variety $V \subset \mathbb{A}^n(k)$ of dimension $n - 1$ is a hypersurface. *Hint:* It suffices to prove that any prime ideal $I \subset k[x_1, \ldots, x_n]$ with $I \cap k[x_1, \ldots, x_{n-1}] = 0$ is principal.

7.8 EXERCISES

Assuming $I \neq 0$, we can express $I = \langle f_1, \ldots, f_r \rangle$ where each f_i is irreducible and does not divide any of the f_j, $j \neq i$. Suppose $r > 1$ and consider f_1, f_2 in $L[x_n]$ where $L = k(x_1, \ldots, x_{n-1})$. The resultant $\text{Res}(f_1, f_2)$ is defined; verify that $\text{Res}(f_1, f_2) \in k[x_1, \ldots, x_{n-1}] \cap I$ and thus is zero. Deduce the existence of an irreducible polynomial in $L[x_n]$ dividing both f_1 and f_2. On clearing denominators and applying Gauss' Lemma, we obtain an irreducible $h \in k[x_1, \ldots, x_n]$ with $h | f_1, f_2$.

8 Primary decomposition

We have shown that every variety is a union of irreducible subvarieties

$$V = V_1 \cup V_2 \cup \ldots \cup V_r, \quad V_i \not\subset V_j, i \neq j.$$

Our goal here is to find an algebraic analog of this decomposition, applicable to *ideals*.

Example 8.1 Let $I = \langle f \rangle \subset k[x_1, \ldots, x_n]$ be principal and decompose the generator into irreducible elements

$$f = f_1^{e_1} \ldots f_r^{e_r},$$

where no two of the f_j are proportional. Then we can write

$$\begin{aligned} I &= \langle f_1^{e_1} \rangle \cap \langle f_2^{e_2} \rangle \ldots \langle f_r^{e_r} \rangle \\ &= P_1^{e_1} \cap \ldots \cap P_r^{e_r}, \quad P_i = \langle f_i \rangle. \end{aligned}$$

Note that P_i is prime by Proposition A.11.

A warning is in order: decomposing even a univariate polynomial into irreducible components can be tricky in practice; the decomposition is very sensitive to the base field k. For example, given a finite extension $L/\mathbb{Q}$ and $f \in \mathbb{Q}[t]$, factoring f into irreducible polynomials over L is really a number-theoretic problem rather than a geometric one. This makes it challenging to implement primary decomposition on a computer, although there are algorithms for extracting some information about the decomposition [10].

8.1 Irreducible ideals

Here we emulate the decomposition into irreducible components described in Theorem 6.4. Keep in mind that unions of varieties correspond to intersections of ideals (cf. Propositions 3.6 and 3.12).

8.1 IRREDUCIBLE IDEALS

Definition 8.2 An ideal $I \subset R$ is *reducible* if can be expressed as the intersection of two larger ideals in R

$$I = J_1 \cap J_2, \quad I \subsetneq J_1, J_2.$$

An ideal is *irreducible* if it is not reducible.

Proposition 8.3 *Let R be Noetherian. Then any ideal $I \subset R$ can be written as an intersection of irreducible ideals*

$$I = I_1 \cap I_2 \ldots \cap I_m.$$

We say that the decomposition is *weakly irredundant* if none of the I_j can be left out, i.e.,

$$I_j \not\supset I_1 \cap \ldots I_{j-1} \cap I_{j+1} \ldots \cap I_r.$$

Proof Suppose this is not the case, so we get an infinite sequence of decompositions $I = I[1] \cap I[1]'$, with $I[1], I[1]' \supsetneq I$, $I[1] = I[2] \cap I'[2]$, with $I[2], I'[2] \supsetneq I[1]$, etc. Thus we obtain an infinite ascending sequence of ideals

$$I \subsetneq I[1] \subsetneq I[2] \ldots ,$$

violating the fact that R is Noetherian. $\square$

A variety V is irreducible precisely when $I(V)$ is prime (Theorem 6.5). This connection persists in our algebraic formulation:

Proposition 8.4 *Any prime ideal is irreducible.*

Proof Suppose that I is prime and $I = J_1 \cap J_2$ with $I \subsetneq J_1$. Pick $f \in J_1$ with $f \notin I$. Given $g \in J_2$ we have

$$fg \in J_1 J_2 \subset J_1 \cap J_2 = I.$$

Since I is prime, it follows that $g \in I$. We conclude that $I = J_2$. $\square$

Example 8.5 If $I = \langle xy, x^2 \rangle \subset k[x, y]$ then we can write

$$I = \langle x \rangle \cap \langle x^2, y \rangle.$$

The first term is prime, hence irreducible, so we focus on the second term. Suppose there were a decomposition

$$\langle x^2, y \rangle = J_1 \cap J_2$$

with $I \subsetneq J_1, J_2$. The quotient ring $R = k[x,y]/\langle x^2, y\rangle$ has dimension $\dim_k R = 2$, so $\dim_k k[x,y]/J_i = 1$ and each J_i is a maximal ideal containing $\langle x^2, y\rangle$. The only possibility is $\langle x, y\rangle$, which contradicts our assumptions.

Theorem 6.4 also asserts the uniqueness of the decomposition into irreducible components. This is conspicuously lacking from Proposition 8.3. Unfortunately, uniqueness fails in the algebraic situation:

Example 8.6   Consider $I = \langle x^2, xy, y^2\rangle \subset k[x,y]$. We have

$$I = \langle y, x^2\rangle \cap \langle y^2, x\rangle = \langle y+x, x^2\rangle \cap \langle x, (y+x)^2\rangle.$$

The analysis in Example 8.5 shows that the ideals involved are all irreducible.

For the rest of this chapter, we will develop *partial* uniqueness results for representations of ideals as intersections of irreducible ideals.

8.2 Quotient ideals

Definition 8.7   Given ideals I, J in a ring R, the *quotient ideal* is defined

$$I : J = \{r \in R : rs \in I \text{ for each } s \in J\}.$$

For any ideals $I, J, K \subset R$, $IJ \subset K$ if and only if $I \subset K : J$. This explains our choice of terminology.

Proposition 8.8   For ideals $I, J \subset k[x_1, \ldots, x_n]$ we have:

1. $I : J \subset I(V(I) \setminus V(J))$;
2. $V(I : J) \supset \overline{V(I) \setminus V(J)}$;
3. $I(V) : I(W) = I(V \setminus W)$.

Proof   Given $f \in I : J$, we have $fg \in I$ for each $g \in J$. For $x \in V(I) \setminus V(J)$, there exists a $g \in J$ such that $g(x) \neq 0$. Since $(fg)(x) = 0$ we have $f(x) = 0$, which proves the first assertion. The second assertion follows by taking varieties associated to the ideals in the first assertion. The inclusion $I(V) : I(W) \subset I(V \setminus W)$ follows from the first assertion; set $I = I(V)$ and $J = I(W)$. To prove the reverse inclusion, note that if $f \in I(V \setminus W)$ and $g \in I(W)$ then $fg \in I(V)$ and thus $f \in I(V) : I(W)$. □

We develop algorithms for computing the quotient by reducing its computation to the computation of intersections:

Proposition 8.9   Let $I \subset R$ be an ideal and $g \in k[x_1, \ldots, x_n]$. If

$$I \cap \langle g\rangle = \langle h_1, \ldots, h_s\rangle$$

8.3 PRIMARY IDEALS

then

$$I : \langle g \rangle = \langle h_1/g, h_2/g, \ldots, h_s/g \rangle.$$

Given an ideal $J = \langle g_1, \ldots, g_r \rangle$ *we have*

$$I : J = (I : \langle g_1 \rangle) \cap (I : \langle g_2 \rangle) \cap \ldots \cap (I : \langle g_r \rangle).$$

Proof Each $h_i/g \in I : \langle g \rangle$, so the inclusion $\supset$ is clear. Given $f \in I : \langle g \rangle$ we have $gf \in I, \langle g \rangle$, i.e., $gf \in I \cap \langle g \rangle$. The last assertion is an immediate consequence of the definitions. $\square$

Algorithm 8.10 *To compute the quotient* $I : J$ *of ideals* $I, J \subset k[x_1, \ldots, x_n]$ *with* $J = \langle g_1, \ldots, g_r \rangle$, *we carry out the following steps:*

1. *compute the intersections* $I \cap \langle g_i \rangle$ *using Proposition 6.19;*
2. *using the first part of Proposition 8.9, write out generators for* $I : \langle g_i \rangle$ *in terms of the generators of* $I \cap \langle g_i \rangle$;
3. *compute the intersection*

$$(I : \langle g_1 \rangle) \cap (I : \langle g_2 \rangle) \cap \ldots \cap (I : \langle g_r \rangle),$$

which is the desired quotient $I : J$.

Remark 8.11 To compute the intersection of a finite collection of ideals

$$J_1, \ldots, J_m \subset k[x_1, \ldots, x_n],$$

the following formula is useful:

$$J_1 \cap \ldots \cap J_m = (s_1 J_1 + \cdots + s_m J_m + \langle s_1 + \cdots + s_m - 1 \rangle) \cap k[x_1, x_2, \ldots, x_n].$$

8.3 Primary ideals

Definition 8.12 An ideal I in a ring R is *primary* if $fg \in I$ implies $f \in I$ or $g^m \in I$ for some m.

Example 8.13

1. If $P = \langle f \rangle \subset k[x_1, \ldots, x_n]$ is principal and prime then P^M is primary.
2. If $\mathfrak{m} = \langle x_1, \ldots, x_n \rangle \subset k[x_1, \ldots, x_n]$ and $I \subset k[x_1, \ldots, x_n]$ is an ideal with $\mathfrak{m}^M \subset I$ for some $M > 0$ then I is primary.

We prove the second assertion. Suppose that $fg \in I$ and $g^m \notin I$ for any m. It follows that $g \notin \mathfrak{m}$ and $g(0, \ldots, 0) \neq 0$. We claim the multiplication

$$\mu_g : k[x_1, \ldots, x_n]/\mathfrak{m}^M \to k[x_1, \ldots, x_n]/\mathfrak{m}^M$$
$$h \mapsto hg$$

is injective: if h has Taylor expansion about the origin

$$h = h_d + \text{higher-order terms}, \quad h_d \neq 0,$$

then gh has expansion

$$gh = g(0, \ldots, 0)h_d + \text{higher-order terms},$$

with the first term nonzero. The quotient $k[x_1, \ldots, x_n]/\mathfrak{m}^M$ is finite-dimensional, so μ_g is also surjective. Since $\mu_g(I) \subset I$, we must have $\mu_g(I) \equiv I \pmod{\mathfrak{m}^M}$. In particular, $fg \in I$ implies $f \in I + \mathfrak{m}^M = I$.

Remark 8.14 It is *not* generally true that if $P \subset k[x_1, \ldots, x_n]$ is prime then P^m is primary! A counterexample is given in Example 8.18.

Proposition 8.15 *Let Q be a primary ideal. Then $\sqrt{Q}$ is a prime ideal P, called* the *associated prime of Q.*

Proof Let $fg \in \sqrt{Q}$. Then $f^M g^M \in Q$ and either $f^M \in Q$ or $g^{Mm} \in Q$, and thus $f \in \sqrt{Q}$ or $g \in \sqrt{Q}$. $\square$

Proposition 8.16 *Any irreducible ideal I in a Noetherian ring R is primary.*

The converse does not hold – see Exercise 8.10 for a counterexample.

Proof Assume that I is irreducible and suppose that $fg \in I$. Consider the sequence of ideals

$$I \subset I : \langle g \rangle \subset I : \langle g^2 \rangle \subset \ldots,$$

which eventually terminates, so that

$$I : \langle g^N \rangle = I : \langle g^{N+1} \rangle$$

for some N. We claim that

$$(I + \langle g^N \rangle) \cap (I + \langle f \rangle) = I.$$

The inclusion $\supset$ is clear. Conversely, given an element $h \in (I + \langle g^N \rangle) \cap (I + \langle f \rangle)$, we can write

$$h = F_1 + H_1 g^N = F_2 + H_2 f, \quad F_1, F_2 \in I.$$

Since $fg \in I$ we have $g^{N+1} H_1 \in I$ and thus $g^N H_1 \in I$ and $h \in I$. Since I is irreducible, the claim implies either $f \in I$ or $g^N \in I$. $\square$

8.3 PRIMARY IDEALS

Combining this with Proposition 8.3, we deduce the following result.

Theorem 8.17 (Existence of primary decompositions) *For any ideal I in a Noetherian ring R, we can write*

$$I = Q_1 \cap Q_2 \cap \ldots \cap Q_r$$

where each Q_j is primary in R. This is called a primary decomposition *of I.*

Example 8.18 Consider the ideal $P \subset k[A, B, C, D, E, F]$ with generators

$$AD - B^2, \quad AF - C^2, \quad DF - E^2, \quad AE - BC, \quad BE - CD, \quad BF - CE,$$

i.e., the two-by-two minors of the symmetric matrix

$$M = \begin{pmatrix} A & B & C \\ B & D & E \\ C & E & F \end{pmatrix}.$$

This ideal is prime: it consists of the equations vanishing on the image V of the map

$$\mathbb{A}^3(k) \to \mathbb{A}^6(k)$$
$$(s, t, u) \to (s^2, st, su, t^2, tu, u^2).$$

We claim that P^2 is not primary. Consider the determinant

$$\det(M) = ADF - AE^2 - B^2F + 2BCE - C^2D,$$

a cubic polynomial contained in P. However, $\det(M) \notin P^2$ because P^2 is generated by polynomials of degree 4.

We have the following relations

$$A \det(M) = (AD - B^2)(AF - C^2) - (AE - BC)^2$$
$$B \det(M) = (AD - B^2)(BF - CE) - (AE - BC)(BE - CD)$$
$$C \det(M) = (BF - CE)(AE - BC) - (AF - C^2)(BE - CD)$$
$$D \det(M) = (AD - B^2)(DF - E^2) - (BE - CD)^2$$
$$E \det(M) = (AE - sBC)(DF - E^2) - (BF - CE)(BE - CD)$$
$$F \det(M) = (AF - C^2)(DF - E^2) - (BF - CE)^2.$$

If P^2 were primary then some power of each of the variables would be contained in P^2, i.e., for some m

$$\langle A, B, C, D, E, F \rangle^m \subset P^2.$$

In particular, $V = V(P^2) \subset \{(0, 0, 0, 0, 0, 0)\}$, a contradiction.

8.4 Uniqueness of primary decomposition

A primary decomposition

$$I = Q_1 \cap Q_2 \cap \ldots \cap Q_r,$$

is *weakly irredundant* if none of the Q_j is superfluous, i.e.,

$$Q_j \not\supseteq Q_1 \cap \ldots Q_{j-1} \cap Q_{j+1} \ldots \cap Q_r$$

for any j. For a weakly irredundant primary decomposition, the prime ideals $P_j = \sqrt{Q_j}$ are called the *associated primes* of I.

Unfortunately, even weakly irredundant primary decompositions are not unique in general:

Example 8.19 Consider two weakly irredundant primary decompositions

$$\begin{aligned} I &= \langle x^2, xy \rangle \\ &= \langle x \rangle \cap \langle y, x^2 \rangle = Q_1 \cap Q_2 \\ &= \langle x \rangle \cap \langle y^2, x^2, xy \rangle = Q_1 \cap Q_2'. \end{aligned}$$

Here $Q_1 = P_1$ is prime but Q_2' and Q_2 are not prime. They are primary with associated prime

$$P_2 = \sqrt{Q_2} = \sqrt{Q_2'} = \langle x, y \rangle.$$

When the ideal is geometric in origin, e.g., $I = I(V)$ for some $V \subset \mathbb{A}^n(k)$, the primary components reflect the geometry:

Example 8.20 Consider the weakly irredundant primary decomposition

$$\begin{aligned} I &= \langle xz - y^2, z - xy \rangle \\ &= \langle xz - y^2, z - xy, y - x^2 \rangle \cap \langle y, z \rangle \\ &= \langle z - xy, y - x^2 \rangle \cap \langle y, z \rangle = Q_1 \cap Q_2 \end{aligned}$$

Both Q_1 and Q_2 are prime: Q_1 consists of the equations vanishing on the image of the morphism

$$t \xrightarrow{\phi} (t, t^2, t^3)$$

and Q_2 consists of the equations for the x-axis. The ideal $I = I(V)$ where $V = \text{image}(\phi) \cup x\text{-axis}$.

See Remark 8.28 for a conceptual explanation of this example.

8.4 UNIQUENESS OF PRIMARY DECOMPOSITION

Proposition 8.21 An ideal $I \subset R$ is primary if and only if each zero divisor $\bar{r} \in R/I$ is nilpotent, i.e., $\bar{r}^m = 0$ for some m. When Q is primary, the nilpotents in R/Q are images of the associated prime $P = \sqrt{Q}$.

Proof By definition, I is primary if and only if

$$fg \in I \Rightarrow f \in I \text{ or } g^m \in I$$

for some m, i.e.,

$$\overline{fg} = 0 \pmod{I} \Rightarrow \overline{f} = 0 \pmod{I} \quad \text{or} \quad \overline{g}^m = 0 \pmod{I},$$

which is the first assertion. Thus when Q is primary, all zero divisors in R/Q come from $P = \sqrt{Q}$. $\square$

Theorem 8.22 (Uniqueness Theorem I) Let R be Noetherian and $I \subset R$ an ideal with weakly irredundant primary decomposition

$$I = Q_1 \cap Q_2 \ldots \cap Q_r.$$

The associated primes are precisely those primes which can be expressed as

$$P = \sqrt{I : \langle f \rangle}$$

for some $f \in R$. In particular, the associated primes are uniquely determined by I.

Proof From the definition of the quotient ideal we see that

$$I : \langle f \rangle = Q_1 : \langle f \rangle \cap \ldots \cap Q_r : \langle f \rangle.$$

Taking radicals and applying Exercise 7.18, we find that

$$\sqrt{I : \langle f \rangle} = \sqrt{Q_1 : \langle f \rangle} \cap \ldots \cap \sqrt{Q_r : \langle f \rangle}.$$

Furthermore, Proposition 8.21 gives

$$\sqrt{Q_i : \langle f \rangle} = \begin{cases} P_i \text{ if } f \notin Q_i \\ R \text{ if } f \in Q_i. \end{cases}$$

By the irredundance assumption, there exists $f \notin Q_j$ with $f \in \cap_{i \neq j} Q_i$. Then

$$\sqrt{I : \langle f \rangle} = \sqrt{Q_j : \langle f \rangle} = P_j,$$

and each associated prime has the desired form.

Conversely, suppose that $P = \sqrt{I : \langle f \rangle}$ for some $f \in R$. Then we have

$$P = \cap_{j : f \notin Q_j} P_j.$$

To finish, we need the following fact:

Lemma 8.23 *Let R be a ring and $P \subset R$ a prime ideal which can be expressed as a finite intersection of primes*

$$P = \cap_{j=1}^{s} P_j.$$

Then $P = P_j$ for some j.

Proof Suppose, on the contrary, that $P \subsetneq P_j$ for each j. Products are contained in intersections, so we have

$$P = \cap_{j=1}^{s} P_j \supset P_1 \ldots P_s.$$

Pick $g_j \in P_j, g_j \notin P$ so the product $g = \prod_{j=1}^{s} g_j \in P$ but each $g_j \notin P$. This contradicts the fact that P is prime. $\square$

Corollary 8.24 *Let $I \subset R$ be an ideal in a Noetherian ring. The zero divisors in R/I are the images of the associated primes for I.*

Proof The proof above shows that every element of an associated prime is a zero divisor. Conversely, suppose we have $ab = 0$ in R/I with $a, b \neq 0$. Then for some j, $a \notin Q_j$ but $ab \in Q_j$. Then b is a zero divisor in Q_j (or is contained in Q_j), and $b \in P_j$ by Proposition 8.21. $\square$

Definition 8.25 Let $I \subset R$ be an ideal in a Noetherian ring. An associated prime is *minimal* if it does not contain any other associated prime; otherwise, it is *embedded*.

Proposition 8.26 *Let $I \subset R$ be an ideal in a Noetherian ring. Then $\sqrt{I}$ is the intersection of the minimal associated primes of I.*

Proof Fix a primary decomposition

$$I = Q_1 \cap \ldots \cap Q_r$$

with associated primes $P_i = \sqrt{Q_i}, i = 1, \ldots, r$. Exercise 7.18 implies

$$\sqrt{I} = \sqrt{Q_1} \cap \ldots \sqrt{Q_r} = P_1 \cap \ldots \cap P_r.$$

Of course, excluding the embedded primes does not change the intersection. $\square$

Example 8.27 We retain the notation of Example 8.18. We showed that P^2 is not primary by computing

$$\langle \det(M) \rangle \mathfrak{m} \subset P^2, \quad \mathfrak{m} := \langle A, B, C, D, E, F \rangle,$$

8.4 UNIQUENESS OF PRIMARY DECOMPOSITION

but $\det(M) \notin P^2$. It follows that

$$\sqrt{P^2 : \langle \det(M) \rangle} = \mathfrak{m}$$

and Uniqueness Theorem I (8.22) implies $\mathfrak{m}$ is an associated prime of P^2.

P is the unique minimal associated prime of P^2: Indeed, for any prime P and integer $m > 0$ we have $P = \sqrt{P^m}$ (cf. Exercise 8.3.) By Proposition 8.26, the minimal associated primes of an ideal are the associated primes of its radical.

The only associated primes of P^2 are $\mathfrak{m}$ and P. The proof uses some Lie theory: Essentially, one classifies prime ideals in $k[A, B, C, D, E, F]$ (the 3×3 symmetric matrices) invariant under similarities, i.e., the equivalence relation $M \sim TMT^t$ where T is a 3×3 invertible matrix. This is carried out in [35, Prop. 4.15]; see also Exercise 8.12.

Remark 8.28 Minimal associated primes have a nice interpretation for ideals arising from geometry. Let $V \subset \mathbb{A}^n(k)$ be a variety with irredundant decomposition into irreducible components

$$V = V_1 \cup \ldots \cup V_r, \quad V_i \not\subset V_j, i \neq j.$$

In Exercise 7.4 we saw that $I(V) \subset k[x_1, \ldots, x_n]$ is radical. Proposition 3.6 gives the decomposition

$$I(V) = I(V_1) \cap \ldots \cap I(V_r), \quad I(V_i) \not\supset I(V_j), i \neq j,$$

and each $I(V_j)$ is prime by Theorem 6.5. Thus the $I(V_j), j = 1, \ldots, r$ are the minimal associated primes of $I(V)$.

We can sharpen this over algebraically closed fields:

Corollary 8.29 *Let k be algebraically closed and $I \subset k[x_1, \ldots, x_n]$ an ideal. The minimal associated primes of I correspond to irreducible components of $V(I)$.*

Proof Proposition 8.26 expresses $\sqrt{I}$ as the intersection of the minimal associated primes of I. The Hilbert Nullstellensatz gives $\sqrt{I} = I(V(I))$, and we know that the minimal associated primes of $I(V)$ correspond to the components of V. □

Based on this geometric intuition, we might expect the primary components associated with *minimal* associated primes to play a special rôle. We shall prove a strong uniqueness theorem for these components.

Proposition 8.30 *Let Q_1 and Q_2 be primary ideals with $\sqrt{Q_1} = \sqrt{Q_2} = P$. Then $Q_1 \cap Q_2$ is also primary with $\sqrt{Q_1 \cap Q_2} = P$.*

Proof Let $fg \in Q_1 \cap Q_2$. Suppose that $f \notin Q_1 \cap Q_2$, and assume that $f \notin Q_1$. Then $g^m \in Q_1$ for some m, and thus $g \in P$. Since $\sqrt{Q_2} = P$, $P^M \subset Q_2$ for some large M and $g^M \in Q_2$. Consequently, $g^{\max(m,M)} \in Q_1 \cap Q_2$. □

Suppose we have a weakly irredundant primary decomposition

$$I = Q_1 \cap Q_2 \cap \ldots \cap Q_r$$

such that two primary components have the *same* associated prime, i.e. $P_i = P_j$ for some $i \neq j$. The proposition allows us to replace Q_i and Q_j with the single *primary* $Q_i \cap Q_j$.

Definition 8.31 A primary decomposition

$$I = Q_1 \cap Q_2 \cap \ldots \cap Q_r$$

is *(strongly) irredundant* if it is weakly irredundant and the associated primes P_i are distinct.

Theorem 8.32 (Uniqueness Theorem II) *Let $I \subset R$ be an ideal in a Noetherian ring with irredundant primary decomposition*

$$I = Q_1 \cap Q_2 \cap \ldots \cap Q_r.$$

Suppose that P_j is a minimal associated prime. Then there exists an element $a \in \cap_{i \neq j} P_i$, $a \notin P_j$, and for each such element we have

$$Q_j = \cup_m (I : \langle a^m \rangle).$$

In particular, the primary components associated to minimal primes are unique.

Proof If there exists no element a with the desired properties then $P_j \supset \cap_{i \neq j} P_i$. Repeating the argument for Lemma 8.23, we find that $P_j \supset P_i$ for some $i \neq j$, which contradicts the minimality of P_j.

Given such an element, for any $r \in R$ there exists an m such that $a^m r \in Q_i, i \neq j$. On the other hand, a is not a zero-divisor modulo Q_j, so if $r \notin Q_j$ then $a^m r \notin Q_j$ for any m. Thus

$$a^m r \in I = Q_1 \cap Q_2 \cap \ldots \cap Q_r$$

for $m \gg 0$ if and only if $r \in Q_j$. □

Remark 8.33 The elements employed in Theorem 8.32 are easy to interpret when the ideal comes from geometry. As in Remark 8.28, suppose that $I = I(V)$ for a

8.4 UNIQUENESS OF PRIMARY DECOMPOSITION

variety $V \subset \mathbb{A}^n(k)$ with irreducible components

$$V = V_1 \cup \ldots \cup V_r, \quad V_i \not\subset V_j, i \neq j.$$

The associated primes of $I(V)$ are $P_j = I(V_j)$, $j = 1, \ldots, r$ and

$$\cap_{i \neq j} P_i = \{a \in k[V] : a \equiv 0 \text{ on each } V_i, i \neq j\}.$$

Theorem 8.32 requires an element $a \in k[V]$ vanishing on V_i for each $i \neq j$ but not on V_j.

Theorem 8.32 is quite useful for computations. Once we have found an element satisfying its hypotheses, we can effectively compute the corresponding primary component. The key tool is the following fact:

Proposition 8.34 *Let R be a ring, $I \subset R$ an ideal, and $a \in R$. Suppose that for some integer $M \geq 0$*

$$I : \langle a^M \rangle = I : \langle a^{M+1} \rangle,$$

where by convention $a^0 = 1$. Then we have

$$\cup_m (I : \langle a^m \rangle) = I : \langle a^M \rangle.$$

Proof It suffices to show that

$$I : \langle a^{M+1} \rangle = I : \langle a^{M+2} \rangle.$$

Indeed, this will imply

$$I : \langle a^M \rangle = I : \langle a^{M+1} \rangle = I : \langle a^{M+2} \rangle = I : \langle a^{M+3} \rangle = \ldots,$$

which is our result.

Pick $r \in (I : \langle a^{M+2} \rangle)$, so that $ra^{M+2} \in I$. We therefore have $ra \in I : \langle a^{M+1} \rangle$ and our hypothesis guarantees then that $ra \in I : \langle a^M \rangle$. We deduce that $(ra)a^M = ra^{M+1} \in I$. $\square$

Example 8.35 We retain the notation of Example 8.27.

What is the distinguished primary ideal $Q_1 \supset P^2$ corresponding to the minimal prime P? We shall show that

$$Q_1 = P^2 + \langle \det(M) \rangle.$$

First we establish that $\det(M) \in Q_1$; if $\det(M) \notin Q_1$ then elements of $\mathfrak{m}$ would be zero divisors (and hence nilpotents) modulo Q_1, whence $\mathfrak{m} \subset \sqrt{Q_1} = P$, a contradiction. To apply Theorem 8.32, we need an element $a \in \mathfrak{m}$, $a \notin P$; we take $a = A$.

Using the algorithms for computing quotient ideals and intersections (Algorithm 8.10 and Proposition 6.19), we compute

$$P^2 : \langle A \rangle = \langle AE^2 - 2BCE - AFD + C^2D + B^2F, D^2F^2 - 2DFE^2 + E^4,$$
$$BF^2D - BFE^2 - CEDF + CE^3, BEDF - BE^3 - CD^2F + CDE^2,$$
$$B^2F^2 - 2BFCE + C^2E^2, BE^2C - AE^3 - CDBF + AEDF,$$
$$B^2E^2 - ADE^2 - B^2DF + AD^2F, C^3E - C^2BF - AFCE + AF^2B,$$
$$BEC^2 - AE^2C - B^2CF + BEAF, CEB^2 - ADCE - B^3F + ADBF,$$
$$DB^2C - AD^2C - EB^3 + ADBE, C^4 - 2AFC^2 + A^2F^2,$$
$$BC^3 - AEC^2 - BCAF + A^2EF, A^2E^2 - 2AEBC + B^2C^2,$$
$$B^3C - ADBC - B^2AE + A^2DE, A^2D^2 - 2ADB^2 + B^4 \rangle$$
$$= \langle \det(M) \rangle + P^2$$

and

$$P^2 : \langle A^2 \rangle = \langle \det(M) \rangle + P^2.$$

Proposition 8.34 implies that

$$Q_1 = \cup_m P^2 : \langle A^m \rangle = \langle \det(M) \rangle + P^2.$$

One last intersection computation implies that our primary decomposition is

$$P^2 = (P^2 + \langle \det(M) \rangle) \cap \mathfrak{m}^4.$$

The component $Q_2 = \mathfrak{m}^4$ is not unique: we could take

$$Q_2' = Q_2 + \langle g \rangle, \quad g \in \mathfrak{m}^3, g \notin \langle \det(M) \rangle;$$

any proper ideal $Q_2' \supset \mathfrak{m}^n$ is $\mathfrak{m}$-primary.

8.5 An application to rational maps

Definition 8.36 Let $\rho : V \dashrightarrow W$ denote a rational map of affine varieties. The *indeterminacy ideal* $I_\rho \subset k[V]$ is defined

$$\{r \in k[V] : r\rho^* f \in k[V] \text{ for each } f \in k[W]\}.$$

We leave it to the reader to check that this is an ideal!

This is compatible with our previous definition of the indeterminacy locus:

Proposition 8.37 *The indeterminacy locus of ρ is equal to $V(I_\rho)$.*

8.5 AN APPLICATION TO RATIONAL MAPS

Proof Choose realizations $V \subset \mathbb{A}^n(k)$ and $W \subset \mathbb{A}^m(k)$, so that ρ is determined by the pull-back homomorphism (see Corolllary 3.46)

$$\rho^* : k[y_1, \ldots, y_m] \to k(V).$$

Suppose first that $v \in V(I_\rho)$. Let

$$\rho' : \mathbb{A}^n(k) \dashrightarrow \mathbb{A}^m(k)$$

be a rational map admissible along V such that $\rho'|_V = \rho$. Express each coordinate $\rho'_j = f_j/g$, with $f_j, g \in k[x_1, \ldots, x_n]$ and g not dividing zero in $k[V]$. We have $g\rho^*k[W] \subset k[V]$ and thus $g \in I_\rho$. Hence $g(v) = 0$ and ρ' is not defined at v. In particular, v is contained in the indeterminacy locus of ρ.

Assume that $v \notin V(I_\rho)$. We claim there exists an element $\bar{g} \in I_\rho$ such that $\bar{g}$ does not divide zero in $k[V]$ and $\bar{g}(v) \neq 0$. It follows that $\bar{g}\rho^*y_j \in k[V]$ for each j. We can then choose polynomials $g, f_1, \ldots, f_m \in k[x_1, \ldots, x_n]$ such that $g \equiv \bar{g}$ and $f_j \equiv \bar{g}\rho^*y_j$ in $k[V]$. The rational map

$$\rho' : \mathbb{A}^n(k) \to \mathbb{A}^m(k)$$
$$(x_1, \ldots, x_n) \mapsto (f_1/g, \ldots, f_m/g)$$

is admissible on V and induces ρ.

To prove the claim, we will require the following

Lemma 8.38 (Prime avoidance) *Let R be a ring and $P_1, \ldots, P_s$ prime ideals in R. If $I \subset R$ is an ideal with $I \subset \cup_{\ell=1}^s P_\ell$ then $I \subset P_\ell$ for some ℓ.*

We first give the application: The zero divisors of $k[V]$ are the union of the primes associated to the zero ideal in $k[V]$; they correspond to functions vanishing on at least one irreducible component of V (see Corollary 8.24 and Remark 8.28.) Let $R = k[V]$, $P_1, \ldots, P_{s-1}$ be the primes associated to zero, and P_s the maximal ideal corresponding to v. Not every element of I_ρ is a zero divisor, as ρ has a representative by a rational map on affine space admissible along V. In particular, $I_\rho \not\subset \cup_{\ell=1}^{s-1} P_\ell$ and $I_\rho \not\subset P_s = I(v)$. The lemma says that $I_\rho \not\subset \cup_{\ell=1}^s P_\ell$, so we can pick $\bar{g} \in I_\rho$ such that $\bar{g}$ neither divides zero nor vanishes at v.

Proof of lemma: The argument is by induction: if $I \subset \cup_{\ell=1}^s P_\ell$ we show that I is contained in the union of some collection of $s - 1$ primes. Suppose, on the contrary, that for each $\ell = 1, \ldots, s$ there exists $r_\ell \in I$ with $r_\ell \notin \cup_{j \neq \ell} P_j$. We must then have $r_\ell \in P_\ell$ for each ℓ. Consider the element

$$t = r_1 + r_2 \ldots r_s \in I \subset P_1 \cup P_2 \cup \ldots \cup P_s.$$

If $t \in P_1$ then $r_2 \ldots r_s \in P_1$ and $r_\ell \in P_1$ for some $\ell \neq 1$. On the other hand, suppose $t \in P_j$ for some $j > 1$; since $r_2 \ldots r_s \in P_j$, we deduce $r_1 \in P_j$. In either case, we reach a contradiction. $\square$

Proposition 8.39 *Retain the notation of Proposition 8.37. The following are equivalent:*

1. *ρ is a morphism;*
2. *$\rho^* k[W] \subset k[V]$;*
3. *$I_\rho = k[V]$.*

Proof The equivalence of the first two conditions follows from Corollaries 3.32 and 3.46: rational maps $V \dashrightarrow W$ (resp. morphisms $V \to W$) correspond to homomorphisms $k[W] \to k(V)$ (resp. $k[W] \to k[V]$). For the third, $I_\rho = k[V]$ precisely when $1 \in I_\rho$, i.e., when $\rho^* f \in k[V]$ for each $f \in k[W]$. $\square$

Our next result sharpens Proposition 8.37 and puts Example 3.49 in a general framework:

Proposition 8.40 *Let k be algebraically closed and consider a rational map of affine varieties $\rho : V \dashrightarrow W$. Then the following are equivalent:*

1. *ρ is a morphism;*
2. *the indeterminacy locus of ρ is empty.*

Proof The first condition obviously defines the second; we prove the converse.

Let $I_\rho \subset k[V]$ denote the indeterminacy ideal; Proposition 8.37 implies $V(I_\rho) = \emptyset$. Realize $V \subset \mathbb{A}^n(k)$ as a closed subset with quotient homomorphism $q : k[x_1, \ldots, x_n] \twoheadrightarrow k[V]$, and set $I = q^* I_\rho$ so that $V(I) = V(I_\rho)$. Since $V(I) = \emptyset$, the Nullstellensatz (Theorem 7.5) implies $I = k[x_1, \ldots, x_n]$ and thus $I_\rho = k[V]$. It follows that ρ is a morphism. $\square$

Given a rational map $\rho : V \dashrightarrow \mathbb{A}^m(k)$, we seek an algorithm for computing the indeterminacy ideal

$$I_\rho = \{r \in k[V] : r\rho^* k[\mathbb{A}^m(k)] \subset k[V]\} \subset k[V],$$

or, more precisely, its preimage in $k[x_1, \ldots, x_n]$

$$J_{V,\rho} := \{h \in k[x_1, \ldots, x_n] : h \pmod{I(V)} \in I_\rho\}.$$

We have

$$\begin{aligned} I_\rho &= \{r \in k[V] : rf_j/g_j \in k[V], j = 1, \ldots, m\} \\ &= \cap_{j=1}^M \{r \in k[V] : rf_j/g_j \in k[V]\} \\ &= \cap_{j=1}^M \{r \in k[V] : rf_j \in g_j k[V]\} \\ &= \cap_{j=1}^M \langle g_j \rangle : \langle f_j \rangle, \end{aligned}$$

which implies

$$J_{V,\rho} = \cap_{j=1}^{m}(\langle g_j \rangle + I(V)) : (\langle f_j \rangle + I(V)). \tag{8.1}$$

The last step is an application of the following general fact

Lemma 8.41 *Let $\psi : R \twoheadrightarrow S$ be a surjective ring homomorphism, $J_1, J_2 \subset S$ ideals with preimages $I_i = \psi^{-1}(J_i) \subset R$. Then*

$$I_1 : I_2 = \psi^{-1}(J_1 : J_2).$$

Proof The surjectivity of ψ implies $\psi(I_i) = J_i$ for $i = 1, 2$. We prove $I_1 : I_2 \subset \psi^{-1}(J_1 : J_2)$ first. Given $r \in R$ with $rI_2 \subset I_1$, applying ψ yields $\psi(r)\psi(I_2) \subset \psi(I_1)$. It follows then that $\psi(r) \in J_1 : J_2$. We turn to the reverse implication. Take $r \in R$ with $\psi(r) \in J_1 : J_2$. For each $w \in I_2$ we have $\psi(r)\psi(w) = \psi(rw) \in J_1$, hence $rw \in \psi^{-1}(J_1) = I_1$. It follows that $r \in I_1 : I_2$. □

Formula 8.1 allows us to compute $J_{V,\rho}$ by iterating our previous algorithms for computing quotients and intersections (Algorithm 8.10 and Proposition 6.19.)

8.6 Exercises

8.1 Flesh out the details in Example 8.6. Prove

$$\langle x^2, xy, y^2 \rangle = \langle y, x^2 \rangle \cap \langle y^2, x \rangle = \langle y + x, x^2 \rangle \cap \langle x, (y + x)^2 \rangle$$

using our algorithm for computing intersections. Verify that the ideals appearing are irreducible.

8.2 Let $C = \{(x_1, x_2) : x_1^3 - x_2^2 = 0\} \subset \mathbb{A}^2(\mathbb{Q})$ and consider the birational parametrization

$$\phi : \mathbb{A}^1(\mathbb{Q}) \to C$$
$$t \mapsto (t^2, t^3).$$

Use the method of §8.5 to compute the indeterminacy ideal $I_{\phi^{-1}} \subset \mathbb{Q}[C]$.

8.3 Let $P \subset R$ be a prime ideal and m a positive integer. Show that $\sqrt{P^m} = P$. In particular, if R is Noetherian then P is a minimal associated prime of P^m. In this case, the primary component of P^m associated to P is called the *mth symbolic power* of P.

8.4 Let R be Noetherian and $\mathfrak{m} \subset R$ a maximal ideal. Let $Q \subset R$ be an ideal such that $\mathfrak{m}^M \subset Q \subset \mathfrak{m}$ for some M. Show that Q is $\mathfrak{m}$-primary.

8.5 (a) Show that the ideal

$$Q = \langle xy - z^2, x \rangle \subset \mathbb{C}[x, y, z]$$

is primary. Describe the associated prime.

(b) Compute an irredundant primary decomposition and associated primes of the ideal
$$I = \langle z^2, yz, xz, y^2 - x^2(x+1) \rangle \subset \mathbb{Q}[x, y, z].$$

(c) Consider the ideal
$$I = \langle xy, xz \rangle \subset \mathbb{Q}[x, y, z].$$

Describe the irreducible components of $V(I)$ and compute the primary decomposition of I.

8.6 Consider the ideals
$$P_1 = \langle x, y \rangle, \quad P_2 = \langle x, z \rangle, \quad P_3 = \langle y, z \rangle \subset \mathbb{Q}[x, y, z].$$

Show that the P_i are prime and compute an irredundant primary decomposition and the associated primes of the product $P_1 P_2 P_3$.

8.7 Consider the ideals
$$I_1 = \langle x, y \rangle, \quad I_2 = \langle y, z \rangle.$$

(a) Compute the intersection $I_1 \cap I_2$, using the Gröbner basis algorithm.
(b) Find a primary decomposition for $I_1 I_2$. Does it have an embedded prime?
(c) Does $I_1 I_2 = I_1 \cap I_2$?

8.8 Let $I_1, I_2 \subset k[x_1, \ldots, x_n]$ be ideals. Show that $I_1 : I_2 = I_1$ if I_2 is not contained in any of the associated primes of I_1.

8.9 Show that a principal ideal in a polynomial ring has no embedded associated primes. On the other hand, if
$$R = k[x_1, x_2, x_3, x_4]/\langle x_1 x_4 - x_2 x_3, x_2^3 - x_1^2 x_3, x_3^3 - x_2 x_4^2 \rangle$$

show that the principal ideal
$$\langle x_2 \rangle \subset R$$

has embedded associated primes.

8.10 Give an example of a primary ideal $Q \subset k[x_1, \ldots, x_n]$ which is not irreducible, i.e., $Q = I_1 \cap I_2, I_1, I_2 \neq Q$.

8.11 (a) Find generators for the ideals $J(X_i)$ for the following closed subsets:
$$X_1 = \{x_0 = x_1 = 0\} \cup \{x_1 = x_2 = 0\} \cup \{x_2 = x_3 = 0\} \cup \{x_3 = x_0 = 0\} \subset \mathbb{P}^3$$
$$X_2 = (\{x_0 = x_1 = 0\} \cup \{x_2 = x_3 = 0\}) \cap \{x_0 + x_1 + x_2 + x_3 = 0\} \subset \mathbb{P}^3$$

(b) Is the ideal $J(X_2)$ equal to
$$J = (\langle x_0, x_1 \rangle \cap \langle x_2, x_3 \rangle) + \langle x_0 + x_1 + x_2 + x_3 \rangle?$$

8.6 EXERCISES

(c) Compute the primary decompositions of $J(X_2)$ and J. *Hint:* What is $\sqrt{J : \langle f \rangle}$ for $f = x_0 + x_1 - x_2 - x_3$?

(d) How do you interpret the embedded prime of J?

8.12 Two symmetric $n \times n$ matrices M, M' are *similar* if there exists an invertible $n \times n$ matrix T with $M' = TMT^t$.

(a) Show that similar matrices have the same rank.

(b) Show that complex matrices of the same rank are similar.

Identify $\mathbb{A}^{n(n+1)/2}(\mathbb{C})$ with the $n \times n$ complex symmetric matrices $M = (m_{ij})$. An ideal $I \subset \mathbb{C}[m_{ij}]$ is *invariant under similarity* if, for every invertible matrix T,

$$f(M) \in I \Rightarrow f(TMT^t) \in I.$$

(c) Let $S_r \subset \mathbb{A}^{n(n+1)/2}(\mathbb{C})$ denote the matrices of rank r. Show that $I(S_r)$ is prime and invariant. *Hint:* Consider the morphism

$$\mathbb{A}^{n^2} \to \mathbb{A}^{n(n+1)/2}$$
$$T \mapsto TMT^t$$

with image $\overline{S_r}$ with $r = \text{rank}(M)$.

(d) Show that each invariant prime ideal $P \subset \mathbb{C}[m_{ij}]$ equals $I(S_r)$ for some r.

9 Projective geometry

Projective geometry arose historically out of plane geometry. It is very fruitful to introduce points 'at infinity' where parallel lines intersect. This leads to a very elegant approach to incidence questions, where points and lines are on an equal and symmetric footing. In the context of algebraic geometry, points at infinity are crucial in the statement of uniform results, like Bezout's Theorem on the intersection of two plane curves.

However, to do projective geometry we must leave the realm of affine varieties. A projective variety is constructed by gluing a number of affine varieties together. There are many subtle issues that arise, especially when the base field is not algebraically closed. These are deferred to the end of this chapter.

Thankfully, there is an extremely concrete approach to projective geometry using the algebra of homogeneous polynomials. This allows us to apply many of the computational techniques developed for affine varieties to projective varieties, with minor modifications. Indeed, concrete problems in affine geometry often become more transparent once they are translated into projective language.

9.1 Introduction to projective space

Projective n-space $\mathbb{P}^n(k)$ is the set of all lines in affine space containing the origin

$$0 \in \ell \subset \mathbb{A}^{n+1}(k).$$

Each such line takes the form

$$\mathrm{span}(a_0, \ldots, a_n) = \lambda(a_0, \ldots, a_n), \quad \lambda \in k,$$

where $(a_0, \ldots, a_n) \in \mathbb{A}^{n+1}(k) - \{0\}$. Two elements $(a_0, \ldots, a_n), (a'_0, \ldots, a'_n) \in \mathbb{A}^{n+1}(k)$ span the same line if

$$(a'_0, \ldots, a'_n) = \lambda(a_0, \ldots, a_n), \lambda \in k^*.$$

Thus we can identify projective space with equivalence classes

$$\mathbb{P}^n(k) = (\mathbb{A}^{n+1}(k) - \{0\})/\sim$$

where

$$(a_0', \ldots, a_n') \sim (a_0, \ldots, a_n) \quad \text{if } (a_0', \ldots, a_n') = \lambda(a_0, \ldots, a_n), \quad \lambda \in k^*.$$

We'll use the notation $[a_0, \ldots, a_n]$ to denote one of the resulting equivalence classes.

There is a natural way to parametrize 'most' lines in $\mathbb{A}^{n+1}(k)$ by an affine space of dimension n. For each $i = 0, \ldots, n$, consider the lines of the form

$$U_i = \{[a_0, \ldots, a_n] : a_i \neq 0\} \subset \mathbb{P}^n(k).$$

The vanishing (or nonvanishing) of a_i is compatible with the following equivalence relation: when $[a_0, \ldots, a_n] \sim [a_0', \ldots, a_n']$ $a_i = 0$ if and only if $a_i' = 0$. We have a function

$$\psi_i : U_i \to \mathbb{A}^n(k)$$
$$[a_0, \ldots, a_n] \mapsto (a_0/a_i, \ldots, a_{i-1}/a_i, a_{i+1}/a_i, \ldots, a_n/a_i)$$

with inverse

$$\psi_i^{-1} : \mathbb{A}^n(k) \to U_i$$
$$(b_0, \ldots, b_{i-1}, b_{i+1}, \ldots, b_n) \mapsto [b_0, \ldots, b_{i-1}, \underbrace{1}_{i\text{th place}}, b_{i+1}, \ldots, b_n].$$

The function ψ_i identifies lines in U_i with points in the affine space $\mathbb{A}^n(k)$ with coordinates $b_0, \ldots, b_{i-1}, b_{i+1}, \ldots, b_n$. For any point in $\mathbb{P}^n(k)$ some $a_i \neq 0$, so we can express

$$\mathbb{P}^n(k) = U_0 \cup U_1 \cup \ldots \cup U_n, \quad U_i \simeq \mathbb{A}^n(k). \tag{9.1}$$

Example 9.1 The line $\mathrm{span}(a_0, a_1) \subset \mathbb{A}^2(k)$ is defined by the linear equation

$$a_1 x_0 - a_0 x_1 = 0.$$

When $a_0 \neq 0$ we can divide through by a_0 to get

$$x_1 = b_1 x_0, \quad b_1 = a_1/a_0;$$

these are the lines in U_0. When $a_1 \neq 0$ we get

$$x_0 = b_0 x_1, \quad b_0 = a_0/a_1,$$

corresponding to the lines in U_1.

What are the lines 'left out' by our distinguished subsets $U_i \subset \mathbb{P}^n(k)$? The complement

$$H_i := \mathbb{P}^n(k) \backslash U_i = \{[a_0, \ldots, a_{i-1}, 0, a_{i+1}, \ldots, a_n] : (a_0, \ldots, a_{i-1}, a_{i+1}, \ldots, a_n) \neq 0\}$$

PROJECTIVE GEOMETRY

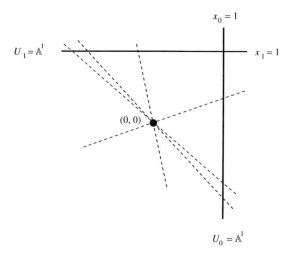

Figure 9.1 Two $\mathbb{A}^1$'s parametrizing lines in $\mathbb{A}^2$.

consists of the lines through the origin in the affine subspace

$$0 \in \ell \subset \mathbb{A}^n(k) = \{a_i = 0\} \subset \mathbb{A}^{n+1}(k).$$

We therefore may express

$$\mathbb{P}^n(k) = U_i \cup H_i = \mathbb{A}^n(k) \cup \mathbb{P}^{n-1}(k),$$

where we interpret $\mathbb{P}^0(k)$ as a point.

How do the identifications $\psi_i : U_i \simeq \mathbb{A}^n(k)$ fit together? Restrict to the intersections

$$\begin{array}{c} U_i \cap U_j \\ {}^{\psi_j}\swarrow \quad \searrow^{\psi_i} \\ \mathbb{A}^n(k) \qquad \mathbb{A}^n(k) \end{array}$$

with $i < j$ for notational simplicity. We consider the compositions $\rho_{ij} = \psi_i \circ \psi_j^{-1}$

$$\begin{array}{ccc} \psi_j(U_i \cap U_j) & \xrightarrow{\psi_i \circ \psi_j^{-1}} & \psi_i(U_i \cap U_j) \\ \| & & \| \\ \mathbb{A}^n(k) \setminus \{b_i = 0\} & \xrightarrow{\rho_{ij}} & \mathbb{A}^n(k) \setminus \{b'_j = 0\}. \end{array}$$

Since $(\psi_i \circ \psi_j^{-1}) \circ (\psi_j \circ \psi_\ell^{-1}) = \psi_i \circ \psi_\ell^{-1}$ we have the *compatibility condition*

$$\rho_{ij} \circ \rho_{j\ell} = \rho_{i\ell}$$

for all i, j, ℓ. We explicitly compute ρ_{ij}

$$(b_0, \ldots, b_{j-1}, b_{j+1}, \ldots, b_n) \xrightarrow{\psi_j^{-1}} [b_0, \ldots, b_{j-1}, 1, b_{j+1}, \ldots, b_n]$$
$$= [b_0/b_i, \ldots, b_{i-1}/b_i, 1, b_{i+1}/b_i, \ldots, b_{j-1}/b_i, 1/b_i, b_{j+1}/b_i, \ldots, b_n/b_i]$$
$$\xrightarrow{\psi_i} (b_0/b_i, \ldots, b_{i-1}/b_i, b_{i+1}/b_i, \ldots, b_{j-1}/b_i, 1/b_i, b_{j+1}/b_i, \ldots, b_n/b_i)$$

which defines a birational map $\mathbb{A}^n(k) \dashrightarrow \mathbb{A}^n(k)$. Eliminating indeterminacy using Proposition 3.47, we get an isomorphism of affine varieties

$$\begin{array}{ccc} \mathbb{A}^n(k) \setminus \{b_i = 0\} & \xrightarrow{\rho_{ij}} & \mathbb{A}^n(k) \setminus \{b'_j = 0\} \\ \| & & \| \\ \mathbb{A}^n(k)_{b_i} & \xrightarrow{\sim} & \mathbb{A}^n(k)_{b'_j}. \end{array}$$

Example 9.2 For $\mathbb{P}^1(k)$ we get a single birational map

$$\rho_{01} : \mathbb{A}^1(k) \dashrightarrow \mathbb{A}^1$$
$$b_0 \mapsto 1/b_0.$$

For $\mathbb{P}^2(k)$ we get

$$\rho_{01} : \mathbb{A}^2(k) \dashrightarrow \mathbb{A}^2(k)$$
$$(b_0, b_2) \mapsto (b_0^{-1}, b_2/b_0),$$
$$\rho_{02} : \mathbb{A}^2(k) \dashrightarrow \mathbb{A}^2(k)$$
$$(b'_0, b'_1) \mapsto (b'_1/b'_0, 1/b'_0),$$
$$\rho_{12} : \mathbb{A}^2(k) \dashrightarrow \mathbb{A}^2(k)$$
$$(b'_0, b'_1) \mapsto (b'_0/b'_1, 1/b'_1)$$

satisfying $\rho_{02} = \rho_{01} \circ \rho_{12}$.

Remark 9.3 (Coordinate-free approach) Let V be a finite-dimensional k-vector space of dimension $n + 1$. The *projectivization of V* is defined

$$\mathbb{P}(V) = \{\text{one-dimensional vector subspaces } \ell \subset V\}.$$

While this is just $\mathbb{P}^n(k)$, for some constructions it is useful to keep track of the underlying vector-space structure. For example, $\mathbb{P}(k[x_0, \ldots, x_n]_d)$ denotes the projective space modeled on the homogeneous forms of degree d.

9.2 Homogenization and dehomogenization

Each polynomial $f \in k[x_0, \ldots, x_n]$ can be decomposed into *homogeneous pieces*

$$f = F_0 + F_1 + \cdots + F_d, \quad d = \deg(f),$$

i.e., each F_j is homogeneous of degree j in $x_0, \ldots, x_n$. An ideal $J \subset k[x_0, \ldots, x_n]$ is *homogeneous* if it admits a collection of homogeneous generators. Equivalently, if a polynomial is in a homogenous ideal then each of its homogeneous pieces is in that ideal (see Exercise 9.1).

Dehomogenization with respect to x_i is defined as the homomorphism

$$\mu_i : k[x_0, \ldots, x_n] \to k[y_0, \ldots, y_{i-1}, y_{i+1}, \ldots, y_n]$$
$$x_i \to 1$$
$$x_j \to y_j, \quad j \neq 1.$$

For $f \in k[y_0, \ldots, y_{i-1}, y_i, \ldots, y_n]$, the preimage $\mu_i^{-1}(f)$ contains

$$\left\{ x_i^D f(x_0/x_i, \ldots, x_{i-1}/x_i, x_{i+1}/x_i, \ldots, x_n/x_i) : D \geq \deg(f) \right\}$$

and equals the affine span of these polynomials. The *homogenization of f with respect to* x_i is defined

$$F(x_0, \ldots, x_n) := x_i^{\deg(f)} f(x_0/x_i, \ldots, x_{i-1}/x_i, x_{i+1}/x_i, \ldots, x_n/x_i).$$

The homogenization of an ideal $I \subset k[y_0, \ldots, y_{i-1}, y_{i+1}, \ldots, y_n]$ is the ideal generated by the homogenizations of each $f \in I$.

Given an ideal $I = \langle f_1, \ldots, f_r \rangle$, the homogenization J need not be generated by the homogenizations of the elements, i.e.,

$$J \neq \left\langle x_i^{\deg(f_j)} f_j(x_0/x_i, \ldots, x_{i-1}/x_i, x_{i+1}/x_i, \ldots, x_n/x_i) \right\rangle_{j=1,\ldots,r}$$

in general.

Example 9.4 Consider

$$I = \langle y_2 - y_1^2, y_3 - y_1 y_2 \rangle = \langle f_1, f_2 \rangle$$

and dehomogenize with respect to x_0

$$\mu_0 : k[x_0, x_1, x_2, x_3] \to k[y_1, y_2, y_3].$$

The homogenization of the elements f_1, f_2 gives an ideal

$$\langle x_2 x_0 - x_1^2, x_3 x_0 - x_1 x_2 \rangle \subsetneq J.$$

The polynomial $h = x_2^2 - x_1 x_3 \in J$ because $y_2^2 - y_1 y_3 = y_2 f_1 - y_1 f_2$, but h is not contained in the ideal generated by the homogenizations of f_1 and f_2.

Definition 9.5 A monomial order on $k[y_1, \ldots, y_n]$ is *graded* if it is compatible with the partial order induced by degree, i.e., $y^\alpha > y^\beta$ whenever $|\alpha| > |\beta|$.

For instance, lexicographic order is not graded, because small degree monomials in the first variable precede large degree monomials in the subsequent variables.

9.2 HOMOGENIZATION AND DEHOMOGENIZATION

Theorem 9.6 Let $I \subset k[y_1, \ldots, y_n]$ be an ideal and $J \subset k[x_0, \ldots, x_n]$ its homogenization with respect to x_0. Suppose that $f_1, \ldots, f_r$ is a Gröbner basis for I with respect to some graded order $>$. Then the homogenizations $F_1, \ldots, F_r$ of $f_1, \ldots, f_r$ generate J.

Proof We first introduce an order $>_x$ on the x-monomials derived from $>$. We define

$$x_0^{\alpha_0} x_1^{\alpha_1} \ldots x_n^{\alpha_n} >_x x_0^{\beta_0} x_1^{\beta_1} \ldots x_n^{\beta_n} \Leftrightarrow \begin{cases} \text{if } y_1^{\alpha_1} \ldots y_n^{\alpha_n} > y_1^{\beta_1} \ldots y_n^{\beta_n}, \\ \text{or } y_1^{\alpha_1} \ldots y_n^{\alpha_n} = y_1^{\beta_1} \ldots y_n^{\beta_n} \text{ and } \alpha_0 > \beta_0. \end{cases}$$

We leave it as an exercise to verify that this defines a monomial order.

The theorem will follow once we show that $F_1, \ldots, F_r$ form a Gröbner basis for J with respect to $>_x$; Corollary 2.14 guarantees they generate J.

Lemma 9.7 Let $G \in k[x_0, \ldots, x_n]$ be homogeneous with dehomogenization $g = \mu_0(G)$. If $\text{LT}_>(g) = c y_1^{\alpha_1} \ldots y_n^{\alpha_n}$ then $\text{LT}_{>_x}(G) = c x_0^{\deg(G)-\deg(g)} x_1^{\alpha_1} \ldots x_n^{\alpha_n}$; in particular, $\text{LT}_>(g) = \mu_0(\text{LT}_{>_x}(G))$.

Let G' denote the homogenenization of g with respect to x_0, so that $G = x_0^{\deg(G)-\deg(g)} G'$. It suffices to show that $\text{LT}_{>_x}(G') = c x_1^{\alpha_1} \ldots x_n^{\alpha_n}$. Since $>$ is graded, $\text{LT}_>(g) = c y_1^{\alpha_1} \ldots y_n^{\alpha_n}$ has degree equal to $\deg(g)$. Thus $c x_1^{\alpha_1} \ldots x_n^{\alpha_n}$ is a term of G'. Consider terms in G' in which only $x_1, x_2, \ldots, x_n$ appear; the leading term of G' is one of these. Indeed, terms containing x_0 dehomogenize to terms in $y_1, \ldots, y_n$ of degree $< \deg(g)$, and thus are smaller than monomials of degree $\deg(g)$ in $x_1, \ldots, x_n$. The order induced on monomials in $y_1, \ldots, y_n$ by $>$ coincides with the order induced on monomials in $x_1, \ldots, x_n$ by $>_x$, so the leading terms of g and G' coincide. This completes the proof of the lemma.

Lemma 9.8 $\mu_0(J) \subset I$.

Choose a homogeneous $H \in J$ and express

$$H = \sum_j A_j g_j(x_1/x_0, \ldots, x_n/x_0) x_0^{\deg g_j}, \quad g_j \in I, A_j \in k[x_0, \ldots, x_n].$$

Dehomogenizing with respect to x_0, we obtain

$$\mu_0(H) = \sum_j A_j(1, y_1, \ldots, y_n) g_j,$$

so $\mu_0(H) \in I$ and the lemma is proven.

Suppose that H is a homogeneous polynomial in J. It suffices to prove that $\text{LT}_{>_x}(H)$ is divisible by $\text{LT}_{>_x}(F_j)$ for some j. By the second lemma above $h = \mu_0(H) \in I$. Since $f_1, \ldots, f_r$ are a Gröbner basis for I we have $\text{LT}_>(f_j)|\text{LT}_>(h)$ for some j. Applying the first lemma twice, we conclude $\text{LT}_{>_x}(F_j)|\text{LT}_{>_x}(H)$. □

9.3 Projective varieties

Definition 9.9   A *projective variety* $X \subset \mathbb{P}^n(k)$ is a subset such that, for each distinguished $U_i \simeq \mathbb{A}^n(k)$, $i = 0, \ldots, n$, the intersection $U_i \cap X \subset U_i$ is affine.

Definition 9.10 $X \subset \mathbb{P}^n(k)$ is *Zariski closed* if $X \cap U_i$ is closed in each distinguished U_i. For any subset $S \subset \mathbb{P}^n(k)$, the *projective closure* $\overline{S} \subset \mathbb{P}^n(k)$ is defined as the smallest closed subset containing S.

Definition 9.11   A projective variety $X \subset \mathbb{P}^n(k)$ is *reducible* if it can be expressed as a union of two closed proper subsets

$$X = X_1 \cup X_2, \quad X_1, X_2 \subsetneq X.$$

It is *irreducible* if there is no such representation.

We describe a natural way to get large numbers of projective varieties:

Proposition 9.12   *Let $F \in k[x_0, \ldots, x_n]$ be homogeneous of degree d. Then there is a projective variety*

$$X(F) := \{[a_0, \ldots, a_n] : F(a_0, \ldots, a_n) = 0\} \subset \mathbb{P}^n(k),$$

called the hypersurface defined by F. More generally, given a homogeneous ideal $J \subset k[x_0, \ldots, x_n]$, we define

$$X(J) := \{[a_0, \ldots, a_n] : F(a_0, \ldots, a_n) = 0 \text{ for each homogeneous } F \in J\},$$

the projective variety defined by J.

Proof   Note that F does not yield a well-defined function on $\mathbb{P}^n(k)$: If $(a_0', \ldots, a_n') = \lambda(a_0, \ldots, a_n)$ then

$$F(a_0', \ldots, a_n') = \lambda^d F(a_0, \ldots, a_n).$$

However, we *can* make sense of the locus $X(F)$ where F vanishes, because $F(a_0', \ldots, a_n') = 0$ if and only if $F(a_0, \ldots, a_n) = 0$.

We check this is closed. On U_i we have $x_i \neq 0$, so $F = 0$ if and only if $x_i^{-d} F = 0$. However,

$$f := x_i^{-d} F = F(x_0/x_i, \ldots, x_{i-1}/x_i, 1, x_{i+1}/x_i, \ldots, x_n/x_i),$$

is a well-defined polynomial on $\mathbb{A}^n(k)$. Hence

$$U_i \cap X(F) = V(f) \subset U_i \simeq \mathbb{A}^n(k)$$

is affine for each i. The final assertion is obtained by intersecting the $X(F)$ for homogeneous $F \in J$. □

Example 9.13

1. The 2×2 minors of

$$\begin{pmatrix} A & B & C \\ B & D & E \\ C & E & F \end{pmatrix}$$

define a closed subset of $\mathbb{P}^5(k)$ (which we'll show is isomorphic to $\mathbb{P}^2(k)$).

2. The 2×2 minors of

$$\begin{pmatrix} A & B & C \\ B & C & D \end{pmatrix}$$

define a closed subset of $\mathbb{P}^3(k)$ (which we'll show is isomorphic to $\mathbb{P}^1(k)$).

Given $S \subset \mathbb{P}^n(k)$, the *homogeneous ideal vanishing along S* is defined

$$J(S) = \langle F \in k[x_0, \ldots, x_n] \text{ homogeneous } : F(s) = 0 \text{ for each } s \in S \rangle.$$

9.4 Equations for projective varieties

Our goal is to prove that Proposition 9.12 is robust enough to produce *every* projective variety:

Theorem 9.14 *Let $X \subset \mathbb{P}^n(k)$ be a projective variety. Then there exists a homogeneous ideal J such that $X = X(J)$.*

Our argument will yield an effective algorithm for computing J from the ideals $I(X \cap U_i)$.

Proposition 9.15 *If $S \subset \mathbb{P}^n(k)$ then $X(J(S)) = \overline{S}$.*

Proof It is clear that $S \subset X(J(S))$, so we have $\overline{S} \subset X(J(S))$. We prove the reverse inclusion. Suppose that $p \notin \overline{S}$. There exists a distinguished open subset $U_i \subset \mathbb{P}^n(k)$ such that $p \in U_i$ and $x_i(p) \neq 0$. Since $U_i \cap \overline{S}$ is closed, there exists a polynomial $f \in I(U_i \cap \overline{S})$ that does not vanish at p. Let F be the homogenization of f; we still have $F(p) \neq 0$. Note that F vanishes at all the points of $\overline{S} \cap U_i$ and x_i vanishes at each point of $\overline{S}$ not contained in U_i. Thus $x_i F \in J(S)$ and $(x_i F)(p) \neq 0$, so $p \notin X(J(S))$. □

Proposition 9.16 Let $V \subset \mathbb{A}^n(k) \simeq U_0 \subset \mathbb{P}^n$ be an affine variety with ideal $I(V) \subset k[y_1, \ldots, y_n]$. Let $J \subset k[x_0, \ldots, x_n]$ denote the homogenization of $I(V)$. Then $J(V) = J$ and $X(J) = \overline{V}$.

Proof Once we prove $J(V) = J$, Proposition 9.15 implies $X(J) = \overline{V}$.

We prove $J(V) \supset J$. For each homogeneous $G \in J$, $g := \mu_0(G) \in I(V)$ by Lemma 9.8. Thus G vanishes on $[1, a_1, \ldots, a_n]$ whenever $(a_1, \ldots, a_n) \in V$, i.e., $G \in J(V)$.

Conversely, suppose we are given a homogeneous $H \in J(V)$, so that $H(1, b_1, \ldots, b_n) = 0$ for each $(b_1, \ldots, b_n) \in V$. Writing $h = \mu_0(H)$, we find that $h \in I(V)$. We can write

$$H = x_0^{\deg(H)} h(x_1/x_0, \ldots, x_n/x_0) = x_0^{\deg(H)-\deg(h)} H'$$

where H' is the homogenization of h. In particular, H is contained in the homogenization of $I(V)$. □

Remark 9.17 In this situation, the hyperplane

$$H_0 := \{x_0 = 0\} = \mathbb{P}^n(k) - U_0$$

is often called the *hyperplane at infinity*. We have $H_0 \supset (\overline{V} \setminus V)$.

Example 9.18 Note that, in Proposition 9.16, we used the ideal of *all* functions vanishing on V rather than an arbitrary ideal vanishing on V. This is actually necessary; using any smaller ideal can lead to unwanted components at infinity.

For instance, consider the ideal

$$I = \langle y_1^2 + y_2^4 \rangle \subset \mathbb{R}[y_1, y_2]$$

which defines the origin in $\mathbb{R}^2$. The homogenization is

$$J = \langle x_0^2 x_1^2 + x_2^4 \rangle,$$

which defines a variety

$$X(J) = \{x_1 = x_2 = 0\} \cup \{x_0 = x_2 = 0\} \subset \mathbb{P}^2(\mathbb{R}).$$

This is strictly larger than the closure $\overline{V(I)}$; it contains an extra point at infinity.

The reason for this pathology is that we are working over a nonclosed field. Over the complex numbers, $\{x_0^2 x_1^2 + x_2^4 = 0\} \subset \mathbb{P}^2(\mathbb{C})$ is a plane curve C with two singularities, $[0, 1, 0]$ and $[1, 0, 0]$. These singularties are the only real points of C.

The following result reduces the computation of the equations of a projective variety to the computation of the equations for affine varieties:

9.4 EQUATIONS FOR PROJECTIVE VARIETIES

Proposition 9.19 Let $X \subset \mathbb{P}^n$ be closed, $I_i = I(U_i \cap X)$ where $U_i \subset \mathbb{P}^n(k)$ is one of the distinguished open subsets, and $J_i \subset k[x_0, \ldots, x_n]$ the homogenization of I_i. Then we have

$$J(X) = J_0 \cap J_1 \cap \ldots \cap J_n.$$

Proof For each $i = 0, \ldots, n$ we have

$$X \supset \overline{U_i \cap X}$$

and thus

$$J(X) \subset J(U_i \cap X) = J_i,$$

where the last equality follows from Proposition 9.16. In particular, $J(X) \subset \cap_{i=0}^n J_i$. Conversely, each $x \in X$ is contained in some distinguished U_i, so $X \subset \cup_{i=0}^n (X \cap U_i)$ and

$$J(X) \supset \cap_{i=0}^n J_i. \qquad \square$$

Example 9.20 Consider

$$X = \{x_0^2 + x_1^2 = x_2^2\} \cup \{x_0 = 0\} \cup \{x_1 = 0\} \cup \{x_2 = 0\} \subset \mathbb{P}^2(\mathbb{C}).$$

We have

$$U_0 \cap X = \{1 + y_1^2 = y_2^2\} \cup \{y_1 = 0\} \cup \{y_2 = 0\}$$

and hence

$$I(U_0 \cap X) = \langle y_1 y_2 (1 + y_1^2 - y_2^2) \rangle$$

with homogenization

$$J_0 = \langle x_1 x_2 (x_0^2 + x_1^2 - x_2^2) \rangle.$$

Similarly,

$$J_1 = \langle x_0 x_2 (x_0^2 + x_1^2 - x_2^2) \rangle, \quad J_2 = \langle x_0 x_1 (x_0^2 + x_1^2 - x_2^2) \rangle$$

and therefore

$$J(X) = J_0 \cap J_1 \cap J_2 = \langle x_0 x_1 x_2 (x_0^2 + x_1^2 - x_2^2) \rangle.$$

Example 9.21 Consider

$$X = \{(x_0, x_1, x_2) : x_0^2 = x_0 x_1 = x_0 x_2 = x_1^2 = x_1 x_2 = x_2^2 = 0\} \subset \mathbb{P}^2(k)$$

so that

$$I(U_0 \cap X) = \langle 1 \rangle, \quad I(U_1 \cap X) = \langle 1 \rangle, \quad I(U_2 \cap X) = \langle 1 \rangle$$

and $J_0 = J_1 = J_2 = J = \langle 1 \rangle$. In particular, X is empty! This shows that we sometimes get additional equations by looking carefully in each of the distinguished open neighborhoods U_i.

9.5 Projective Nullstellensatz

In the previous section, we gave an example of a homogeneous ideal

$$J \subsetneq k[x_0, \ldots, x_n]$$

where $X(J) = \emptyset$ and $J(X(J)) = k[x_0, \ldots, x_n]$. Now we describe systematically when this happens.

Definition 9.22 A homogeneous ideal $J \subset k[x_0, \ldots, x_n]$ is *irrelevant* if $J \supset \langle x_0^N, \ldots, x_n^N \rangle$ for some N.

Proposition 9.23 Fix $\mathfrak{m} = \langle x_0, \ldots, x_n \rangle$. A homogeneous ideal J is irrelevant if and only if either of the following two equivalent conditions holds

1. $J \supset \mathfrak{m}^M$ for some M;
2. J is $\mathfrak{m}$-primary.

Proof The analysis of $\mathfrak{m}$-primary ideals in Example 8.13 shows the equivalence of the two conditions. It is clear that if the first condition holds then J is irrelevant. Conversely, if $x_0^N, x_1^N, \ldots, x_n^N \in J$ then every monomial in $x_0, \ldots, x_n$ of degree at least $(n+1)N$ is contained in J. $\square$

Proposition 9.24 If $J \subset k[x_0, \ldots, x_n]$ is irrelevant then $X(J) = \emptyset$.

Proof By Proposition 9.23, we know that $x_0^N, \ldots, x_n^N \in J$ for some large N, hence

$$X(J) \subset \{x_0^N = x_1^N = \ldots = x_n^N = 0\} = \{x_0 = x_1 = \ldots = x_n = 0\} = \emptyset. \quad \square$$

Theorem 9.25 (Projective Nullstellensatz) Let k be algebraically closed and $J \subset k[x_0, \ldots, x_n]$ be a homogeneous ideal. Then $X(J) = \emptyset$ if and only if J is irrelevant.

Proof The implication $\Leftarrow$ is Proposition 9.24. To prove $\Rightarrow$, assume that $X(J) = \emptyset$. Let I_i be the dehomogenization of J with respect to x_i, i.e., $I_i = \mu_i(J)$. For any $g \in I_i$ with homogenization G, we have

$$x_i^e x_i^{\deg(g)} g(x_0/x_i, \ldots, x_n/x_i) = x_i^e G \in J$$

for some $e > 0$. Indeed, g is the dehomogenization of some homogeneous $H \in J$ and we have $H = x_i^e G$. We deduce $U_i \cap X = V(I_i)$ because on $U_i \cap X$, g vanishes exactly where G vanishes. Since $U_i \cap X = \emptyset$, the affine Nullstellensatz II (Theorem 7.5) implies $I_i = \langle 1 \rangle$, and it follows that $x_i^e \in J$. We conclude that J is irrelevant. □

Theorem 9.26 (Projective Hilbert Nullstellensatz) *Let k be algebraically closed and $J \subset k[x_0, \ldots, x_n]$ be homogeneous with variety $X = X(J)$. If $H \in k[x_0, \ldots, x_n]$ is homogeneous and vanishes on X then $H^N \in J$ for some N.*

Proof Retain the notation introduced in the proof of the last theorem. If $h_i = \mu_i(H)$ is the dehomogenization of H with respect to x_i then $h_i^{N_i} \in I_i$ for some N_i by the affine Hilbert Nullstellensatz (Theorem 7.3). Rehomogenizing, we get $x_i^{e_i} H^{N_i} \in J$ for suitable $e_i > 0$. Each monomial of sufficiently large degree is contained in $\langle x_0^{e_0}, \ldots, x_n^{e_n} \rangle$, so we have $H^M \in \langle x_0^{e_0}, \ldots, x_n^{e_n} \rangle$ for some $M \gg 0$. It follows that

$$H^M H^{\max\{N_0, \ldots, N_n\}} \in J.$$

□

9.6 Morphisms of projective varieties

Recall that a morphism of affine spaces is a polynomial map

$$\phi : \mathbb{A}^n(k) \to \mathbb{A}^m(k)$$
$$(x_1, \ldots, x_n) \mapsto (\phi_1, \ldots, \phi_m), \quad \phi_j \in k[x_1, \ldots, x_n].$$

Naively, we might try to define a morphism $\phi : \mathbb{P}^n(k) \to \mathbb{P}^m(k)$ analogously

$$\phi : \mathbb{P}^n(k) \to \mathbb{P}^m(k)$$
$$[x_0, \ldots, x_n] \mapsto [\phi_0, \ldots, \phi_m], \quad \phi_j \in k[x_0, \ldots, x_n].$$

This only makes sense when the ϕ_j are all homogeneous of degree d, in which case

$$\phi[\lambda x_0, \ldots, \lambda x_n] = [\lambda^d \phi_0, \ldots, \lambda^d \phi_m] = [\phi_0, \ldots, \phi_m] = \phi[x_0, \ldots, x_n]$$

and ϕ is well-defined on equivalence classes.

Definition 9.27 A *polynomial (rational) map*

$$\phi : \mathbb{P}^n(k) \dashrightarrow \mathbb{P}^m(k)$$

is given by a rule

$$[x_0, \ldots, x_n] \mapsto [\phi_0, \ldots, \phi_m],$$

where the $\phi_j \in k[x_0, \ldots, x_n]$ are all homogeneous of degree $d \geq 0$.

Of course, ϕ is defined everywhere on $\mathbb{P}^n(k)$ provided the ϕ_j have only trivial common zeros, i.e., if

$$\phi_0(a_0, \ldots, a_n) = \ldots = \phi_m(a_0, \ldots, a_n) = 0 \quad \Rightarrow \quad a_0 = \ldots = a_n = 0. \quad (\dagger)$$

In other words, $\phi(a_0, \ldots, a_n)$ determines a point in $\mathbb{P}^m(k)$ for each $[a_0, \ldots, a_n] \in \mathbb{P}^n(k)$. The following example suggests why we might want to ask for a bit more:

Example 9.28   Consider the polynomial map

$$\phi : \mathbb{P}^2(\mathbb{R}) \dashrightarrow \mathbb{P}^1(\mathbb{R})$$
$$[x_0, x_1, x_2] \to [x_0^2 + x_1^2, x_1^2 + x_2^2],$$

which satisfies both of our naive conditions. However, the same rule does *not* define a function

$$\phi : \mathbb{P}^2(\mathbb{C}) \to \mathbb{P}^1(\mathbb{C}).$$

Over $\mathbb{C}$, ϕ_0 and ϕ_1 have a common solution.

Thus condition ($\dagger$) does not behave well as the coefficient field is varied. We faced the same issue for the indeterminacy of rational maps over nonclosed fields (see Example 3.49 and Proposition 8.40.)

We can avoid this problem by insisting that $\langle \phi_0, \ldots, \phi_m \rangle$ be irrelevant. Proposition 9.24 then guarantees that ϕ is well-defined over *any* extension:

Proposition 9.29   *A polynomial map*

$$\phi : \mathbb{P}^n(k) \dashrightarrow \mathbb{P}^m(k)$$

defines a morphism provided the ideal $\langle \phi_0, \ldots, \phi_m \rangle$ is irrelevant.

The irrelevance is *not* a necessary condition for the polynomial map to induce a morphism (see Exercise 9.10).

This generalizes readily to arbitrary projective varieties $X \subset \mathbb{P}^n(k)$, with a slight twist. For a polynomial map ϕ to specify a morphism

$$\phi : X \to \mathbb{P}^m(k)$$

it is not necessary that the ϕ_j have no common solution anywhere on $\mathbb{P}^n(k)$. We don't care whether ϕ makes sense on the complement $\mathbb{P}^n(k) \setminus X$:

Proposition 9.30   *Let $X \subset \mathbb{P}^m(k)$ be a projective variety. A polynomial map $\phi : \mathbb{P}^n(k) \dashrightarrow \mathbb{P}^m(k)$ restricts to a morphism $X \to \mathbb{P}^m(k)$ provided $J(X) + \langle \phi_0, \ldots, \phi_m \rangle$ is irrelevant.*

9.6 MORPHISMS OF PROJECTIVE VARIETIES

Remark 9.31 This situation is in marked contrast to the affine case discussed in Chapter 3: given affine $V \subset \mathbb{A}^n(k)$ and $W \subset \mathbb{A}^m(k)$, a morphism $\phi : V \to W$ always extends to a morphism $\phi' : \mathbb{A}^n(k) \to \mathbb{A}^m(k)$.

The proofs of Propositions 9.29 and 9.30 make reference to the general definitions of abstract varieties and morphisms in §9.8. Readers who wish to avoid this machinery might take Proposition 9.30 as a working definition of a morphism from a projective variety to projective space.

9.6.1 Projection and linear maps

We generalize the distinguished open affine subsets U_i introduced in the definition of projective space:

Proposition 9.32 *Let $L = \sum_{i=0}^{n} c_i x_i \in k[x_0, \ldots, x_n]$ be a nonzero homogeneous linear form. Then the open subset*

$$U_L = \{[a_0, \ldots, a_n] : L(a_0, \ldots, a_n) \neq 0\} \subset \mathbb{P}^n(k)$$

is naturally an affine variety, isomorphic to $\mathbb{A}^n(k)$.

This expanded inventory of affine open subsets will be useful in describing morphisms from $\mathbb{P}^n(k)$.

Proof We describe the identification with affine space. Consider the map

$$\Phi : U_L \to \mathbb{A}^{n+1}(k)$$
$$[a_0, \ldots, a_n] \mapsto (a_0/L(a_0, \ldots, a_n), \ldots, a_n/L(a_0, \ldots, a_n))$$

and write $b_0, \ldots, b_n$ for the coordinates on affine space. The image of Φ is the affine hyperplane

$$V = \{(b_0, \ldots, b_n) : c_0 b_0 + \cdots + c_n b_n = 1\} \subset \mathbb{A}^{n+1}(k).$$

Note that $V \simeq \mathbb{A}^n(k)$: If $c_i \neq 0$ then the projection

$$\Pi : V \to \mathbb{A}^n(k)$$
$$(b_0, \ldots, b_n) \mapsto (b_0, \ldots, b_{i-1}, b_{i+1}, \ldots, b_n)$$

has inverse

$$(b_0, \ldots, b_{i-1}, b_{i+1}, \ldots, b_n) \mapsto \left(b_0, \ldots, b_{i-1}, \left(1 - \sum_{j \neq i} c_j b_j\right)/c_i, b_{i+1}, \ldots, b_n\right).$$

Thus we obtain a bijective map

$$\psi_L = (\Pi \circ \Phi) : U_L \to \mathbb{A}^n(k)$$
$$[a_0, \ldots, a_n] \mapsto (a_0/L, \ldots, a_{i-1}/L, a_{i+1}/L, \ldots, a_n/L).$$

This discussion might leave a nagging doubt in the reader's mind: is ψ_L compatible with the identifications $\psi_i : U_i \to \mathbb{A}^n(k)$ used in the definition of projective space? Consider how these fit together over the intersections

$$\begin{array}{c} U_L \cap U_j \\ \psi_j \swarrow \qquad \searrow \psi_L \\ \mathbb{A}^n(k) \qquad\qquad \mathbb{A}^n(k) \end{array}$$

where we assume $i < j$ for notational simplicity. The composition

$$\varphi_j(U_L \cap U_j) \xrightarrow{\psi_L \circ \psi_j^{-1}} \psi_L(U_L \cap U_j)$$
$$\| \qquad\qquad\qquad\qquad \|$$
$$\mathbb{A}^n(k) \setminus \{L = 0\} \xrightarrow{\rho_{ij}} \mathbb{A}^n(k) \setminus \{b_j = 0\},$$

where $L = L(b'_0, \ldots, b'_{j-1}, 1, b'_{j+1}, \ldots, b'_n)$, is given by

$$(b'_0/L, \ldots, b'_{i-1}/L, b'_{i+1}/L, \ldots, b'_{j-1}/L, 1/L, b'_{j+1}/L, \ldots, b'_n/L).$$

These are birational, just as in the case of the distinguished affine open subsets. $\square$

Remark 9.33 A more formal approach, explicitly using the language of abstract varieties, is sketched in Exercise 9.22.

Corollary 9.34 *Let $X \subset \mathbb{P}^n(k)$ be projective and $L \in k[x_0, \ldots, x_n]$ be a linear form such that $L \notin J(X)$. Then*

$$U_L \cap X \subset U_L \simeq \mathbb{A}^n(k)$$

is naturally an affine variety.

Definition 9.35 A *linear map* $\phi : \mathbb{P}^n(k) \dashrightarrow \mathbb{P}^m(k)$ is a polynomial map induced by a linear transformation, i.e., there exists an $(m+1) \times (n+1)$ matrix A with entries in k such that

$$\phi[x_0, \ldots, x_n] = [a_{00}x_0 + \cdots + a_{0n}x_n, \ldots, a_{m0}x_0 + \cdots + a_{mn}x_n].$$

Proposition 9.36 (Linear case) *Fix a linear map $\tau : \mathbb{P}^n(k) \dashrightarrow \mathbb{P}^m(k)$ with matrix A. It defines a morphism $\mathbb{P}^n(k) \to \mathbb{P}^m(k)$ if A has trivial kernel. If $X \subset \mathbb{P}^n(k)$ is projective then τ defines a morphism $X \to \mathbb{P}^m(k)$ if*

$$J(X) + \langle L_i := a_{i0}x_0 + \cdots + a_{in}x_n, \ i = 0, \ldots, m \rangle$$

is irrelevant.

9.6 MORPHISMS OF PROJECTIVE VARIETIES

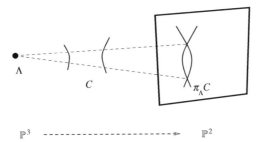

$\mathbb{P}^3 \dashrightarrow \mathbb{P}^2$

Figure 9.2 Projection from a point in $\mathbb{P}^3$.

A morphism $\tau : \mathbb{P}^n(k) \to \mathbb{P}^n(k)$ induced by an invertible $(n+1) \times (n+1)$ matrix is called a *projectivity*.

Proof It suffices to prove the second assertion, which subsumes the first. Let $V_0, \ldots, V_m$ be the distinguished affine open subsets of $\mathbb{P}^m(k)$ and $U_{L_i} \subset \mathbb{P}^n(k)$ the corresponding affine open subsets. We tacitly ignore rows of A which are zero.

Using the identifications of Proposition 9.32, τ induces a sequence of morphisms of affine varieties

$$U_{L_i} \xrightarrow{\tau} V_i, i = 0, \ldots, m$$
$$\| \quad \quad \|$$
$$\mathbb{A}^n(k) \xrightarrow{\tau_i} \mathbb{A}^m(k).$$

Assuming that the entry $a_{i,\ell} \neq 0$, we can compute τ_i:

$$(b_0, \ldots, b_{\ell-1}, b_{\ell+1}, \ldots, b_n) \xrightarrow{\psi_{L_i,\mathbb{P}^n}^{-1}} [b_0, \ldots, b_{\ell-1}, (1 - \sum_{j \neq \ell} a_{ij}b_j)/a_{i\ell}, b_{\ell+1}, \ldots, b_n]$$
$$\xrightarrow{\tau} [L'_0, \ldots, L'_{i-1}, 1, L'_{i+1}, \ldots, L'_m] \xrightarrow{\psi_{i,\mathbb{P}^m}} (L'_0, \ldots, L'_{i-1}, L'_{i+1}, \ldots, L'm)$$

where

$$L'_r = a_{r0}b_0 + \cdots + a_{r\ell-1}b_{\ell-1} + a_{r\ell}\left(1 - \sum_{j \neq \ell} a_{ij}b_j\right)/a_{i\ell} + a_{r\ell+1}b_{\ell+1} + \cdots a_{rn}b_n$$

for $r \neq i$. These are morphisms of affine spaces.

These together define a morphism $X \to \mathbb{P}^m(k)$ provided the U_{L_i}, $i = 0, \ldots, m$ cover X. In the language of Section 9.8, $\{U_{L_i} \cap X\}_{i=0,\ldots,m}$ is an affine open covering of X. This is guaranteed by our irrelevance hypothesis: the linear equations

$$L_0 = L_1 = \ldots = L_m = 0$$

have no nontrivial solutions along X–even over extensions of k. Thus the U_{L_i} contain all of X. $\square$

PROJECTIVE GEOMETRY

Definition 9.37 Let $\Lambda \subset \mathbb{P}^n(k)$ be a *linear subspace*, i.e., the locus defined by the vanishing of linearly independent forms $L_0, \ldots, L_m$ for $m < n$. The polynomial map

$$\pi_\Lambda : \mathbb{P}^n(k) \dashrightarrow \mathbb{P}^m(k)$$
$$[x_0, \ldots, x_n] \mapsto [L_0, \ldots, L_m]$$

is called *projection from the subspace* Λ.

9.6.2 Veronese morphisms

We give another special instance of Proposition 9.29. Because of the importance of this example, we offer a precise description of its image:

Proposition 9.38 (Veronese embedding) *The polynomial map*

$$v(d) : \mathbb{P}^n(k) \to \mathbb{P}(k[x_0, x_1, \ldots, x_n]_d) = \mathbb{P}^{\binom{n+d}{d}-1}(k)$$
$$[x_0, \ldots, x_n] \mapsto \underbrace{\left[x^\alpha = x_0^{\alpha_0} \ldots x_n^{\alpha_n}\right]}_{\text{all monomials with } |\alpha|=d}$$

is a morphism. Its image is closed and defined by the equations

$$z_\alpha z_\beta = z_\gamma z_\delta \qquad (9.2)$$

for all indices with $\alpha + \beta = \gamma + \delta$. Moreover, $v(d)$ is an isomorphism onto its image.

This is called the *Veronese embedding*, after Giuseppe Veronese (1854–1917).

Proof The equations correspond to the identities

$$x^\alpha x^\beta = x^\gamma x^\delta, \quad \alpha + \beta = \gamma + \delta.$$

We show that $v(d)$ defines a morphism by analyzing it over each of the distinguished affine open subset $U_i \subset \mathbb{P}^n(k)$. For notational simplicity we assume $i = 0$. Now U_0 is mapped into the distinguished affine open subset

$$U_{d0\ldots 0} = \{z_{d0\ldots 0} \neq 0\} \subset \mathbb{P}^{\binom{n+d}{d}-1}(k),$$

so we must show that the induced map of affine spaces $v(d)_0 : U_0 \to U_{d0\ldots 0}$ is a morphism.

Let $y_1, \ldots, y_n$ and $w_{\alpha_1 \ldots \alpha_n}$ be coordinates on U_0 and $U_{d0\ldots 0}$ respectively. The second set of variables is indexed by nonconstant monomials of degree $\leq d$ in n variables, i.e., the dehomogenizations of the x^α other than x_0^d. The map $v(d)_0$ is given by

$$(y_1, \ldots, y_d) \mapsto (y^\alpha)_{0 < \alpha_1 + \cdots + \alpha_n \leq d}.$$

This corresponds to the homomorphism

$$\nu(d)_0^* : k[w_\alpha] \to k[y_1, \ldots, y_n]$$
$$w_{\alpha_1 \ldots \alpha_n} \mapsto y_1^{\alpha_1} \cdots y_n^{\alpha_n}$$

so $\nu(d)_0$ is a morphism of affine varieties.

The image of $\nu(d)_0$ is easy to characterize in terms of polynomial equations: For each $i = 1, \ldots, n$, write

$$\hat{\imath} = (\underbrace{0, \ldots, 0}_{i-1 \text{ times}}, 1, \underbrace{0, \ldots, 0}_{n-i \text{ times}})$$

and $w_{\hat{\imath}}$ the corresponding variable, i.e., $\nu(d)_0^* w_{\hat{\imath}} = y_i$. For each $(\beta_1, \ldots, \beta_n)$ with $\beta_1 + \cdots + \beta_n < d$, we have the relation (cf. Exercise 1.3):

$$q_\beta := w_{\hat{\imath}} w_{\beta_1 \ldots \beta_i \ldots \beta_n} - w_{\beta_1 \ldots \beta_{i-1} \beta_i + 1 \beta_{i+1} \ldots \beta_n} = 0.$$

Thus all the w_α are determined from the $w_{\hat{\imath}}$ and we have

$$k[w_\alpha] / \underbrace{\langle q_\beta \rangle}_{\text{all } \beta} \simeq k[w_{\hat{1}}, \ldots, w_{\hat{n}}].$$

This proves that image($\nu(d)_0$) is closed and isomorphic to affine n-space with coordinates $w_{\hat{1}}, \ldots, w_{\hat{n}}$; moreover, $\nu(d)_0 : U_0 \to$ image($\nu(d)_0$) is an isomorphism. Finally, the homogenization of q_β is

$$z_{d-10\ldots 010\ldots 0} z_{\beta_0 \beta_1 \ldots \beta_i \ldots \beta_n} - z_{\beta_0 - 1 \beta_1 \ldots \beta_i + 1 \ldots \beta_n} z_{d0\ldots 0} = 0$$

where the '1' in the first subscript occurs in the ith position and $\beta_0 = d - (\beta_1 + \cdots + \beta_n) > 0$. These are instances of the equations (9.2).

To complete the argument, we check that the projective variety defined by the equations (9.2) lies in the union of the $n + 1$ distinguished affine open subsets

$$U_{d0\ldots 0} \cup U_{0d0\ldots 0} \cup \ldots U_{0\ldots 0d}.$$

(Then the local analysis above implies $\nu(d)$ is an isomorphism onto its image.) Suppose we have

$$z_{d0\ldots 0} = z_{0d0\ldots 0} = \ldots = z_{0\ldots 0d} = 0.$$

Iteratively applying the equations (9.2), we obtain

$$z_{\alpha_0 \ldots \alpha_n}^d = z_{d0\ldots 0}^{\alpha_0} z_{0d0\ldots 0}^{\alpha_1} \cdots z_{0\ldots 0d}^{\alpha_n}$$

for each α. (This combinatorial deduction is left as an exercise.) It follows that $z_{\alpha_0 \ldots \alpha_n} = 0$ for each α. $\square$

We pause to flesh out some important special cases of Proposition 9.38:

Example 9.39 The Veronese embeddings of $\mathbb{P}^1$ take the form

$$\phi : \mathbb{P}^1(k) \to \mathbb{P}^m(k)$$
$$[x_0, x_1] \to [x_0^m, x_0^{m-1}x_1, \ldots, x_1^m].$$

The image is called the *rational normal curve of degree m*. It is defined by the 2×2 minors of the $2 \times m$ matrix

$$\begin{pmatrix} z_m & z_{m-1} & \cdots & z_1 \\ z_{m-1} & z_{m-2} & \cdots & z_0 \end{pmatrix}.$$

Example 9.40 The degree-2 Veronese embedding takes the form

$$\phi : \mathbb{P}^n(k) \to \mathbb{P}^{\binom{n+2}{2}-1}(k)$$
$$[x_0, x_1, \ldots, x_n] \to \underbrace{[x_0^2, x_0x_1, x_0x_2 \ldots, x_{n-1}x_n, x_n^2]}_{\text{all monomials of degree 2}}.$$

The image is defined by the 2×2 minors of the symmetric $(n+1) \times (n+1)$ matrix M, where $M_{ij} = M_{ji}$ is the coordinate function mapped to $x_i x_j$.

Proposition 9.38 gives a large number of affine open subsets in $\mathbb{P}^n(k)$.

Proposition 9.41 *Let $F \in k[x_0, \ldots, x_n]$ be homogeneous of degree $d > 0$. The open subset*

$$V_F = \{(x_0, \ldots, x_n) \in \mathbb{P}^n(k) : F(x_0, \ldots, x_n) \neq 0\} \subset \mathbb{P}^n(k)$$

naturally carries the structure of an affine variety.

Proof Express

$$F = \sum_{|\alpha|=d} c_\alpha x^\alpha$$

and write

$$H_F := \{[z_\alpha] : L_F = \sum_\alpha c_\alpha z_\alpha = 0\} \subset \mathbb{P}^{\binom{n+d}{d}-1}(k)$$

for the corresponding hyperplane with complement

$$U_F := \left\{ [z_\alpha] : \sum_\alpha c_\alpha z_\alpha \neq 0 \right\}.$$

By Proposition 9.32, U_F is naturally affine space with coordinate ring

$$k[b_\alpha]/\langle \sum_\alpha c_\alpha b_\alpha - 1 \rangle.$$

9.6 MORPHISMS OF PROJECTIVE VARIETIES

The Veronese map $v(d)$ identifies V_F with the intersection $U_F \cap v(d)(\mathbb{P}^n)$. Proposition 9.38 implies $v(d)(\mathbb{P}^n)$ is cut out by homogeneous equations, so

$$U_F \cap v(d)(\mathbb{P}^n) \subset U_F \simeq \mathbb{A}^{\binom{n+d}{d}-1}$$

is defined by polynomial equations and thus is affine. □

Remark 9.42 The coordinate ring of U_F has an alternate description

$$k[U_F] \simeq k[z_\alpha][L_F^{-1}]_{\text{degree } 0},$$

the quotients G/L_F^N with G homogeneous of degree N in the z_α. With a bit more work, we could compute analogously

$$k[V_F] = (k[x_0, \ldots, x_n][1/F])_{\text{degree } 0},$$

the homogeneous fractions in $k[x_0, \ldots, x_n][1/F]$ of degree zero.

Example 9.43 Let $F = x_1^2 + 4x_1x_2 - 3x_2^2$. Then we can realize $V_F \subset \mathbb{P}^2(k)$ as the affine variety

$$\{(y_{02}, y_{11}, y_{20}) : y_{11}^2 = y_{02}y_{20}, y_{20} + 4y_{11} - 3y_{02} = 1\}.$$

Example 9.44 Let $X = \{(x_0, x_1, x_2) : x_0^3 + x_1^3 + x_2^3 = 0\} \subset \mathbb{P}^2(\mathbb{C})$ be a plane curve. The rule from Example 9.28 does define a morphism on X

$$\phi : X \to \mathbb{P}^1(\mathbb{C})$$
$$[x_0, x_1, x_2] \mapsto [x_0^2 + x_1^2, x_1^2 + x_2^2].$$

We have

$$\phi^{-1}[0, 0] = \{[1, i, 1], [1, -i, 1], [-1, i, 1], [-1, -i, 1]\} \subset \mathbb{P}^2(k),$$

which is disjoint from X.

9.6.3 Proof of Propositions 9.29 and 9.30

Every polynomial rational map

$$\phi : \mathbb{P}^n(k) \dashrightarrow \mathbb{P}^m(k)$$

can be factored

$$\mathbb{P}^n(k) \xrightarrow{v(d)} \mathbb{P}^{\binom{n+d}{d}-1}(k) \xdashrightarrow{\tau} \mathbb{P}^m,$$

where $v(d)$ is the Veronese embedding and τ is linear. Indeed, writing each coordinate function

$$\phi_i = \sum_\alpha c_{i;\alpha} x^\alpha, \quad \deg(\alpha) = d,$$

τ arises from the $(m+1) \times \binom{n+d}{d}$ matrix $(c_{i;\alpha})$. Our results follow from Propositions 9.38 and 9.36.

9.7 Products

Proposition 9.45 (Segre embedding) *The product $\mathbb{P}^m(k) \times \mathbb{P}^n(k)$ is a projective variety. It can be realized as the locus in $\mathbb{P}^{mn+m+n}(k)$ where the 2×2 minors of*

$$\begin{pmatrix} z_{00} & z_{01} & \cdots & z_{0n} \\ z_{10} & z_{11} & \cdots & z_{1n} \\ \cdots & \cdots & \cdots & \cdots \\ z_{m0} & z_{m1} & \cdots & z_{mn} \end{pmatrix} \tag{9.3}$$

vanish. Here we regard $\mathbb{P}^{mn+m+n}(k)$ as the projective space of all $(m+1) \times (n+1)$ matrices (z_{ij}).

This is called the *Segre embedding* of $\mathbb{P}^m(k) \times \mathbb{P}^n(k)$ into $\mathbb{P}^{mn+m+n}(k)$, after Corrado Segre (1863–1924).

Proof Our analysis here presumes the abstract variety structure on the product $\mathbb{P}^m(k) \times \mathbb{P}^n(k)$ (cf. Proposition 9.52).

Our first task is to construct a morphism

$$\phi : \mathbb{P}^m(k) \times \mathbb{P}^n(k) \to \mathbb{P}^{(m+1)(n+1)-1}(k).$$

This is given by the rule

$$[x_0, \ldots, x_m] \times [y_0, \ldots, y_n] \to [x_0 y_0, x_0 y_1, \ldots, x_0 y_n, x_1 y_0, \ldots, x_1 y_n, x_2 y_0, \ldots],$$

which is well-defined on projective equivalence classes.

Consider the distinguished affine open subsets $V_i = \{[x_0, \ldots, x_m] : x_i \neq 0\} \subset \mathbb{P}^m(k)$, $W_j = \{[y_0, \ldots, y_n] : y_j \neq 0\} \subset \mathbb{P}^n(k)$, and $U_{ij} = \{[z_{00}, \ldots, z_{mn}] : z_{ij} \neq 0\} \subset \mathbb{P}^{mn+m+n}(k)$. Note that $\phi(V_i \times W_j) \subset U_{ij}$ for each i and j.

The restriction of ϕ to $V_0 \times W_0$ takes the form

$$V_0 \times W_0 \to U_{00}$$
$$[1, x_1/x_0, \ldots, x_m/x_0], [1, y_1/y_0, \ldots, y_n/y_0] \mapsto [x_i y_j / x_0 y_0].$$

If $a_1, \ldots, a_m$ and $b_1, \ldots, b_n$ are coordinates on $V_0 \simeq \mathbb{A}^m(k)$ and $W_0 \simeq \mathbb{A}^n(k)$, ϕ can be expressed

$$\phi_{00} : \mathbb{A}^m(k) \times \mathbb{A}^n(k) \to \mathbb{A}^{mn+m+n}(k)$$
$$(a_1, \ldots, a_m, b_1, \ldots, b_n) \mapsto (b_1, b_2, \ldots, b_n, a_1, a_1 b_1, \ldots, a_1 b_n, a_2, a_2 b_1, \ldots, a_m b_n).$$

This is a morphism of affine varieties.

9.7 PRODUCTS

We claim that image(ϕ) is closed and

$$\mathbb{P}^m(k) \times \mathbb{P}^n(k) \xrightarrow{\cong} \text{image}(\phi).$$

Fix coordinates on U_{00}:

$$c_{01}, \ldots, c_{0n}, c_{11}, \ldots, c_{1n}, \ldots, c_{mn}$$

such that

$$\phi_{00}^* c_{0j} = b_j, \quad \phi_{00}^* c_{i0} = a_i, \quad \phi_{00}^* c_{ij} = a_i b_j, \quad i = 1, \ldots, m, \ j = 1, \ldots, n.$$

The image of ϕ_{00} is the closed set defined by the relations $c_{ij} = c_{i0} c_{0j}$, which is isomorphic to affine space $\mathbb{A}^{m+n}(k)$ with coordinates $c_{10}, \ldots, c_{m0}, c_{01}, \ldots, c_{0n}$. (Indeed, ϕ_{00} has a left inverse: project onto the variables corresponding to the a_j and b_i.) Repeating this over each of the distinguished affine opens, we conclude that each local realization $\phi_{ij} : V_i \times W_j \to U_{ij}$ has closed image and is an isomorphism onto its image.

It remains to extract homogeneous equations for the image. Homogenizing $c_{ij} = c_{i0} c_{j0}$ yields $z_{ij} z_{00} = z_{i0} z_{0j}$. Working over all the distinguished affine open sets, we obtain the relations

$$z_{ij} z_{h\ell} = z_{i\ell} z_{hj}, \quad i, h = 0, \ldots, m, \ j, \ell = 0, \ldots, n,$$

which are the 2×2 minors of our matrix. $\square$

Remark 9.46 Identify the $(m+1) \times (n+1)$ matrices (z_{ij}), up to scalar multiplication, with $\mathbb{P}^{mn+m+n}(k)$. The matrices of rank 1 form a closed subset $R_1 \subset \mathbb{P}^{mn+m+n}(k)$, defined by the 2×2 minors of (z_{ij}). The image of a rank-1 matrix (z_{ij}) is a one-dimensional subspace of k^{m+1}, thus a point in $\mathbb{P}^m(k)$. The kernel is a codimension-1 linear subspace of k^{n+1}, defined by a linear form

$$\ker(z_{ij}) = \{(w_0, \ldots, w_n) \in k^{n+1} : a_0 w_0 + \cdots + a_n w_n = 0\},$$

where $(a_0, \ldots, a_n)$ is proportional to any of the nonzero rows of (z_{ij}). We regard $[a_0, \ldots, a_n] \in \mathbb{P}^n(k)$, the projective space associated to the vector space dual to k^{n+1}.

This gives us a function

$$R_1 \to \mathbb{P}^m(k) \times \mathbb{P}^n(k)$$
$$(z_{ij}) \mapsto (\text{image}(z_{ij}), \text{ linear form vanishing on } \ker(z_{ij})).$$

Since there is a unique rank-1 matrix with prescribed kernel and image, this is a bijection.

We can now address one complication that arises in the study of morphisms of projective varieties. Consider projection onto the first factor

$$\pi_1 : \mathbb{P}^1(k) \times \mathbb{P}^1(k) \to \mathbb{P}^1(k)$$
$$([x_0, x_1], [y_0, y_1]) \mapsto [x_0, x_1],$$

which is definitely a morphism. And we have just seen how $\mathbb{P}^1(k) \times \mathbb{P}^1(k)$ can be realized as the surface

$$\{[z_{00}, z_{01}, z_{10}, z_{11}] : z_{00}z_{11} = z_{01}z_{10}\} \subset \mathbb{P}^3(k).$$

Does π_1 come from a polynomial map $\mathbb{P}^3(k) \to \mathbb{P}^1(k)$? Definitely not! Any degree-d polynomial in the z_{ij} corresponds to a polynomial homogeneous of degree d in both $\{x_0, x_1\}$ and $\{y_0, y_1\}$; the degree in the x_i equals the degree in the y_i. However, π_1 cannot be represented with such polynomials.

The polynomial rational maps $\mathbb{P}^3(k) \dashrightarrow \mathbb{P}^1(k)$ best approximating π_1 are

$$\phi : \mathbb{P}^3(k) \dashrightarrow \mathbb{P}^1(k)$$
$$[z_{00}, z_{01}, z_{10}, z_{11}] \mapsto [z_{00}, z_{10}],$$
$$\phi' : \mathbb{P}^3(k) \dashrightarrow \mathbb{P}^1(k)$$
$$[z_{00}, z_{01}, z_{10}, z_{11}] \mapsto [z_{01}, z_{11}].$$

Note that ϕ agrees with π_1 away from the line

$$\{z_{00} = z_{10} = 0\} = \mathbb{P}^1(k) \times [0, 1] \subset \mathbb{P}^1(k) \times \mathbb{P}^1(k) \subset \mathbb{P}^3(k);$$

ϕ' agrees with π_1 away from the line

$$\{z_{01} = z_{11} = 0\} = \mathbb{P}^1(k) \times [1, 0] \subset \mathbb{P}^1(k) \times \mathbb{P}^1(k) \subset \mathbb{P}^3(k).$$

9.8 Abstract varieties

We would like to endow $\mathbb{P}^n(k)$ with the structure of a variety, compatible with our decomposition (9.1) as a union of affine spaces. To achieve this we will need a more flexible definition of algebraic varieties, going beyond the affine examples we have studied up to this point. This section can be omitted on first reading.

9.8.1 Graphs of birational maps

Let $\rho : V \dashrightarrow W$ be a birational map and $\Gamma_\rho \subset V \times W$ the closure of its graph. Let $p_V : \Gamma_\rho \to V$ and $p_W : \Gamma_\rho \to W$ be the morphisms induced by the projections. Let $I_\rho \subset k[V]$ denote the indeterminacy ideal of ρ and $I_{\rho^{-1}} \subset k[W]$ the indeterminacy ideal of its inverse. Consider the intersection of the graphs of ρ and ρ^{-1}

$$U_\rho = \{(v, w) : \rho \text{ defined at } v, \rho^{-1} \text{ defined at } w\}.$$

We can regard this as an open subset of both V and W. By Proposition 8.37, we have

$$U_\rho = \Gamma_\rho \setminus \left(p_V^{-1}V(I_\rho) \cup p_W^{-1}V(I_{\rho^{-1}})\right).$$

We are interested in whether $U_\rho = \Gamma_\rho$, i.e., whether the indeterminacy $p_V^{-1}V(I_\rho)$ or $p_W^{-1}V(I_{\rho^{-1}})$ actually meets Γ_ρ. This might be sensitive to the field – indeterminacy may be apparent only after base extension (see Example 3.49.) However, if k is

9.8 ABSTRACT VARIETIES

algebraically closed then Proposition 8.40 (or a direct argument using the Nullstellensatz) implies that $U_\rho = \Gamma_\rho$ if and only if

$$I(\Gamma_\rho) + (p_V^* I_\rho \cap p_W^* I_{\rho^{-1}}) = k[V \times W].$$

We say then that $\rho : V \dashrightarrow W$ satisfies the *closed graph condition*. In this case, U_ρ is naturally an affine variety.

Example 9.47   The birational map

$$\rho : \mathbb{A}^1(k) \dashrightarrow \mathbb{A}^1(k)$$
$$x \mapsto 1/x$$

satisfies the closed graph condition. We have $I(\Gamma_\rho) = \langle xz - 1 \rangle$, $I_\rho = \langle x \rangle$, and $I_{\rho^{-1}} = \langle z \rangle$, so that

$$I(\Gamma_\rho) + (k[x, z]I_\rho \cap k[x, z]I_{\rho^{-1}}) = \langle 1 - xz, xz \rangle = k[x, z].$$

The birational morphism

$$\rho : \mathbb{A}^2(k) \dashrightarrow \mathbb{A}^2(k)$$
$$(x_1, x_2) \mapsto (x_1, x_1 x_2)$$

does not satisfy the closed graph condition. Here $I(\Gamma_\rho) = \langle y_1 - x_1, y_2 - x_1 x_2 \rangle$, $I_\rho = k[x_1, x_2]$, and $I_{\rho^{-1}} = \langle y_1 \rangle$, so we have

$$\langle y_1 - x_1, y_2 - x_1 x_2, y_1 \rangle \neq k[x_1, x_2, y_1, y_2].$$

9.8.2 Coverings of affine varieties

For pedagogical reasons, we introduce the ideas behind abstract varieties in the familiar context of affine varieties, where they are more transparent.

Let V be an affine variety. Choose $g_0, \ldots, g_n \in k[V]$ and write $U_i = \{v \in V : g_i(v) \neq 0\}$ for each i. Suppose that $\langle g_0, \ldots, g_n \rangle = k[V]$, so there exist $h_0, \ldots, h_n \in k[V]$ with

$$h_0 g_0 + \cdots + h_n g_n = 1.$$

In particular, for each $v \in V$ we have $g_i(v) \neq 0$ for some i and

$$V = U_0 \cup U_1 \ldots \cup U_n.$$

If k is algebraically closed then the fact that V can be expressed as a union of the U_is implies that $\langle g_0, \ldots, g_n \rangle = k[V]$; this follows from the Nullstellensatz (Theorem 7.5).

As in the proof of Proposition 3.47, we endow each open subset U_i with the structure of an affine variety. Let

$$V_{g_i} := \{(v, z) : g_i(v)z = 1\} \subset V \times \mathbb{A}^1(k)$$

so that projection onto V maps V_{g_i} bijectively onto U_i. The collection $\{V_{g_i}\}_{i=0,\ldots,n}$ is called an *affine open covering* of V. The intersections $U_{i_1} \cap \ldots \cap U_{i_r}$ also are naturally affine varieties; they correspond to the affine open subsets $V_{g_{i_1}\cdots g_{i_r}}$.

Here is the crucial property of affine open coverings:

Theorem 9.48 *Let V and W be affine varieties. A morphism $\phi : V \to W$ is equivalent to the data of an affine open covering $\{V_{g_i}\}_{i=0,\ldots,n}$ and a collection of morphisms $\{\phi_i : V_{g_i} \to W, i = 0, \ldots, n\}$ satisfying the compatibility $\phi_i|V_{g_i g_j} = \phi_j|V_{g_i g_j}$.*

Proof Given $\phi : V \to W$, we obtain ϕ_i by composing

$$V_{g_i} \hookrightarrow V \to W$$

where the first arrow is the standard projection morphism. These are compatible on the intersections $V_{g_i g_j}$ by construction.

Conversely, suppose we are given a collection $\{\phi_i\}$ as above. By Proposition 3.31, we get a collection of k-algebra homomorphisms

$$\phi_i^* : k[W] \to k[V][g_i^{-1}].$$

The morphism $V_{g_i g_j} \hookrightarrow V_{g_i}$ corresponds to the localization homomorphism $k[V][g_i^{-1}] \to k[V][(g_i g_j)^{-1}]$. By assumption the compositions

$$k[W] \xrightarrow{\phi_i^*} k[V][g_i^{-1}] \to k[V][(g_i g_j)^{-1}] \qquad k[W] \xrightarrow{\phi_j^*} k[V][g_j^{-1}] \to k[V][(g_i g_j)^{-1}]$$

are equal.

We construct a unique k-algebra homomorphism

$$\psi : k[W] \to k[V]$$

such that for each $f \in k[W]$ and $i = 0, \ldots, n$ we have $\psi(f) = \phi_i^* f$ in $k[V][g_i^{-1}]$. Our desired $\phi : V \to W$ is the unique morphism with $\phi^* = \psi$.

Lemma 9.49 *Let R be a ring and $g_0, \ldots, g_n \in R$ such that*

$$h_0 g_0 + h_1 g_1 + \cdots + h_n g_n = 1$$

for some $h_0, \ldots, h_n \in R$. Then for each $N > 0$ there exist $h_0', \ldots, h_n'$ such that

$$h_0' g_0^N + h_1' g_1^N + \cdots + h_n' g_n^N = 1.$$

Proof of lemma: Taking $N(n+1)$ powers and expanding gives

$$\sum_{e_0+\cdots+e_n=N(n+1)} P_{e_0\ldots e_n}(h_0, \ldots, h_n) g_0^{e_0} \cdots g_n^{e_n} = 1$$

9.8 ABSTRACT VARIETIES

where $P_{e_0\ldots e_n}(h_0, \ldots, h_n)$ is a suitable polynomial in the h_i. For each term, one of the $e_i \geq N$, i.e., each summand is divisible by g_i^N for some i. Regrouping terms in the summation gives the result. □

First, if $\psi(f)$ exists it must be unique: Suppose we have another $\hat{\psi}(f) \in k[V]$ with $\hat{\psi}(f) = \psi(f) \in k[V][g_i^{-1}]$ for each i. Then there exists an N such that $g_i^N(\hat{\psi}(f) - \psi(f)) = 0$ for each i. It follows that

$$\hat{\psi}(f) - \psi(f) = \sum_{i=0}^{n} h'_i g_i^N ((\hat{\psi}(f) - \psi(f)) = 0.$$

Second, observe that this uniqueness forces ψ to be a homomorphism: Given $f_1, f_2 \in k[W]$, $\psi(f_1) = \phi_i^*(f_1)$ and $\psi(f_2) = \phi_i^*(f_2)$ in $k[V][g_i^{-1}]$ imply that

$$\psi(f_1) + \psi(f_2) = \phi_i^*(f_1 + f_2).$$

Thus $\psi(f_1 + f_2) = \psi(f_1) + \psi(f_2)$. A similar argument proves $\psi(f_1 f_2) = \psi(f_1)\psi(f_2)$.

It suffices then to write a formula for ψ and verify the required properties. Given $f \in k[W]$, choose N such that $g_i^N \phi_i^*(f) \in k[V]$ for each i and $(g_i g_j)^N \phi_i^*(f) = (g_i g_j)^N \phi_j^*(f)$ for each i, j. Write

$$\psi(f) = \sum_{j=0}^{n} h'_j \left(g_j^N \phi_j^*(f) \right)$$

so that

$$\begin{aligned} g_i^N \psi(f) &= \sum_{j=0}^{n} h'_j (g_j g_i)^N \phi_j^*(f) \\ &= \sum_{j=0}^{n} h'_j (g_j g_i)^N \phi_i^*(f) \\ &= \left(\sum_{j=0}^{n} h'_j g_j^N \right) g_i^N \phi_i^*(f) \\ &= 1 \cdot g_i^N \phi_i^*(f) \end{aligned}$$

which means that $\psi(f) = \phi_i^*(f)$ in $k[V][g_i^{-1}]$. □

How are the various affine open subsets V_{g_i} related? Suppose that $g_i, g_j \in k[V]$ are not zero divisors so that

$$k\left(V_{g_i}\right) = k(V) = k\left(V_{g_j}\right).$$

In particular, V_{g_i} and V_{g_j} are birational, and Corollary 3.46 yields inverse rational maps

$$\rho_{ij} : V_{g_j} \dashrightarrow V_{g_i}, \quad \rho_{ji} : V_{g_i} \dashrightarrow V_{g_j}.$$

The birational maps ρ_{ij} are called the *gluing maps*.

The birational maps ρ_{ij} satisfy the closed-graph property. Since
$$V_{g_i g_j} \subset \Gamma_{\rho_{ij}} \subset V_{g_i} \times V_{g_j}$$
it suffices to show that $V_{g_i g_j}$ is closed in the product. If $\pi_i : V_{g_i} \to V$ and $\pi_j : V_{g_j} \to V$ are the standard projections then
$$(\pi_i, \pi_j)^{-1}\Delta_V = \{(v_1, v_2) : \pi_i(v_1) = \pi_j(v_2)\} \subset V_{g_i} \times V_{g_j} \subset (V \times \mathbb{A}^1(k))^2$$
is closed by Exercise 3.10. This is equal to
$$\{(v, z_i, z_j) \in V \times \mathbb{A}^1(k) \times \mathbb{A}^1(k) : z_i g_i(v) = z_j g_j(v) = 1\},$$
i.e., the affine variety with coordinate ring $k[V][g_i^{-1}, g_j^{-1}]$. The isomorphism of k-algebras
$$k[V][g_i^{-1}, g_j^{-1}] \xrightarrow{\sim} k[V][(g_i g_j)^{-1}]$$
induces an isomorphism of affine varieties
$$V_{g_i g_j} \xrightarrow{\sim} (\pi_i, \pi_j)^{-1}\Delta_V.$$

9.8.3 Definition of abstract varieties

An *abstract variety* X consists of a collection of affine varieties $U_0, \ldots, U_n$ and birational maps
$$\rho_{ij} : U_j \dashrightarrow U_i$$
satisfying the closed-graph property and the compatibility conditions
$$\rho_{ij} \circ \rho_{j\ell} = \rho_{i\ell}$$
for all $i, j, \ell \in \{0, \ldots, n\}$.

We regard X as the quotient of the disjoint union
$$U_0 \sqcup \ldots \sqcup U_n$$
under the equivalence relation $\approx$ generated by the following 'gluing data': Given $u_i \in U_i$ and $u_j \in U_j$, $u_i \approx u_j$ if ρ_{ij} is defined at u_j and $u_i = \rho_{ij}(u_j)$. The ρ_{ij} are called the *gluing maps*. The intersection $U_i \cap U_j \subset X$ therefore has the structure of the affine variety $U_{ij} := \Gamma_{\rho_{ij}}$.

The collection $\{U_i\}_{i=0,\ldots,n}$ is called an *affine open covering* of X. We say that $Z \subset X$ is *closed* if each intersection $Z \cap U_i$ is closed in U_i.

Example 9.50 Let V be affine and $g_0, \ldots, g_n \in k[V]$ as in § 9.8.2. Then the collection of open affines
$$U_i = \{v \in V : g_i(v) \neq 0\}$$
makes V into an abstract variety.

9.8 ABSTRACT VARIETIES

A *refinement* $\{W_{i;\ell}\}$ of a covering $\{U_i\}$ is a union of open coverings $\{W_{i;\ell}\}_{\ell=1,\ldots,m_i}$ of each U_i, as described in §9.8.2. There are induced birational gluing maps among the $W_{i;\ell}$.

Let X be an abstract variety and W an affine variety. A *morphism* $\phi : X \to W$ is specified over an affine open covering $\{U_i\}$ of X by a collection of morphisms

$$\phi_i : U_i \to W$$

compatible with the gluing maps $\rho_{ij} : U_j \to U_i$, i.e., $\phi_j = \phi_i \circ \rho_{ij}$, for each i, j. The compatibility condition is equivalent to stipulating that $\phi_i|U_{ij} = \phi_j|U_{ij}$, for each i, j. Theorem 9.48 shows that our new definition is consistent with our orginal definition when $X = V$ is affine and $\{U_i = V_{g_i}\}$ arises as in § 9.8.2. It also guarantees our definition respects refinements of affine open coverings. We do not distinguish between a morphism specified over a covering and the morphisms arising from refinements of that covering.

A *morphism of abstract varieties* $\phi : X \to Y$ is specified by the following data: affine open coverings $\{U_i\}_{i=0,\ldots,n}$ and $\{V_j\}_{j=0,\ldots,m}$ with gluing maps $\rho_{ia} : U_a \dashrightarrow U_i$ and $\xi_{jb} : V_b \dashrightarrow V_j$ for X and Y respectively, and a collection of morphisms

$$\{\phi_{j(i),i} : U_i \to V_{j(i)}\}_{i=0,\ldots,n},$$

with $j(i) \in \{0, \ldots, m\}$, such that the compatibility condition

$$\phi_{j(i),i} \circ \rho_{ia} = \xi_{j(i)j(a)} \circ \phi_{j(a),a}$$

holds for each $i, a \in \{0, \ldots, n\}$. Equivalently

$$\phi_{j(i),i}|U_{ia} = \left(\xi_{j(i),j(a)} \circ \phi_{j(a),a}\right)|U_{ia}$$

holds on the interesections $U_{ia} = U_i \cap U_a$.

Remark 9.51   If we *fix* coverings $\{U_i\}$ and $\{V_j\}$ of X and Y then most $\phi : X \to Y$ cannot be realized by compatible collections of morphisms $\{\phi_{j(i),i} : U_i \to V_{j(i)}\}$. As we saw in our analysis of morphisms of projective spaces, the choice of covering must take into account the geometry of the morphism.

We define rational maps of abstract varieties analogously.

The formalism of abstract varieties is quite useful for constructing products:

Proposition 9.52   Let X and Y be abstract varieties. Then the product $X \times Y$ has a natural structure as an abstract variety and admits natural projection morphisms $\pi_1 : X \times Y \to X$ and $\pi_2 : X \times Y \to Y$.

Sketch proof   Let $\{U_i\}_{i=0,\ldots,m}$ and $\{V_j\}_{j=0,\ldots,n}$ be coverings of X and Y respectively by affine open subsets, and $\rho_{ia} : U_a \dashrightarrow U_i$ and $\xi_{jb} : V_b \dashrightarrow V_j$ the birational gluing maps. The product $X \times Y$ is covered by the affine varieties $\{U_i \times V_j\}$ with gluing maps

$$(\rho_{ia}, \xi_{jb}) : U_a \times V_b \dashrightarrow U_i \times V_j.$$

The projections

$$\pi_1 : U_i \times V_j \to U_i, \quad \pi_2 : U_i \times V_j \to V_j$$

are clearly compatible with the gluing maps, and thus define the desired morphisms. $\square$

9.9 Exercises

9.1 Show that an ideal $J \subset k[x_0, \ldots, x_n]$ is homogeneous if, and only if, for each $f \in J$ the homogeneous pieces of f are all in J.

9.2 For $a = [a_0, \ldots, a_n] \in \mathbb{P}^n(k)$, show that

$$J(a) = \langle x_i a_j - x_j a_i \rangle_{i,j=0,\ldots,n}.$$

9.3 Let $Z = \mathbb{P}^1(k) \sqcup \mathbb{P}^1(k) \sqcup \mathbb{P}^1(k)$ be the disjoint union of three copies of $\mathbb{P}^1(k)$. Show that $Z \simeq X \times \mathbb{P}^1(k)$ where $X = \{p_0, p_1, p_2\} \subset \mathbb{P}^1(k)$. Realize Z explicitly as a projective variety and find its homogeneous equations. Consider the disjoint union

$$W = \underbrace{\mathbb{P}^1(k) \sqcup \mathbb{P}^1(k) \ldots \sqcup \mathbb{P}^1(k)}_{n \text{ times}}.$$

Explain how W can be realized as a closed subset of $\mathbb{P}^3(k)$.

9.4 Let $I, J \subset k[x_0, \ldots, x_n]$ be irrelevant ideals. Show that $I \cap J$ and $I + J$ are also irrelevant.

9.5 Consider the polynomial map

$$[x_0, x_1, x_2, x_3] \mapsto [x_0 x_1, x_0 x_2, x_0 x_3, x_1 x_2, x_1 x_3, x_2 x_3].$$

(a) Does this induce a well-defined morphism

$$\phi : \mathbb{P}^3(\mathbb{Q}) \to \mathbb{P}^5(\mathbb{Q})?$$

(b) Consider the hypersurface

$$X = \{[x_0, x_1, x_2, x_3] : x_0^2 + x_1^2 + x_2^2 + x_3^2 = 0\} \subset \mathbb{P}^3(\mathbb{C}).$$

Does our polynomial map induce a well-defined morphism $X \to \mathbb{P}^5(\mathbb{C})$?

9.6 Consider the homogeneous ideal

$$J = \langle x_0^2 x_1, x_1^3, x_1 x_2^2 \rangle.$$

9.9 EXERCISES

(a) Let I_i, $i = 0, 1, 2$ denote the dehomogenization of J with respect to x_i. Compute each I_i.

(b) Let J_i be the homogenization of I_i with respect to x_i. Compute each J_i.

(c) Compute the intersection $J' := J_0 \cap J_1 \cap J_2$. Show that $J \subsetneq J'$.

(d) Show that $(J : \langle x_0, x_1, x_2 \rangle) \neq J$.

9.7 Describe the irreducible components of the projective variety

$$X = \{[x_0, x_1, x_2, x_3] : x_0 x_1 - x_2 x_3 = x_0 x_2 - x_1 x_3 = 0\} \subset \mathbb{P}^3(k).$$

Hint: You may find it easier to work first in the distinguished affine neighborhoods.

9.8 Consider the ideal

$$I = \langle y_2 - y_1^2, y_3 - y_1 y_2, \ldots, y_n - y_1 y_{n-1} \rangle = \langle y_{i+1} - y_1 y_i, i = 1, \ldots, n-1 \rangle.$$

Show that the homogenization of I is generated by the 2×2 minors of the matrix

$$\begin{pmatrix} x_0 & x_1 & \cdots & x_{n-1} \\ x_1 & x_2 & \cdots & x_n \end{pmatrix}.$$

Compare this with Examples 1.5 and 9.39 and Exercise 1.3.

9.9 Consider the rule $\phi : [s, t] \to [s^4, s^2 t^2, t^4]$.

(a) Show that ϕ defines a morphism $\mathbb{P}^1(k) \to \mathbb{P}^2(k)$.

(b) Compute the image $\phi(\mathbb{P}^1(k)) \subset \mathbb{P}^2(k)$ and its closure $\overline{\phi(\mathbb{P}^1(k))}$.

(c) Show that ϕ is not dominant.

9.10 Consider the polynomial rational map

$$\phi : \mathbb{P}^1(k) \dashrightarrow \mathbb{P}^2(k)$$
$$[x_0, x_1] \mapsto [x_0^3, x_0^2 x_1, x_0 x_1^2].$$

Show this is well-defined on the distinguished open set $U_0 \subset \mathbb{P}^1(k)$ and there is a morphism

$$\hat{\phi} : \mathbb{P}^1(k) \to \mathbb{P}^2(k)$$

such that $\hat{\phi}|_{U_0} = \phi|_{U_0}$.

9.11 Let $f = a_d x^d + \cdots + a_0$ and $g = b_e x^e + \cdots + b_0$ be nonzero polynomials in x. Consider the corresponding points $[f] = [a_0, a_1, \ldots, a_d] \in \mathbb{P}^d(k)$ and $[g] = [b_0, b_1, \ldots, b_e] \in \mathbb{P}^e(k)$. Show that

$$\{([f], [g]) : \text{Res}(f, g) = 0\} \subset \mathbb{P}^d(k) \times \mathbb{P}^e(k)$$

is a well-defined closed subset.

9.12 Let $X \subset \mathbb{P}^n(k)$ be closed and $\Sigma \subset \mathbb{P}^n(k)$ a finite set disjoint from X. Show there exists an affine open subset V with

$$\Sigma \subset V \subset (\mathbb{P}^n(k) \setminus X).$$

9.13 Consider the complex plane curves

$$C_1 = \{[x_0, x_1, x_2] : x_0^2 x_2 = x_1^3\}$$

and

$$C_2 = \{[x_0, x_1, x_2] : x_0 x_2^2 = x_0 x_1^2 + x_1^3\}.$$

Show that each of these is the image of the rational normal cubic curve

$$X = \nu(3)(\mathbb{P}^1) \subset \mathbb{P}^3(\mathbb{C})$$

under projection from a point $\Lambda \in \mathbb{P}^3(\mathbb{C})$ not on that curve.

9.14 Our discussion of projective geometry ignores its historical origins in the study of parallelism. Here we partially fill this gap.

A line $L \subset \mathbb{P}^2(k)$ is defined $L = \{[x_0, x_1, x_2] : p_0 x_0 + p_1 x_1 + p_2 x_2 = 0\}$ where $(p_0, p_1, p_2) \neq 0$. Note that (p_0, p_1, p_2) and $\lambda(p_0, p_1, p_2)$, $\lambda \in k^*$, define the same line. Thus the lines in $\mathbb{P}^2(k)$ are also parametrized by a projective plane, the *dual projective plane* $\check{\mathbb{P}}^2(k)$.

(a) Show that the *incidence correspondence*

$$W = \{[x_0, x_1, x_2], [p_0, p_1, p_2] : p_0 x_0 + p_1 x_1 + p_2 x_2 = 0\} \subset \mathbb{P}^2(k) \times \check{\mathbb{P}}^2(k)$$

is closed. Check that the projections

$$p : W \to \mathbb{P}^2(k), \quad \check{p} : W \to \check{\mathbb{P}}^2(k)$$

have projective lines as fibers, i.e., $p^{-1}([a_0, a_1, a_2]) \simeq \mathbb{P}^1(k)$. (Translation: the lines through a point in the plane form a projective line.)

(b) Show that the open subset $U = \check{\mathbb{P}}^2 \setminus \{[1, 0, 0]\}$ corresponds to lines in the affine plane

$$\mathbb{A}^2(k) = U_0 = \{x_0 \neq 0\} \subset \mathbb{P}^2(k).$$

(c) Show that

$$Y = \{[x_0, x_1, x_2], [p_0, p_1, p_2], [q_0, q_1, q_2] : p_0 x_0 + p_1 x_1 + p_2 x_2 = 0,$$
$$q_0 x_0 + q_1 x_1 + q_2 x_2 = 0\} \subset \mathbb{P}^2(k) \times \check{\mathbb{P}}^2(k) \times \check{\mathbb{P}}^2(k)$$

is closed.

(d) Show that the open subset

$$V = \{[p_0, p_1, p_2], [q_0, q_1, q_2] : p_1 q_2 - p_2 q_1 \neq 0\} \subset U \times U$$

parametrizes pairs of non-parallel lines in the affine plane.

(e) Let $q : Y \to \mathbb{P}^2(k)$ and $\check{q} : Y \to \check{\mathbb{P}}^2(k) \times \check{\mathbb{P}}^2(k)$ be the projection morphisms. Show that $\check{q}(q^{-1}(U_0))$ contains V. (Translation: any two non-parallel lines in the affine plane intersect somewhere in the affine plane.)

(f) Show that $\check{q}(Y) = \check{\mathbb{P}}^2(k) \times \check{\mathbb{P}}^2(k)$. (Translation: any two lines in the projective plane intersect.)

(g) Let $\ell_1, \ldots, \ell_r \subset \mathbb{A}^2(k)$ be a collection of parallel lines and $L_1, \ldots, L_r$ their closures in $\mathbb{P}^2(k)$. Show that $L_1, \ldots, L_r$ share a common point $s \in H_0 = \{x_0 = 0\}$, the line at infinity. Moreover, each point $s \in H_0$ arises in this way. (Translation: the points on the line at infinity correspond to equivalence classes of parallel lines in the affine plane.)

9.15 Let X be a projective variety. Show that X is isomorphic to a variety $Y \subset \mathbb{P}^N(k)$ defined by quadratic equations. *Hint:* Given $X \subset \mathbb{P}^n$ defined by homogeneous equations of degree $\leq d$, analyze the equations satisfied by $Y = \nu(d)(X)$ in $\mathbb{P}^{\binom{n+d}{d}-1}(k)$.

9.16 It is quite fruitful to consider real morphisms

$$\phi : \mathbb{P}^n(\mathbb{R}) \to \mathbb{P}^m(\mathbb{R})$$

whose image happens to lie in a distinguished open subset.

Consider the following version of the 2-Veronese morphism:

$$\varphi : \mathbb{P}^1(\mathbb{R}) \to \mathbb{P}^2(\mathbb{R})$$
$$[x_0, x_1] \mapsto [x_0^2 + x_1^2, x_0^2 - x_1^2, 2x_0 x_1].$$

Show that the image satisfies the equation

$$z_1^2 + z_2^2 = z_0^2$$

and verify that $\varphi(\mathbb{P}^1(\mathbb{R})) \subset U_0$. Conclude that there is a bijection between the real projective line and the unit circle

$$C = \{(y_1, y_2) : y_1^2 + y_2^2 = 1\}.$$

Challenge: Show that $\mathbb{P}^1(\mathbb{R})$ and C are not isomorphic as varieties, by proving there exists no non-constant morphism $\mathbb{P}^1(\mathbb{R}) \to \mathbb{A}^1(\mathbb{R})$.

9.17 (Steiner Roman surface) This is a continuation of Exercise 9.16. Consider the polynomial map

$$\phi : \mathbb{P}^2(\mathbb{R}) \dashrightarrow \mathbb{P}^3(\mathbb{R})$$
$$[x_0, x_1, x_2] \mapsto [x_0^2 + x_1^2 + x_2^2, x_1 x_2, x_0 x_2, x_0 x_1].$$

(a) Show that ϕ is a morphism and $\phi(\mathbb{P}^2(\mathbb{R})) \subset U_0$.

(b) Verify that image(ϕ) satisfies the quartic

$$z_1^2 z_2^2 + z_1^2 z_3^2 + z_2^2 z_3^2 = z_0 z_1 z_2 z_3.$$

(c) Writing

$$V = \{(y_1, y_2, y_3) : y_1^2 y_2^2 + y_1^2 y_3^2 + y_2^2 y_3^2 = y_1 y_2 y_3\} \subset \mathbb{A}^3(\mathbb{R})$$

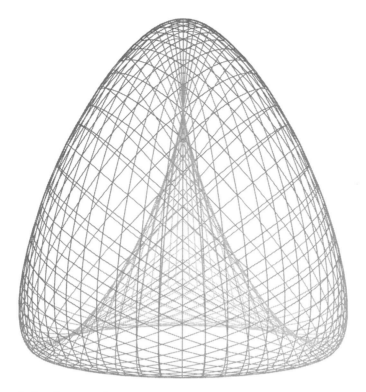

Figure 9.3 Steiner Roman surface.

we have a well-defined function

$$\varphi : \mathbb{P}^2(\mathbb{R}) \to V.$$

For each point $v \in V$, determine $\#\{p \in \mathbb{P}^2(\mathbb{R}) : \varphi(p) = v\}$. *Hint:* Analyze carefully what happens along the coordinate axes.

The surface $V \subset \mathbb{R}^3$ is known as the *Steiner Roman surface*, after Jakob Steiner (1796–1863). It is a useful tool for visualizing the real projective plane.

9.18 (a) Working from the definitions, show that $\mathbb{P}^1(k) \times \mathbb{P}^1(k) \times \mathbb{P}^1(k)$ is an abstract variety.

(b) Verify that the rule

$$\phi : \mathbb{P}^1(k) \times \mathbb{P}^1(k) \times \mathbb{P}^1(k) \to \mathbb{P}^7(k)$$

taking $([s_0, s_1], [t_0, t_1], [u_0, u_1])$ to

$$[s_0 t_0 u_0, s_0 t_0 u_1, s_0 t_1 u_0, s_0 t_1 u_1, s_1 t_0 u_0, s_1 t_0 u_1, s_1 t_1 u_0, s_1 t_1 u_1]$$

gives a well-defined morphism of varieties.

(c) Show that the image of ϕ is closed in $\mathbb{P}^7(k)$ and write down equations for the image.

9.9 EXERCISES

(d) Check that ϕ is an isomorphism onto its image and conclude that $\mathbb{P}^1(k) \times \mathbb{P}^1(k) \times \mathbb{P}^1(k)$ is projective.

9.19 Let X be a projective variety.
(a) Show that the *diagonal map*
$$\Delta : X \to X \times X$$
$$x \mapsto (x, x)$$
is a morphism. Its image is denoted Δ_X.

(b) Let $X = \mathbb{P}^n(k)$ and fix coordinates $x_0, x_1, \ldots, x_n$ and $y_0, y_1, \ldots, y_n$ on $\mathbb{P}^n(k) \times \mathbb{P}^n(k)$. Show that
$$I(\Delta_{\mathbb{P}^n(k)}) = \langle x_i y_j - x_j y_i \rangle_{i,j=0,\ldots,n}.$$

(c) For general X, show that Δ_X is closed in $X \times X$ and isomorphic to X.

(d) Repeat parts (a) and (c) for an arbitrary abstract variety. *Hint:* Let $\{U_i\}$ be a covering for X with gluing maps $\rho_{ij} : U_j \dashrightarrow U_i$. Verify that $\Delta_X \cap (U_i \times U_j)$ corresponds to the graph of ρ_{ij} and apply the closed graph property.

9.20 For each $m, n > 0$, show that there exists no constant morphism
$$\phi : \mathbb{P}^n(\mathbb{C}) \to \mathbb{A}^m(\mathbb{C}).$$

Hints:

(a) Reduce to the $m = 1$ case: Suppose that for each projection $\pi_i : \mathbb{A}^m(\mathbb{C}) \to \mathbb{A}^1(\mathbb{C})$ the composition $\pi_i \circ \phi$ is constant. Then ϕ is constant.

(b) Reduce to the $n = 1$ case: Suppose that $\phi : \mathbb{P}^n(\mathbb{C}) \to \mathbb{A}^1(\mathbb{C})$ is constant along each line $\ell \subset \mathbb{P}^n(\mathbb{C})$. Then ϕ is constant.

(c) Express $\mathbb{C}[\mathbb{A}^1] = \mathbb{C}[t]$. Show there exists a collection of homogeneous forms
$$\{F_0, \ldots, F_m\} \subset \mathbb{C}[x_0, x_1]_{\text{degree } d}$$
such that $\langle F_0, \ldots, F_m \rangle$ is irrelevant and
$$\phi^* t|_{V_{F_i}} \in \mathbb{C}[x_0, x_1][F_i^{-1}]_{\text{degree } 0}$$
for each i (cf. Proposition 9.41).

(d) Suppose $F_1, F_2 \in \mathbb{C}[x_0, x_1]$ are homogeneous of degree d with greatest common divisor G. Suppose that $h \in \mathbb{C}(x_0, x_1)$ is a rational function expressible in two different forms
$$h = H_1/F_1^N, \quad h = H_2/F_2^N,$$
where $N > 0$ and H_1 and H_2 are homogeneous of degree Nd in $\mathbb{C}[x_0, x_1]$. Show that $h = H/G$ for some homogeneous H with $\deg(H) = \deg(G)$.

9.21 Let V be an affine variety over an algebraically closed field, and $\{g_i\}_{i \in I} \in k[V]$ a possibly infinite collection of elements such that
$$V = \cup_{i \in I} U_i, \quad U_i = \{v \in V : g_i(v) \neq 0\}.$$

PROJECTIVE GEOMETRY

Show that there exists a finite collection of indices $i(0), \ldots, i(n)$ such that $\{V_{g_{i(j)}}\}_{j=0,\ldots,n}$ is an affine open covering of V. In other words, every affine open covering admits a finite subcovering; this property is known as *quasi-compactness*.

9.22 Recall the notation of Proposition 9.32. Consider the map

$$\phi_L^{-1} : \mathbb{A}^n(k) \to U_L \subset \mathbb{P}^n(k).$$

Show this defines a morphism $\mathbb{A}^n(k) \to \mathbb{P}^n(k)$, using Theorem 9.48 and the definition of a morphism to an abstract variety.

9.23 Recall that the birational morphism

$$U_0 := \mathbb{A}^2(k) \xrightarrow{\rho} U_1 := \mathbb{A}^2(k)$$
$$(x_1, x_2) \mapsto (x_1, x_1 x_2)$$

with inverse

$$\rho^{-1}(y_1, y_2) = (y_1, y_2/y_1)$$

does not satisfy the closed graph condition.

Show that the intersection U_{01} of the graphs of ρ and ρ^{-1} is nonetheless an affine variety with coordinate ring

$$k[U_{01}] = k[y_1, y_2]\left[y_1^{-1}\right] \simeq k[x_1, x_2]\left[x_1^{-1}\right].$$

Observe that U_{01} is naturally an affine open subset of both U_0 and U_1.

Describe the quotient of $U_0 \sqcup U_1$ obtained by gluing along U_{01}: given $u_0 \in U_0$ and $u_1 \in U_1$, we have $u_0 \approx u_1$ if $u_0, u_1 \in U_{01}$ and $\rho(u_0) = u_1$.

Is this an affine variety? A projective variety?

10 Projective elimination theory

What is projective geometry good for? We focus on a key property of projective varieties that differentiates them from affine varieties: the image of a projective variety under a morphism is always closed.

Recall that the image of a morphism of affine varieties is not closed. For example, consider the variety

$$V = \{(x, y) : xy = 1\} \subset \mathbb{A}^2(k)$$

and the projection map

$$\pi_2 : \mathbb{A}^2(k) \to \mathbb{A}^1(k)$$
$$(x, y) \mapsto y.$$

The image

$$\pi_2(V) = \{y \in \mathbb{A}^1(k) : y \neq 0\}$$

is not closed. This phenomenon complicates description of the image of a morphism. Of course, we have developed algorithms for computing the *closure* of the image of a map of affine varieties in Chapter 4.

Projective geometry allows us to change the problem slightly, so that the image of the morphism becomes closed. We introduce the method for our archetypal example. Regard

$$V \subset \mathbb{A}^1(k) \times \mathbb{A}^1(k) \subset \mathbb{P}^1(k) \times \mathbb{A}^1(k),$$

where the affine factor corresponding to the variable x is completed to the projective line. The projective closure

$$\overline{V} \subset \mathbb{P}^1(k) \times \mathbb{A}^1(k)$$

is obtained by homogenization

$$\overline{V} = \{([x_0, x_1], y) : x_1 y = x_0\}.$$

We still have a projection map

$$\pi_2 : \mathbb{P}^1(k) \times \mathbb{A}^1(k) \to \mathbb{A}^1(k)$$
$$([x_0, x_1], y) \mapsto y,$$

but now $\pi_2(\overline{V}) = \mathbb{A}^1(k)$. Indeed, the point added at infinity is mapped to the origin,

$$\pi_2(([0, 1], 0)) = 0.$$

10.1 Homogeneous equations revisited

A polynomial $F \in k[x_0, \ldots, x_n, y_1, \ldots, y_m]$ is homogeneous of degree d in $x_0, \ldots, x_n$ if

$$F = \sum_{|\alpha|=d} x^\alpha h_\alpha(y_1, \ldots, y_m),$$

where the $h_\alpha \in k[y_1, \ldots, y_m]$. Regard $f \in k[w_1, \ldots, w_n, y_1, \ldots, y_m]$ as a polynomial in $w_1, \ldots, w_n$ with coefficients in $k[y_1, \ldots, y_m]$ of degree d. The homogenization of f relative to $w_1, \ldots, w_n$ is defined as

$$F(x_0, \ldots, x_n, y_1, \ldots, y_m) = x_0^d f(x_1/x_0, \ldots, x_n/x_0, y_1, \ldots, y_m).$$

An ideal $J \subset k[x_0, \ldots, x_n, y_1, \ldots, y_m]$ is homogeneous in $x_0, \ldots, x_n$ if it can be generated by polynomials that are homogeneous in $x_0, \ldots, x_n$. Given $I \subset k[w_1, \ldots, w_n, y_1, \ldots, y_m]$, the homogenization of I relative to $w_1, \ldots, w_n$ is the ideal $J \subset k[x_0, \ldots, x_n, y_1, \ldots, y_m]$ generated by homogenizations of elements in I.

A monomial order $>$ on $k[w_1, \ldots, w_n, y_1, \ldots, y_m]$ is graded relative to $w_1, \ldots, w_n$ if it is compatible with the partial order induced by degree in the w-variables, i.e., $w^\alpha y^\gamma > w^\beta y^\delta$ whenever $|\alpha| > |\beta|$. We have the following straightforward generalization of Theorem 9.6:

Proposition 10.1 *Let $I \subset k[w_1, \ldots, w_n, y_1, \ldots, y_m]$ be an ideal and $J \subset k[x_0, \ldots, x_n, y_1, \ldots, y_m]$ its homogenization relative to $w_1, \ldots, w_n$. Suppose that $f_1, \ldots, f_r$ is a Gröbner basis for I with respect to some order $>$ graded relative to $w_1, \ldots, w_n$. Then the homogenizations of $f_1, \ldots, f_r$ relative to $w_1, \ldots, w_n$ generate J.*

For each $F \in k[x_0, \ldots, x_n, y_1, \ldots, y_m]$ homogeneous in $x_0, \ldots, x_n$ of degree d, we have

$$F(\lambda x_0, \ldots, \lambda x_n, y_1, \ldots, y_m) = \lambda^d F(x_0, \ldots, y_m)$$

so the locus where F vanishes is well-defined in $\mathbb{P}^n(k) \times \mathbb{A}^m(k)$. Let

$$X(F) := \{([x_0, \ldots, x_n], y_1, \ldots, y_m) \in \mathbb{P}^n(k) \times \mathbb{A}^m(k) : F(x_0, \ldots, y_m) = 0\}$$

denote the corresponding hypersurface. More generally, for any ideal $J \subset k[x_0, \ldots, x_n, y_1, \ldots, y_m]$ homogeneous in $x_0, \ldots, x_n$, we have the closed subset

$$X(J) \subset \mathbb{P}^n(k) \times \mathbb{A}^m(k).$$

For each subset $S \subset \mathbb{P}^n(k) \times \mathbb{A}^m(k)$, write

$$J(S) = \{F \in k[x_0, \ldots, x_n, y_1, \ldots, y_m] : F(s) = 0 \text{ for each } s \in S\}.$$

We would like to find equations for closed subsets of $\mathbb{P}^n(k) \times \mathbb{A}^m(k)$. When $m = 0$ we have already addressed this problem: Theorem 9.14 implies each closed $X \subset \mathbb{P}^n(k)$ is given by a homogeneous ideal. The general picture is very similar:

Proposition 10.2 *If $S \subset \mathbb{P}^n(k) \times \mathbb{A}^m(k)$ then*

$$J(S) \subset k[x_0, \ldots, x_n, y_1, \ldots, y_m]$$

is homogeneous and $X(J(S)) = \overline{S}$. Moreover, $J(S)$ can be computed effectively from the local affine equations vanishing along S. Consider the distinguished subsets of $\mathbb{P}^n(k) \times \mathbb{A}^m(k)$

$$U_i = \{([x_0, \ldots, x_n], y_1, \ldots, y_m) : x_i \neq 0\} \simeq \mathbb{A}^n(k) \times \mathbb{A}^m(k)$$

and the ideals

$$I_i = I(U_i \cap S) \subset k[w_0, \ldots, w_{i-1}, w_{i+1}, \ldots, w_n, y_1, \ldots, y_m].$$

If J_i denotes the homogenization of I_i relative to $w_0, \ldots, w_{i-1}, w_{i+1}, \ldots, w_n$ (using x_i as the homogenizing variable) then

$$J(S) = \cap_{i=0}^n J_i.$$

The proof proceeds just as in Section 9.4.

10.2 Projective elimination ideals

Given a closed subset $X \subset \mathbb{P}^n(k) \times \mathbb{A}^m(k)$, we would like algorithms for computing the image $\pi_2(X) \subset \mathbb{A}^m(k)$.

Definition 10.3 Let $J \subset k[x_0, \ldots, x_n]$ be homogeneous and $\mathfrak{m} = \langle x_0, \ldots, x_n \rangle$. The *saturation* of J is defined

$$\tilde{J} = \{F \in k[x_0, \ldots, x_n] : \mathfrak{m}^N F \subset J \text{ for some } N \gg 0\}.$$

An ideal is *saturated* if it is equal to its saturation.

Note that an ideal is irrelevant if and only if its saturation is $k[x_0, \ldots, x_n]$. Generally, we have $X(\tilde{J}) = X(J)$ (see Exercise 10.2.)

Definition 10.4 Given an ideal $J \subset k[x_0, \ldots, x_n, y_1, \ldots, y_m]$, homogeneous in $x_0, \ldots, x_n$, the *projective elimination ideal* is defined

$$\hat{J} = \{f \in k[y_1, \ldots, y_m] : \mathfrak{m}^N \langle f \rangle \subset J \text{ for some } N \gg 0\}, \quad \mathfrak{m} = \langle x_0, \ldots, x_n \rangle.$$

The reason for the fudge-factor $\mathfrak{m}^N$ is that we want any irrelevant ideal to have elimination ideal $\langle 1 \rangle$: The image of the closed subset defined by an irrelevant ideal (i.e., the empty set) is empty!

Proposition 10.5 *Consider an ideal homogeneous in $x_0, \ldots, x_n$,*

$$J \subset k[x_0, \ldots, x_n, y_1, \ldots, y_m],$$

defining a closed subset $X(J) \subset \mathbb{P}^n(k) \times \mathbb{A}^m(k)$. Then we have

$$\pi_2(X(J)) \subset V(\hat{J}).$$

Proof Suppose we have $(b_1, \ldots, b_m) \in \pi_2(X(J))$, corresponding to

$$([a_0, \ldots, a_n], b_1, \ldots, b_m) \in X(J).$$

For each $F \in J$ we have $F(a_0, \ldots, a_n, b_1, \ldots, b_m) = 0$. For some i, $a_i \neq 0$, say, $a_0 \neq 0$. If $h(y_1, \ldots, y_m) \in \hat{J}$ then $x_0^N h \in F$ for some $N \gg 0$, so

$$a_0^N h(b_1, \ldots, b_m) = 0$$

and hence $h(b_1, \ldots, b_m) = 0$. $\square$

Theorem 10.6 (Projective Elimination) *Assume k is algebraically closed. Let $J \subset k[x_0, \ldots, x_n, y_1, \ldots, y_m]$ be homogeneous in $x_0, \ldots, x_n$, $X = X(J) \subset \mathbb{P}^n(k) \times \mathbb{A}^m(k)$, and $\hat{J} \subset k[y_1, \ldots, y_m]$ the projective elimination ideal. Then*

$$\pi_2(X(J)) = V(\hat{J}).$$

Proof It suffices to show $V(\hat{J}) \subset \pi_2(X)$: the reverse implication has already been proven.

Suppose we are given $c = (c_1, \ldots, c_m) \in V(\hat{J})$, but $c \notin \pi_2(X)$. Write $J = \langle F_1, \ldots, F_r \rangle$, with $F_1, \ldots, F_r$ homogenous in $x_0, \ldots, x_n$, and set

$$F_i(x_0, \ldots, x_n, c) = F_i(x_0, \ldots, x_n, c_1, \ldots, c_m).$$

Since the equations

$$F_1(x_0, \ldots, x_n, c) = \ldots = F_r(x_0, \ldots, x_n, c) = 0$$

have no common solutions, Projective Nullstellensatz gives

$$\mathfrak{m}^N \subset \langle F_i(x_0, \ldots, x_n, c) \rangle_{i=1,\ldots,r}$$

10.2 PROJECTIVE ELIMINATION IDEALS

for some $N \gg 0$. In particular, for each monomial x^α with $|\alpha| = N$ we can write

$$x^\alpha = \sum_{i=1}^{r} F_i(x_0, \ldots, x_n, c) H_{i,\alpha}(x_0, \ldots, x_n),$$

where the $H_{i,\alpha}$ are homogeneous.

We may choose monomials x^{β_j} and indices $i_j \in \{1, \ldots, r\}$ for $j = 1, \ldots, \binom{N+n}{N}$ such that

$$\{x^{\beta_j} F_{i_j}(x_0, \ldots, x_n, c)\}_{j=1,\ldots,\binom{N+n}{N}}$$

forms a basis for homogeneous forms of degree N in $x_0, \ldots, x_n$. Consider the corresponding polynomials in $x_0, \ldots, x_n, y_1, \ldots, y_m$:

$$G_j(x_0, \ldots, x_n, y_1, \ldots, y_m) := x^{\beta_j} F_{i_j}(x_0, \ldots, x_n, y_1, \ldots, y_m).$$

Express

$$G_j = \sum_{|\alpha|=N} x^\alpha A_{j,\alpha}(y_1, \ldots, y_m)$$

so that $A = (A_{j,\alpha})$ is an $\binom{N+n}{N} \times \binom{N+n}{N}$ matrix of polynomials in $y_1, \ldots, y_m$. The determinant

$$D(y_1, \ldots, y_m) = \det(A)$$

is hence also a polynomial in $y_1, \ldots, y_m$, and $D(c_1, \ldots, c_m) \neq 0$ by hypothesis. By Cramer's rule

$$D(y_1, \ldots, y_m) x^\alpha = \sum_{j=1}^{\binom{N+n}{N}} B_{j,\alpha}(y_1, \ldots, y_m) G_j(x_0, \ldots, x_m, y_1, \ldots, y_m)$$

for a suitable matrix $B = (B_{j,\alpha})$, with entries polynomials in $y_1, \ldots, y_m$. It follows that

$$D(y_1, \ldots, y_m) x^\alpha \in \langle F_1, \ldots, F_r \rangle$$

and thus

$$D(y_1, \ldots, y_m) \in \hat{J}.$$

This contradicts the assumption that $(c_1, \ldots, c_m) \in V(\hat{J})$. □

Remark 10.7 This is strikingly similar to our analysis of resultants, especially Theorems 5.5 and 5.8. We encourage the reader to apply the argument for Theorem 10.6 to

$$J = \langle a_d x_1^d + a_{d-1} x_1^{d-1} x_0 + \cdots + a_0 x_0^d, b_e x_1^e + b_{e-1} x_1^{e-1} x_0 + \cdots + b_0 x_0^e \rangle,$$

which defines a closed subset $X \subset \mathbb{P}^1(k) \times \mathbb{A}^{m+n+2}(k)$. Here $a_0, \ldots, a_d, b_0, \ldots, b_e$ play the role of the y-variables in our argument; the elimination ideal is generated by resultant of the two generators. One advantage of this special case is the good control we have over N, which makes possible the elegant determinantal form of the resultant.

10.2.1 Caveats for nonclosed fields

The Projective Elimination Theorem needs the assumption that k is algebraically closed. For general fields, at least two problems arise.

First, the ideal $J(X(J))$ may be much larger than J, and may include equations that are not algebraic consequences of the equations in J. This is the case if $X(J)$ happens to have few points with coordinates in k, i.e., when there are polynomials vanishing on $X(J)$ that are not consequences of polynomials in J. See Example 3.17 for concrete examples and further discussion.

The second problem arises even when $X(J)$ does have lots of k-rational points. Consider the real variety

$$V = \{([x_0, x_1], y) : x_0^2 = y x_1^2\} \subset \mathbb{P}^1(\mathbb{R}) \times \mathbb{A}^1(\mathbb{R}).$$

The image,

$$\pi_2(V) = \{y \in \mathbb{A}^1(\mathbb{R}) : y \geq 0\},$$

is not a Zariski-closed subset of $\mathbb{A}^1(\mathbb{R})$.

10.3 Computing the projective elimination ideal

Let $J \subset k[x_0, \ldots, x_n, y_1, \ldots, y_m]$ be an ideal, homogeneous in $x_0, \ldots, x_n$. How do we compute the projective elimination ideal $\hat{J}$?

Method I By definition,

$$\hat{J} = (\cup_N J : \mathfrak{m}^N) \cap k[y_1, \ldots, y_m]$$

for $\mathfrak{m} = \langle x_0, \ldots, x_n \rangle$. For any particular N, Algorithm 8.10 computes the quotient $J : \mathfrak{m}^N$. To compute the union, we use the ascending chain condition and a generalization of Proposition 8.34:

Proposition 10.8   *Let R be a ring and $I, J \subset R$ ideals. Suppose that for some integer $M \geq 0$*

$$I : J^M = I : J^{M+1},$$

where $J^0 = R$ by convention. Then we have

$$I : J^M = I : J^{M+1} = I : J^{M+2} = I : J^{M+3} = \ldots.$$

Method II   For each $i = 0, \ldots, n$, let I_i denote the dehomogenization of J with respect to x_i. Then

$$\hat{I} := \cap_{i=0}^{n}(I_i \cap k[y_1, \ldots, y_m])$$

is equal to $\hat{J}$.

Proof of equivalence   Suppose $f \in \hat{J}$, so that for some $N \gg 0$ we have

$$x_0^N f, x_1^N f, \ldots, x_n^N f \in J.$$

It follows that f is contained in each dehomogenization I_i, and thus in $\hat{I}$.

We prove the reverse inclusion. Recall that if F is homogeneous in $x_0, \ldots, x_n$, f its dehomogenization with respect to x_i, and F' the rehomogenization of f, then $F = x_i^e F'$ for some e. Hence if J is a homogeneous ideal, I_i its dehomogenization with respect to x_i, and J_i the rehomogenization with respect to x_i, then

$$J_i = \left(J : \langle x_i^N \rangle\right), \text{ for all } N \gg 0.$$

We have

$$I_i \cap k[y_1, \ldots, y_m] = J_i \cap k[y_1, \ldots, y_m]$$

and hence

$$\hat{I} = \cap_{i=0}^{n}(J_i \cap k[y_1, \ldots, y_m]).$$

On the other hand,

$$\cap_{i=0}^{n} J_i = (J : \mathfrak{m}^N), N \gg 0,$$

so $\hat{I} = \hat{J}$. □

10.4 Images of projective varieties are closed

Theorem 10.9   Assume k is algebraically closed. Let $X \subset \mathbb{P}^n(k)$ be a projective variety, Y an abstract variety, $\phi : X \to Y$ a morphism. Then $\phi(X)$ is closed.

Proof   Let

$$\Gamma_\phi \subset X \times Y \subset \mathbb{P}^n(k) \times Y$$

denote the graph of ϕ. We claim that this is closed. Consider the induced morphism

$$(\phi, \text{Id}) : X \times Y \to Y \times Y,$$

with the second factor the identity. We have

$$\Gamma_\phi = \{(x, y) \in X \times Y : (f, \text{Id})(x, y) = (y, y)\} = (\phi, \text{Id})^{-1}(\Delta_Y),$$

where Δ_Y is the diagonal. Since Δ_Y is closed in $Y \times Y$ (see Exercise 9.19), Γ_ϕ is closed as well.

Choose an affine open covering $\{V_j\}$ for Y; realize $V_j \subset \mathbb{A}^m(k)$ as a closed subset. The intersection

$$\Gamma_\phi \cap \pi_2^{-1}(V_j) \subset \mathbb{P}^n(k) \times \mathbb{A}^m(k)$$

is also closed, and the Projective Elimination Theorem (10.6) implies

$$\pi_2\left(\Gamma_\phi \cap \pi_2^{-1}(V_j)\right)$$

is closed, which is the equal to $\pi_2(\Gamma_\phi) \cap V_j = \phi(X) \cap V_j$. Thus the intersection of $\phi(X)$ with each affine open neighborhood is closed in that neighborhood, and $\phi(X)$ is closed in Y. □

Proposition 10.10   Assume k is algebraically closed. Let $V \subset \mathbb{A}^n(k)$ be an affine variety, $\phi : V \to \mathbb{A}^m(k)$ a morphism with graph Γ_ϕ. Let

$$\overline{\Gamma_\phi} \subset \mathbb{P}^n(k) \times \mathbb{A}^m(k)$$

be the projective closure. Then $\pi_2(\overline{\Gamma_\phi}) = \overline{\phi(V)}$.

Proof   The Projective Elimination Theorem (10.6) implies $\pi_2(\overline{\Gamma_\phi})$ is closed, and it contains the image $\phi(V)$, so it also contains $\overline{\phi(V)}$. On the other hand, $\pi_2^{-1}(\overline{\phi(V)})$ contains Γ_ϕ, and also $\overline{\Gamma_\phi}$; applying ϕ_2 gives the desired result. □

Remark 10.11   Let $V \subset \mathbb{A}^n(k)$ be affine and $\phi : V \to \mathbb{A}^m(k)$ a morphism. If its image fails to be closed, the extra points in the closure come from points 'at infinity', i.e.,

$$\overline{\phi(V)} - \phi(V) \subset \pi_2(\overline{\Gamma_\phi} \cap X(x_0)).$$

Here we are identifying $\mathbb{A}^n(k)$ with $U_0 = \{x_0 \neq 0\} \subset \mathbb{P}^n(k)$; $X(x_0) = (\mathbb{P}^n(k) \times \mathbb{A}^m(k)) \setminus (U_0 \times \mathbb{A}^m(k))$ is the hyperplane at infinity.

10.5 Further elimination results

For the remainder of this section, we discuss how elimination results for affine varieties can be applied in projective contexts.

Proposition 10.12   Let $X \subset \mathbb{P}^n(k)$ be a projective variety and $\phi : X \dashrightarrow \mathbb{P}^m(k)$ a rational map induced by the polynomial map

$$\mathbb{P}^n(k) \dashrightarrow \mathbb{P}^m(k)$$
$$[x_0, \ldots, x_n] \mapsto [\phi_0, \ldots, \phi_m].$$

Writing

$$I = \langle y_0 - \phi_0, \ldots, y_m - \phi_m \rangle + k[x_0, \ldots, x_n, y_0, \ldots, y_m] J(X)$$
$$J = I \cap k[y_0, \ldots, y_m].$$

then $X(J)$ is the closure of the image of ϕ.

Of course, if k is algebraically closed and ϕ is a morphism then $\phi(X)$ is automatically closed.

Proof Let $C(X) \subset \mathbb{A}^{n+1}(k)$ be the *cone* over X, i.e., the subvariety defined by the equations in $J(X)$, or the union of lines

$$\{\lambda(x_0, \ldots, x_n) : \lambda \in k\} \subset \mathbb{A}^{n+1}(k)$$

for $[x_0, \ldots, x_n] \in X$. The polynomials defining ϕ also determine a map of affine varieties

$$\varphi : C(X) \to \mathbb{A}^{m+1}(k),$$

whose graph is defined by the equations in I. It is easy to see that

$$C(\phi(X)) = \varphi(C(X)),$$

which has equations that may be analyzed with affine elimination theory (see §4.1). In particular, the image is given by the ideal J. □

10.6 Exercises

10.1 Prove Proposition 10.8.

10.2 Let $J \subset k[x_0, \ldots, x_n]$ be a homogeneous ideal with saturation $\tilde{J}$.
 (a) Show that $X(J) = X(\tilde{J})$.
 (b) For each $i = 0, \ldots, n$, let $I_i \subset k[y_0, \ldots, y_{i-1}, y_{i+1}, \ldots, y_n]$ denote the dehomogenization of J and $J_i \subset k[x_0, \ldots, x_n]$ the homogenization of I_i. Show that

$$\tilde{J} = J_0 \cap J_1 \cap \ldots \cap J_n.$$

 (c) Show that $J = \tilde{J}$ if and only if $\mathfrak{m} = \langle x_0, \ldots, x_n \rangle$ is not an associated prime of J.

10.3 Consider the ideal

$$J = \langle x_0 y_0 + x_1 y_1, x_0 y_1 + x_1 y_0 \rangle \subset \mathbb{C}[x_0, x_1, y_0, y_1],$$

which is homogeneous in x_0, x_1.
 (a) Compute the intersection $J \cap \mathbb{C}[y_0, y_1]$.
 (b) Compute the projective elimination ideal $\hat{J}$.

(c) Compute the image $\pi_2(X(J))$, where $X(J) \subset \mathbb{P}^1(\mathbb{C}) \times \mathbb{A}^2(\mathbb{C})$.

(d) Note that J is also homogeneous in y_0 and y_1 and thus determines a closed subset $X \subset \mathbb{P}^1(\mathbb{C}) \times \mathbb{P}^1(\mathbb{C})$. Describe X.

10.4 (Cayley cubic surface) Consider the rational map

$$\rho : \mathbb{P}^2(k) \dashrightarrow \mathbb{P}^3(k)$$

taking $[x_0, x_1, x_2]$ to

$$[-x_0 x_1 x_2, x_0 x_1 (x_0 + x_1 + x_2), x_0 x_2 (x_0 + x_1 + x_2), x_1 x_2 (x_0 + x_1 + x_2)].$$

Describe the indeterminacy of ρ in $\mathbb{P}^2(k)$ and the equations of the image in $\mathbb{P}^3(k)$. *Optional:* Show that $\mathbb{P}^2(k) \dashrightarrow \overline{\text{image}(\rho)}$ is birational.

10.5 $F \in k[x_0, \ldots, x_n, y_0, \ldots, y_m]$ is *bihomogeneous of bidegree* (d, e) if it is homogeneous in $x_0, \ldots, x_n$ of degree d and homogeneous in $y_0, \ldots, y_m$ of degree e.

(a) Show that if F is bihomogeneous then

$$Y(F) = \{([a_0, \ldots, a_n], [b_0, \ldots, b_m]) : F(a_0, \ldots, a_n, b_0, \ldots, b_m) = 0\}$$

is a well-defined closed subset of $\mathbb{P}^n(k) \times \mathbb{P}^m(k)$.

(b) Using the Segre embedding, express $Y(F)$ as the locus in $\mathbb{P}^{mn+m+n}(k)$ where a collection of homogeneous forms vanish. *Hint:* If $d \geq e$ express $y^\beta F$, for each β with $|\beta| = d - e$, as a polynomial in the products $x_i y_j$.

(c) For $F = x_0^2 y_0 + x_1^2 y_1 + x_2^2 y_2$ write down explicit equations for $Y(F) \subset \mathbb{P}^8(k)$.

(d) If $J \subset k[x_0, \ldots, x_n, y_0, \ldots, y_m]$ is a bihomogeneous ideal show that

$$Y(J) := \cap_{F \in J} Y(F) \subset \mathbb{P}^n(k) \times \mathbb{P}^m(k)$$

is projective.

(e) Let $X \subset \mathbb{P}^m(k)$ and $Y \subset \mathbb{P}^n(k)$ be projective varieties. Show that $X \times Y$ is projective.

10.6 Given a polynomial morphism

$$\phi : \mathbb{P}^n(k) \to \mathbb{P}^m(k)$$
$$[x_0, \ldots, x_n] \mapsto [\phi_0, \ldots, \phi_m]$$

it can be tricky to extract equations for the graph.

(a) Show that the graph always contains the bihomogeneous equations

$$y_i \phi_j - y_j \phi_i.$$

(b) For the Veronese morphism

$$\nu(2) : \mathbb{P}^1(k) \to \mathbb{P}^2(k)$$
$$[x_0, x_1] \mapsto [x_0^2, x_0 x_1, x_1^2]$$

show that the equations of the graph are

$$\langle y_0 x_1 - y_1 x_0, y_1 x_1 - y_2 x_0 \rangle.$$

(c) Extract bihomogenous equations for the graph of $v(2) : \mathbb{P}^2(k) \to \mathbb{P}^5(k)$.

10.7 A *binary form of degree d* is a nonzero homogeneous polynomial $F = a_0 x_0^d + a_1 x_0^{d-1} x_1 + \cdots + a_d x_1^d \in k[x_0, x_1]$. Binary forms up to scalar multiples are parametrized by the projective space $\mathbb{P}(k[x_0, x_1]_d) = \{[a_0, \ldots, a_d]\}$.

(a) Show that multiplication induces a morphism

$$\mathbb{P}(k[x_0, x_1]_e) \times \mathbb{P}(k[x_0, x_1]_{d-e}) \to \mathbb{P}(k[x_0, x_1]_d)$$
$$(F, G) \mapsto FG.$$

Suppose k is algebraically closed. We say F has a *root of multiplicity* $\geq e$ if there exists a nonzero linear form $L = l_0 x_0 + l_1 x_1$ with $L^e | F$.

(b) Prove that the binary forms with a root of multiplicity e form a closed subset $R_e \subset \mathbb{P}(k[x_0, x_1]_d)$. *Hint:* Verify that the map

$$\mathbb{P}(k[x_0, x_1]_1) \times \mathbb{P}(k[x_0, x_1]_{d-e}) \to \mathbb{P}(k[x_0, x_1]_d)$$
$$(L, G) \mapsto L^e G$$

is a morphism of projective varieties.

(c) Write down explicit equations for
- $R_1 \subset \mathbb{P}^d(k)$;
- $R_2 \subset \mathbb{P}^3(k), \mathbb{P}^4(k)$.

10.8 Let F be homogeneous of degree d in $k[x_0, \ldots, x_n]$; all such forms (up to scalars) are parametrized by

$$\mathbb{P}^{\binom{n+d}{d}-1}(k) = \mathbb{P}(k[x_0, \ldots, x_n]_d).$$

For $a = [a_0, \ldots, a_n] \in \mathbb{P}^n(k)$, we say that F has *multiplicity* $\geq e$ *at* a (or *vanishes to order e at a*) if (cf. Exercise 9.2):

$$F \in J(a)^e = \langle x_i a_j - x_j a_i, \quad i, j = 0, \ldots, n \rangle^e.$$

(a) Show that F has multiplicity $\geq e$ at $[1, 0, 0, \ldots, 0]$ if and only if the dehomogenization

$$\mu_0(F) \in k[y_1, \ldots, y_n]$$

has no terms of degree $< e$.

Consider the locus

$$Z_e := \{(a, F) : F \text{ has multiplicity} \geq e \text{ at } a \} \subset \mathbb{P}^n(k) \times \mathbb{P}^{\binom{n+d}{d}-1}(k).$$

Assume k is a field of characteristic other than two and $d = 2$.

(b) Show that $Z_2 = \{(a, F) : \partial F/\partial x_i(a_0, \ldots, a_n) = 0, i = 0, \ldots, n\}$.

(c) Write

$$F(x_0, \ldots, x_n) = (x_0 \; x_1 \; \cdots \; x_n) \begin{pmatrix} y_{00} & y_{01} & \cdots & y_{0n} \\ y_{01} & y_{11} & \cdots & y_{1n} \\ \vdots & \vdots & \vdots & \vdots \\ y_{0n} & y_{1n} & \cdots & y_{nn} \end{pmatrix} \begin{pmatrix} x_0 \\ x_1 \\ \vdots \\ x_n \end{pmatrix}$$

for $Y = (y_{ij})$ a symmetric $(n+1) \times (n+1)$ matrix. Compute the projective elimination ideal of

$$\langle \partial F/\partial x_0, \ldots, \partial F/\partial x_n \rangle \subset k[x_0, \ldots, x_n, y_{00}, \ldots, y_{nn}]$$

and show that $\pi_2(Z_2) = \{Y : \det(Y) = 0\}$.

(d) Returning to the general case, show that Z_e is closed.

Conclude that the hypersurfaces with point of multiplicity $\geq e$ are closed in $\mathbb{P}^{\binom{n+d}{d}-1}(k)$.

11 Parametrizing linear subspaces

The vector subspaces of k^N are parametrized by a projective variety called the *Grassmannian*. This is the first instance of a very important principle: algebraic varieties with common properties are often themselves classified by an algebraic variety. Applying the techniques of algebraic geometry to this 'classifying variety' gives rise to rich insights. For example, we write down explicit equations for the Grassmannian in projective space, using the formalism of exterior algebra. Such representations are crucial for many applications.

11.1 Dual projective spaces

Recall that we defined projective space

$$\mathbb{P}^n(k) = \text{space of all lines } 0 \in \ell \subset k^{n+1}$$
$$= \text{one-dimensional subspaces } \ell = \text{span}(x_0, \ldots, x_n).$$

Definition 11.1 The *dual projective space* $\check{\mathbb{P}}^n(k)$ is the space of all n-dimensional vector subspaces $H \subset k^{n+1}$.

It is a basic fact of linear algebra that every n-dimensional subspace can be expressed as

$$H = H(p_0, \ldots, p_n) = \{(x_0, \ldots, x_n) : p_0 x_0 + \cdots + p_n x_n = 0\}$$

for some $(p_0, \ldots, p_n) \neq 0$, where

$$H(p_0, \ldots, p_n) = H(p'_0, \ldots, p'_n) \Leftrightarrow [p_0, \ldots, p_n] = [p'_0, \ldots, p'_n] \in \mathbb{P}^n(k).$$

Thus the map $H(p_0, \ldots, p_n) \to [p_0, \ldots, p_n]$ allows us to identify $\mathbb{P}^n(k)$ with $\check{\mathbb{P}}^n(k)$.

PARAMETRIZING LINEAR SUBSPACES

Example 11.2 (General incidence correspondence) Consider the locus

$$W := \{(\ell, H) : \ell \subset H \subset k^{n+1}\}$$
$$= \{[x_0, \ldots, x_n], [p_0, \ldots, p_n] : x_0 p_0 + \cdots + x_n p_n = 0\}$$
$$\subset \mathbb{P}^n(k) \times \check{\mathbb{P}}^n(k).$$

Recall the Segre embedding

$$\phi : \mathbb{P}^n(k) \times \check{\mathbb{P}}^n(k) \hookrightarrow \mathbb{P}^{n^2+2n}(k)$$
$$k[z_{00}, \ldots, z_{nn}] \to k[x_0, \ldots, x_n] \times k[p_0, \ldots, p_n]$$
$$z_{ij} \to x_i p_j$$

with image given by the vanishing of the 2×2 minors of the matrix

$$Z = \begin{pmatrix} z_{00} & z_{01} & \cdots & z_{0n} \\ z_{10} & z_{11} & \cdots & z_{1n} \\ \vdots & \vdots & \cdots & \cdots \\ z_{n0} & z_{n1} & \cdots & z_{nn} \end{pmatrix}.$$

The locus $W \subset \phi(\mathbb{P}^n(k) \times \check{\mathbb{P}}^n(k))$ is defined by

$$z_{00} + \cdots + z_{nn} = \mathrm{Trace}(Z) = 0,$$

and thus is a projective variety.

11.2 Tangent spaces and dual varieties

Let $V \subset \mathbb{A}^n(k)$ be a hypersurface, i.e., $I(V) = \langle g \rangle$ for some $g \in k[y_1, \ldots, y_n]$. Given $b \in V$, the affine tangent space is defined:

$$T_b V = \left\{ (y_1, \ldots, y_n) : \sum_{j=1}^n \partial g/\partial y_j|_b \cdot (y_j - b_j) = 0 \right\}. \tag{11.1}$$

This is an affine-linear subspace of $\mathbb{A}^n(k)$. It is a hyperplane if $\partial g/\partial y_j|_b \neq 0$ for some index j; in this case, we say that V is *smooth* at b. Otherwise, it is *singular*.

For a general affine $V \subset \mathbb{A}^n(k)$ and $b \in V$, we define the affine tangent space by

$$T_b V = \left\{ (y_1, \ldots, y_n) : \sum_{j=1}^n \partial g/\partial y_j|_b (y_j - b_j) = 0 \text{ for each } g \in I(V) \right\}.$$

We say that V is *smooth* at b if it has a unique irreducible component containing b of dimension $\dim T_b V$. (The dimension of a variety at a point is well-defined when the variety has a unique irreducible component containing the point.)

11.2 TANGENT SPACES AND DUAL VARIETIES

Consider an affine linear subspace expressed as the solutions to a system of linear equations, i.e.,

$$L = \left\{ y = \begin{pmatrix} y_1 \\ \vdots \\ y_n \end{pmatrix} : Ay = b \right\} \subset \mathbb{A}^n(k)$$

where A is an $m \times n$ matrix of maximal rank, $b \in k^m$ is a column vector, and $\dim L = n - m$. Express $\mathbb{A}^n(k)$ as a distinguished open subset $U_0 \subset \mathbb{P}^n(k)$. Using Proposition 9.16, the projective closure of L can be expressed

$$\overline{L} = \left\{ [x_0, \ldots, x_n] \in \mathbb{P}^n(k) : A \begin{pmatrix} x_1 \\ \vdots \\ x_n \end{pmatrix} = bx_0 \right\}.$$

This is a linear subspace of $\mathbb{P}^n(k)$ as well.

Proposition 11.3 *Let $X \subset \mathbb{P}^n(k)$ be a hypersurface with $J(X) = \langle F \rangle$, where $F \in k[x_0, \ldots, x_n]$ is homogeneous of degree d, and $a = [a_0, \ldots, a_n] \in X \cap U_i$. Write $V = X \cap U_i$, $b = (a_0/a_i, \ldots, a_{i-1}/a_i, a_{i+1}/a_i, \ldots, a_n/a_i) \in \mathbb{A}^n(k)$ the corresponding point of affine space, and $f \in k[y_0, \ldots, y_{i-1}, y_{i+1}, \ldots, y_m]$ the dehomogenization of F with respect to x_i. Then the projective closure $\overline{T_b V}$ equals the linear subspace*

$$\mathbb{T}_a X := \left\{ [x_0, \ldots, x_n] : \sum_{i=0}^n (\partial F / \partial x_i)|_a x_i = 0 \right\}.$$

This is called the projective tangent space *of X at a.*

In particular, a hypersurface in $\mathbb{P}^n(k)$ is singular at a point if its projective tangent space there is $\mathbb{P}^n(k)$.

Proof For notational simplicity assume $i = 0$. Let

$$\mu_0 : k[x_0, \ldots, x_n] \to k[y_1, \ldots, y_m]$$

denote the dehomogenization homomorphism. For $i = 1, \ldots, n$ we have

$$\mu_0(\partial F / \partial x_i) = \partial f / \partial y_i.$$

Writing out f as a sum of homogeneous pieces

$$f = f_0 + f_1 + \cdots + f_d,$$

we find

$$\mu_0(\partial F / \partial x_0) = df_0 + (d-1)f_1 + \cdots + f_{d-1}.$$

PARAMETRIZING LINEAR SUBSPACES

We analyze the expression

$$\sum_{i=0}^{n} (\partial F/\partial x_i)|_a x_i. \tag{11.2}$$

This is homogeneous in a of degree $d-1$, i.e.,

$$\sum_{i=0}^{n} (\partial F/\partial x_i)|_{\lambda a} = \lambda^{d-1} \sum_{i=0}^{n} (\partial F/\partial x_i)|_a.$$

Since we are only interested in where this vanishes, we may assume that $a_0 = 1$ and $a_i = b_i, i = 1, \ldots, n$. Dehomogenizing (11.2) therefore yields

$$(df_0 + (d-1)f_1 + \cdots + f_{d-1})|_b + \sum_{j=1}^{n} \partial f/\partial y_j|_b y_j.$$

This is equal to (11.1) provided we can establish

$$-\sum_{j=1}^{n} b_j \partial f/\partial y_j|_b = (df_0 + (d-1)f_1 + \cdots + f_{d-1})|_b.$$

Lemma 11.4 (Euler's Formula) *If $F \in k[x_0, \ldots, x_n]$ is homogeneous then*

$$\deg(F) \cdot F = \sum_{i=0}^{n} x_i \partial F/\partial x_i.$$

Proof of lemma Both sides are linear in F. It suffices then to check the formula for $x^\alpha = x_0^{\alpha_0} \ldots x_n^{\alpha_n}$, $\alpha_0 + \cdots + \alpha_n = d$. In this case we have

$$\sum_{i=0}^{n} x_i \frac{\partial}{\partial x_i} x^\alpha = x^\alpha \sum_{i=0}^{n} \alpha_i = dx^\alpha. \qquad \square$$

Applying μ_0 to Euler's formula and evaluating at $(b_1, \ldots, b_n)$ yields

$$d \cdot f(b_1, \ldots, b_n) = (df_0 + (d-1)f_1 + \cdots + f_{d-1})|_b + \sum_{j=1}^{n} b_j \partial f/\partial y_j|_b.$$

Since $f(b_1, \ldots, b_n) = 0$ we obtain

$$(df_0 + (d-1)f_1 + \cdots + f_{d-1})|_b = -\sum_{j=1}^{n} b_j \partial f/\partial y_j|_b. \qquad \square$$

Corollary 11.5 *Retain the notation of Proposition 11.3.*

1. *If $a \in X \cap U_i$ and $\partial F/\partial x_j|_a = 0$ for each $j \neq i$ then $\partial F/\partial x_i|_a = 0$ as well.*
2. *$a \in X$ is singular if and only if $\partial F/\partial x_i|_a = 0$ for each $i = 0, \ldots, n$.*

3. *Suppose that* $(\operatorname{char}(k), d) = 1$ *or* $\operatorname{char}(k) = 0$. *Singular points* $a \in X$ *are precisely the simultaneous solutions to the equations*

$$\partial F/\partial x_0 = \ldots = \partial F/\partial x_n = 0.$$

Proof Only the last assertion requires proof. Euler's formula gives

$$F(a_0, \ldots, a_n) = 1/d \sum_{i=0}^{n} a_i \partial F/\partial x_i|_{(a_0,\ldots,a_n)},$$

which vanishes whenever the partials all vanish. □

Definition 11.6 Let $X \subset \mathbb{P}^n(k)$ be a projective variety and $a \in X$. The *projective tangent space* of X at a is defined

$$\mathbb{T}_a X = \left\{ [x_0, \ldots, x_n] : \sum_{i=0}^{n} \partial F/\partial x_i|_a x_i = 0 \text{ for each } F \in J(X) \right\} \subset \mathbb{P}^n(k).$$

X is *smooth* at a if it has a unique irreducible component containing a of dimension $\dim \mathbb{T}_a X$.

Definition 11.7 Let $X \subset \mathbb{P}^n(k)$ be an irreducible projective variety. The *dual variety* $\check{X} \subset \check{\mathbb{P}}^n(k)$ is the closure of the locus of all hyperplanes tangent to X at smooth points, i.e.,

$$\check{X} = \overline{\{H \in \check{\mathbb{P}}^n(k) : \mathbb{T}_a X \subset H \text{ for some } a \in X \text{ smooth }\}}.$$

Remark 11.8 Let k be algebraically closed and $X \subset \mathbb{P}^n(k)$ projective. Consider the incidence variety

$$W_X = \{(a, H) : a \subset H \subset \mathbb{T}_a X\}$$
$$\subset X \times \check{\mathbb{P}}^n(k) \subset \mathbb{P}^n(k) \times \check{\mathbb{P}}^n(k),$$

which is contained in the incidence correspondence introduced in Example 11.2. Note that $\pi_2(W_X)$ is closed by Theorem 10.9; $\check{X}$ is a union of irreducible components of this variety.

For many applications, it is important to restrict attention to hypersurfaces without singularities. However, a hypersurface with no singularities over a given field may acquire them after the field is extended. This is just the problem we faced in defining a morphism. The following definition circumvents this difficulty:

Definition 11.9 A hypersurface $X(F) \subset \mathbb{P}^n(k)$ is *smooth* if $\langle F, \partial F/\partial x_0, \ldots, \partial F/\partial x_n \rangle$ is irrelevant.

Proposition 11.10 Let $X \subset \mathbb{P}^n(k)$ be a hypersurface with $J(X) = \langle F \rangle$. The polynomial map

$$\mathbb{P}^n(k) \dashrightarrow \check{\mathbb{P}}^n(k)$$
$$[x_0, \ldots, x_n] \mapsto [\partial F/\partial x_0, \ldots, \partial F/\partial x_n]$$

induces a morphism $X \to \mathbb{P}^n(k)$ when X is smooth. Even when X is singular its image is $\check{X}$, with equations given by

$$\langle F, p_0 - \partial F/\partial x_0, \ldots, p_n - \partial F/\partial x_n \rangle \cap k[p_0, \ldots, p_n].$$

Proof The first statement follows from general properties of polynomial maps of projective spaces (Proposition 9.30). Let $X^s \subset X$ denote the union of the irreducible components containing smooth points. We obtain a rational map $X^s \dashrightarrow \mathbb{P}^n$ with closed image $\check{X}$. The equations are obtained from Proposition 10.12. $\square$

Example 11.11 Assume $\text{char}(k) \neq 3$. Consider the smooth plane curve

$$X = \{(x_0, x_1, x_2) : x_0^3 + x_1^3 + x_2^3 = 0\} \subset \mathbb{P}^2(k).$$

Then the dual is given as

$$\check{X} = \{(p_0, p_1, p_2) : p_0^6 + p_1^6 + p_2^6 - 2p_0^3 p_2^3 - 2p_1^3 p_2^3 - 2p_1^3 p_0^3\} \subset \check{\mathbb{P}}^2(k).$$

11.2.1 Plücker formulas

When X is a smooth hypersurface the dual $\check{X}$ is usually singular, even for plane curves! Figure 11.1 shows two typical cases where the dual curve acquires singularities; for generic X these are only possibilities. In the first case we have an *inflectional tangent*, where the tangent line ℓ meets X at p with multiplicity 3 rather than 2. (Formal definitions can be found in §12.4.2.) X admits an inflectional tangent at p precisely when the differential of the map $X \to \check{\mathbb{P}}^2$ vanishes at p. One can show that the image $\check{X}$ has a *cusp* at ℓ, i.e., a singularity with local normal form $y^2 = x^3$. In the second case, we have a bitangent, i.e. ℓ is tangent to X in two points p_1, p_2. Thus p_1 and p_2 are mapped to the same point of $\check{\mathbb{P}}^2$, so the image $\check{X}$ has two local branches at these points. We say that $\check{X}$ has a *node* at ℓ, with local normal form $y^2 = x^2$.

There are formulas relating the invariants of X and $\check{X}$. Let d and $\check{d}$ be the degrees of X and $\check{X}$, f the number of inflectional tangents X (which equals the number of cusps of $\check{X}$), and b the number of bitangents to X (the number of nodes of $\check{X}$). Then we have the *Plücker formulas*

$$\check{d} = d(d-1)$$
$$f = 3d(d-2)$$
$$b = (\check{d}(\check{d}-1) - d - 3f)/2.$$

We will deduce the second formula from the Bezout Theorem in §12.4.2.

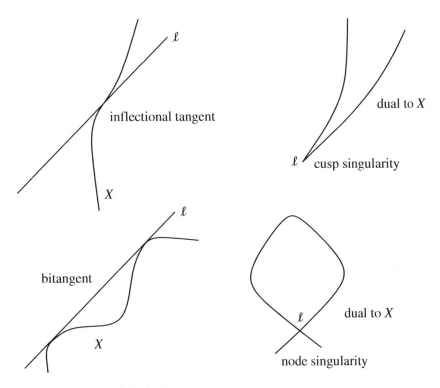

Figure 11.1 Typical singularities of the dual curve.

11.3 Grassmannians: Abstract approach

We have defined the spaces of all lines and hyperplanes in k^{n+1}. Why not consider subspaces of arbitrary dimension? The study of arbitrary finite-dimensional spaces was pioneered by Herman Günter Grassmann (1809–1877) in his 1844 book *Die Lineale Ausdehnungslehre, ein neuer Zweig der Mathematik*. (An English translation is available in [13].)

Definition 11.12   For each $M = 1, 2, \ldots, N-1$, the *Grassmannian* $\mathrm{Gr}(M, N)$ is the set of all vector subspaces $\Lambda \subset k^N$ of dimension M.

In particular, $\mathrm{Gr}(1, N) = \mathbb{P}^{N-1}(k)$ and $\mathrm{Gr}(N-1, N) = \check{\mathbb{P}}^{N-1}(k)$.

Theorem 11.13   *The Grassmannian $\mathrm{Gr}(M, N)$ carries the structure of an abstract variety. It is irreducible and rational of dimension $M(N-M)$.*

For the moment, we only describe the affine open covering and the gluing maps.

Here is the idea of the argument: fix a basis $e_1, \ldots, e_N$ for k^N. Each Λ can be expressed as the row-space of an $M \times N$ matrix W of maximal rank M. This W is not at all unique. Indeed, applying elementary row operations to W does not affect

Λ, so we may replace W by its reduced row echelon form. For 'most' W, we have

$$\mathrm{RREF}(W) = \begin{pmatrix} 1 & 0 & \ldots & b_{1M+1} & \ldots & b_{1N} \\ 0 & \ddots & 0 & \vdots & \ldots & \vdots \\ \vdots & 0 & 1 & b_{MM+1} & \ldots & b_{MN} \end{pmatrix} = (I|B),$$

where I is the identity. Allowing permutations of $e_1, \ldots, e_N$, every subspace admits a reduced row echelon form of this type.

To formalize this, we will need several results from linear algebra:

Lemma 11.14 *Fix a partition*

$$\{1, \ldots, N\} = S \cup T, \quad S = \{s_1, \ldots, s_M\}, T = \{t_1, \ldots, t_{N-M}\}$$

with $s_1 < \ldots < s_M$, $t_1 < \ldots < t_{N-M}$. For each $N \times (M - N)$ matrix $B = (b_{st})$ with rows and columns indexed by S and T respectively, consider the subspace

$$\Lambda(S; B) = \mathrm{span}\left(e_s + \sum_{t \in T} b_{st} e_t : s \in S\right).$$

These satisfy the following:

- $\dim(\Lambda(S; B)) = M$;
- $\Lambda(S; B) = \Lambda(S; C)$ only if $B = C$.

Proof Let $R(S; B)$ be the $M \times N$ matrix with ith row equal to $e_{s_i} + \sum_{t \in T} b_{s_i t} e_t$. For example, when $S = \{1, \ldots, M\}$ we have

$$R(S; B) = \begin{pmatrix} 1 & 0 & \ldots & b_{1M+1} & \ldots & b_{1N} \\ 0 & \ddots & \vdots & \vdots & \ldots & \vdots \\ 0 & \ldots & 1 & b_{MM+1} & \ldots & b_{MN} \end{pmatrix} = (I|B),$$

where I is an $M \times M$ identity matrix. The rows of $R(S; B)$ span $\Lambda(S; B)$; since $R(S; B)$ has rank M, $\dim(\Lambda(S; B)) = M$ and the first assertion follows.

We leave it to the reader to verify the second assertion. $\square$

Lemma 11.15 *Retain the notation of the previous lemma. Each of the distinguished subsets*

$$U_S = \{\Lambda(S; B) : B \text{ is a } M \times (N - M) \text{ matrix}\} \subset \mathrm{Gr}(M, N)$$

admits a natural identification

$$\psi_S : U_S \xrightarrow{\sim} \mathbb{A}^{M(N-M)}(k)$$
$$\Lambda(S; B) \mapsto B.$$

11.3 GRASSMANNIANS: ABSTRACT APPROACH

The Grassmannian is covered by these subsets, i.e.,

$$\mathrm{Gr}(M, N) = \cup_{partitions\ S \sqcup T}\ U_S.$$

Proof Each $\Lambda \in \mathrm{Gr}(M, N)$ can be expressed as $\mathrm{span}(w_1, \ldots, w_M)$ for suitable linearly independent $w_1, \ldots, w_M \in \Lambda \subset k^N$. Let W be the $M \times N$ matrix having these vectors as its rows. Note that multiplication from the left by an invertible matrix does not change the span of the rows, i.e.,

$$\mathrm{row\ span}(W) = \mathrm{row\ span}(AW)$$

provided $\det(A) \neq 0$. Choose indices $s_1, \ldots, s_M$ such that the corresponding columns of W are linearly independent and let W_S be the square matrix built from these columns. The matrix $W_S^{-1} W$ is of the form $R(S; B)$ for some $M \times (N - M)$ matrix B. The natural identification with affine space arises by identifying $M \times (N - M)$ matrices with points in $\mathbb{A}^{M(N-M)}(k)$. □

Lemma 11.16 Given partitions corresponding to $S, S' \subset \{1, \ldots, N\}$, the overlap maps

$$\rho_{S'S} := \psi_{S'} \circ \psi_S^{-1} : \mathbb{A}^{M(N-M)}(k) \dashrightarrow \mathbb{A}^{M(N-M)}(k)$$

are birational, given by the rule

$$B \overset{\psi_S^{-1}}{\mapsto} \Lambda(S; B) = \Lambda(S'; B') \overset{\psi_{S'}}{\mapsto} B'$$
$$R(S; B) \to R(S; B)_{S'}^{-1} R(S; B).$$

Here $R(S; B)_{S'}$ is the $M \times M$ matrix obtained by extracting the columns of $R(S; B)$ indexed by S'; this is invertible provided $B \in \psi_S(U_S \cap U_{S'})$.

Proof Observe that $R(S; B)_{S'}^{-1} R(S; B)$ contains the identity matrix in the columns indexed by S'. It is therefore of the form $R(S'; B')$ for some suitable B'. Again we have

$$\mathrm{row\ span}(R(S; B)) = \mathrm{row\ span}(R(S'; B')),$$

i.e., $\Lambda(S; B)$ and $\Lambda(S'; B')$ coincide. □

What is left to do in the proof of Theorem 11.13? We have not verified that the $\rho_{S'S}$ satisfy the closed-graph condition. In Proposition 11.30, we will establish that the Grassmannian is projective, in a way that is compatible with the proposed abstract variety structure: our distinguished open subsets U_S will be intersections of the Grassmannian with distinguished open subsets of the ambient projective space.

Example 11.17   We give examples of gluing maps for $\mathrm{Gr}(2, 4)$. Let $S = \{1, 2\}$ and $S' = \{1, 3\}$ so that

$$U_S = \mathrm{span}(e_1 + b_{13}e_3 + b_{14}e_4, e_2 + b_{23}e_3 + b_{24}e_4) \simeq \mathbb{A}^4(k),$$
$$U_{S'} = \mathrm{span}(e_1 + b'_{12}e_2 + b'_{14}e_4, e_3 + b'_{32}e_2 + b'_{34}e_4) \simeq \mathbb{A}^4(k).$$

Start with the matrix

$$R(S; B) = \begin{pmatrix} 1 & 0 & b_{13} & b_{14} \\ 0 & 1 & b_{23} & b_{24} \end{pmatrix}$$

and left multiply by

$$R(S; B)_{S'}^{-1} = \begin{pmatrix} 1 & -b_{13}/b_{23} \\ 0 & 1/b_{23} \end{pmatrix}$$

to get

$$R(S'; B') = R(S; B)_{S'}^{-1} = \begin{pmatrix} 1 & -b_{13}/b_{23} & 0 & b_{14} - b_{13}b_{24}/b_{23} \\ 0 & 1/b_{23} & 1 & b_{24}/b_{23} \end{pmatrix}.$$

The gluing map is

$$\rho_{S'S}^* b'_{12} = -b_{13}/b_{23}, \quad \rho_{S'S}^* b'_{14} = b_{14} - b_{13}b_{24}/b_{23}$$
$$\rho_{S'S}^* b'_{32} = 1/b_{23}, \quad \rho_{S'S}^* b'_{34} = b_{24}/b_{23}.$$

We have seen that $\mathbb{P}^{N-1}(k) \simeq \check{\mathbb{P}}^{N-1}(k)$; this is not a coincidence:

Proposition 11.18   *Choose a nondegenerate inner product on k^N, e.g., the standard dot-product*

$$(a_1, \ldots, a_N) \cdot (b_1, \ldots, b_N) = a_1 b_1 + \cdots + a_N b_N.$$

These we have a natural identification

$$\mathrm{Gr}(M, N) \simeq \mathrm{Gr}(N - M, N).$$

Proof   Given a subspace Λ, we define the orthogonal complement

$$\Lambda^\perp = \{x \in k^N : x \cdot v = 0 \text{ for each } v \in \Lambda\}.$$

Since the product is nondegenerate, $\dim \Lambda^\perp = N - \dim \Lambda$ and $(\Lambda^\perp)^\perp = \Lambda$. The association

$$\mathrm{Gr}(M, N) \to \mathrm{Gr}(N - M, N)$$
$$\Lambda \mapsto \Lambda^\perp$$

gives the desired identification.   □

11.4 Exterior algebra

In the last section, we introduced the Grassmannian as an abstract variety. Both for theoretical and practical reasons, it is very useful to have a concrete realization of the Grassmannian in projective space. Here we develop the algebraic formalism needed to write down its homogeneous equations.

We work over a field with characteristic char$(k) \neq 2$.

11.4.1 Basic definitions

Let $V = \{c_1 e_1 + \cdots + c_N e_N\}$ be a finite-dimensional vector space with basis $\{e_1, \ldots, e_N\}$. For each $M = 0, \ldots, N$, the *Mth exterior power* is defined as the vector space

$$\bigwedge^M V = \left\{ \sum_{1 \leq i_1 < i_2 < \ldots < i_M \leq N} c_{i_1 \ldots i_M} e_{i_1} \wedge e_{i_2} \ldots \wedge e_{i_M} \right\},$$

with the convention that $\bigwedge^0 V = \{c_\emptyset 1\}$. The basis for $\bigwedge^M V$ is indexed by subsets

$$\{i_1, \ldots, i_M\} \subset \{1, \ldots, N\}$$

with M elements, so

$$\dim\left(\bigwedge^M V\right) = \binom{\dim V}{M}.$$

The direct sum of all the exterior powers is written:

$$\bigwedge^* V = \bigwedge^0 V \oplus \bigwedge^1 V \oplus \ldots \oplus \bigwedge^N V.$$

We describe an associative but *noncommutative* multiplication operation

$$\bigwedge^* V \times \bigwedge^* V \to \bigwedge^* V$$
$$(\eta, \omega) \mapsto \eta \wedge \omega,$$

called the *wedge product*. It satisfies the following axioms:

1. $\wedge$ is k-linear in each factor, i.e.,

$$(r_1 \omega_1 + r_2 \omega_2) \wedge \eta = r_1(\omega_1 \wedge \eta) + r_2(\omega_2 \wedge \eta)$$
$$\eta \wedge (r_1 \omega_1 + r_2 \omega_2) = r_1(\eta \wedge \omega_1) + r_2(\eta \wedge \omega_2)$$
$$r_1, r_2 \in k, \quad \omega_1, \omega_2, \eta \in \bigwedge^* V;$$

2. $\wedge$ is *graded commutative*:

$$e_i \wedge e_j = -e_j \wedge e_i,$$

so, in particular, $e_i \wedge e_i = 0$.

Example 11.19

$$(e_1 + e_3) \wedge (e_1 \wedge e_2 + e_3 \wedge e_4)$$
$$= e_1 \wedge e_1 \wedge e_2 + e_1 \wedge e_3 \wedge e_4 + e_3 \wedge e_1 \wedge e_2 + e_3 \wedge e_3 \wedge e_4$$
$$= 0 + e_1 \wedge e_3 \wedge e_4 + (-1)^2 e_1 \wedge e_2 \wedge e_3 + 0$$
$$= e_1 \wedge e_3 \wedge e_4 + e_1 \wedge e_2 \wedge e_3.$$
$$(e_1 \wedge e_2 + e_3 \wedge e_4)^2 = e_1 \wedge e_2 \wedge e_1 \wedge e_2$$
$$+ e_1 \wedge e_2 \wedge e_3 \wedge e_4 + e_3 \wedge e_4 \wedge e_1 \wedge e_2 + e_3 \wedge e_4 \wedge e_3 \wedge e_4$$
$$= 2 e_1 \wedge e_2 \wedge e_3 \wedge e_4.$$

Proposition 11.20 *The wedge product is uniquely determined by the two axioms and associativity.*

Proof By linearity, we can compute arbitrary wedge products once we have specified products of basis elements

$$(e_{j_1} \wedge \ldots \wedge e_{j_L}) \wedge (e_{j_{L+1}} \wedge \ldots \wedge e_{j_M}),$$

where $j_1 < j_2 < \ldots < j_L$ and $j_{L+1} < \ldots < j_M$.

To evaluate these, we will use basic facts about permutations

$$\sigma : \{1, \ldots, M\} \xrightarrow{\sim} \{1, \ldots, M\}$$
$$i \mapsto \sigma(i).$$

The *sign* of the permutation satisfies the following:

1. $\mathrm{sign}(\sigma) \in \{\pm 1\}$;
2. $\mathrm{sign}(\sigma \sigma') = \mathrm{sign}(\sigma) \mathrm{sign}(\sigma')$;
3. if σ is a transposition, i.e., a permutation exchanging i and j but fixing all the other elements of $\{1, \ldots, M\}$, then $\mathrm{sign}(\sigma) = -1$.

Every permutation can be represented as a product of transpositions, and any two such representations have the same number of transpositions modulo 2. These two facts guarantee that the sign is well-defined.

Let σ be a permutation of $\{1, \ldots, M\}$ with

$$j_{\sigma(1)} \leq j_{\sigma(2)} \ldots \leq j_{\sigma(M)}$$

and express σ is a product of transpositions. Successively applying the second axiom of wedge products, using the properties of the sign, we can write

$$e_{j_1} \wedge \ldots \wedge e_{j_M} = \mathrm{sign}(\sigma) e_{j_{\sigma(1)}} \wedge \ldots \wedge e_{j_{\sigma(M)}}.$$

However, if any $j_{\sigma(\ell)} = j_{\sigma(\ell+1)}$ then $e_{j_{\sigma(\ell)}} \wedge e_{j_{\sigma(\ell+1)}} = 0$ and thus the whole product is zero. Otherwise, we find

$$(e_{j_1} \wedge \ldots \wedge e_{j_L}) \wedge (e_{j_{L+1}} \wedge \ldots \wedge e_{j_M}) = \mathrm{sign}(\sigma) e_{j_{\sigma(1)}} \wedge \ldots \wedge e_{j_{\sigma(M)}}. \qquad \square$$

11.4 EXTERIOR ALGEBRA

It remains to verify that the multiplication rule arising out of this analysis is a well-defined associative product:

Proposition 11.21 *Consider the following multiplication rule on $\bigwedge^* V$: given basis elements $e_{j_1} \wedge \ldots \wedge e_{j_A} \in \bigwedge^A V$ and $e_{j_{A+1}} \wedge \ldots \wedge e_{j_{A+B}} \in \bigwedge^B V$ with $j_1 < j_2 < \ldots < j_A$ and $j_{A+1} < \ldots < j_{A+B}$, we take*

$$(e_{j_1} \wedge \ldots \wedge e_{j_A}) \wedge (e_{j_{A+1}} \wedge \ldots \wedge e_{j_{A+B}}) = \begin{cases} 0 \text{ if } j_a = j_b \text{ for some } a \neq b, \\ \text{sign}(\sigma) e_{j_{\sigma(1)}} \wedge \ldots \wedge e_{j_{\sigma(A+B)}} \text{ otherwise,} \end{cases}$$

where in the second case σ is the unique permutation of $\{1, \ldots, A+B\}$ with

$$j_{\sigma(1)} < j_{\sigma(2)} < \ldots < j_{\sigma(A+B)}.$$

This defines an associative multiplication on $\bigwedge^* V$.

Proof We need to verify the identity

$$\begin{aligned}&\left((e_{j_1} \wedge \ldots \wedge e_{j_A}) \wedge (e_{j_{A+1}} \wedge \ldots \wedge e_{j_{A+B}})\right) \wedge (e_{j_{A+B+1}} \wedge \ldots \wedge e_{j_{A+B+C}}) \\ &= (e_{j_1} \wedge \ldots \wedge e_{j_A}) \wedge \left((e_{j_{A+1}} \wedge \ldots \wedge e_{j_{A+B}}) \wedge (e_{j_{A+B+1}} \wedge \ldots \wedge e_{j_{A+B+C}})\right),\end{aligned} \quad (11.3)$$

where $j_1 < \ldots < j_A$, $j_{A+1} < \ldots < j_{A+B}$, and $j_{A+B+1} < \ldots < j_{A+B+C}$. Both sides are zero whenever any two of the indices coincide, so it suffices to consider the case where all the indices are distinct.

We introduce four permutations

$$\sigma, \sigma', \tau, \tau' : \{1, \ldots, A+B+C\} \to \{1, \ldots, A+B+C\} :$$

- σ is the unique permutation fixing $A+B+1, \ldots, A+B+C$ with $j_{\sigma(1)} < j_{\sigma(2)} < \ldots < j_{\sigma(A+B)}$;
- σ' is the unique permutation such that $j_{\sigma'(\sigma(1))} < \ldots < j_{\sigma'(\sigma(A+B+C))}$;
- τ is the unique permutation fixing $1, \ldots, A$ with $j_{\tau(A+1)} < j_{\tau(A+2)} < \ldots < j_{\tau(A+B+C)}$;
- τ' is the unique permutation such that $j_{\tau'(\tau(1))} < \ldots < j_{\tau'(\tau(A+B+C))}$.

For each $r = 1, \ldots, A+B+C$, we have $\sigma'(\sigma(r)) = \tau'(\tau(r))$ — we write this index i_r.

The left-hand side of (11.3) is

$$\text{sign}(\sigma')\text{sign}(\sigma) e_{i_1} \wedge \ldots \wedge e_{i_{A+B+C}}$$

and the right-hand side is

$$\text{sign}(\tau')\text{sign}(\tau) e_{i_1} \wedge \ldots \wedge e_{i_{A+B+C}}.$$

The multiplicative property of the sign

$$\operatorname{sign}(\sigma')\operatorname{sign}(\sigma) = \operatorname{sign}(\sigma'\sigma) = \operatorname{sign}(\tau'\tau) = \operatorname{sign}(\tau')\operatorname{sign}(\tau)$$

yields our identity. □

Thus $\bigwedge^* V$ is a graded-commutative k-algebra under wedge product. This is called the *exterior algebra* of V.

11.4.2 Exterior powers and linear transformations

Proposition 11.22 *Let $T : V \to W$ be a linear transformation. Then there is an induced k-algebra homomorphism*

$$\bigwedge T : \bigwedge^* V \to \bigwedge^* W.$$

Suppose that $\{e_j\}$ and $\{f_i\}$ are bases of V and W. If (A_{ij}) is the matrix of T with respect to these bases then

$$\left(\bigwedge T\right)(e_{j_1} \wedge \ldots \wedge e_{j_M}) = \sum_{i_1 < \ldots < i_M} A_{i_1,\ldots,i_M; j_1,\ldots,j_M} f_{i_1} \wedge \ldots \wedge f_{i_M}, \quad (11.4)$$

where $A_{i_1,\ldots,i_M; j_1,\ldots,j_M}$ is the determinant of the $M \times M$ minor of (A_{ij}) obtained from extracting the rows $\{i_1, i_2, \ldots, i_M\}$ and the columns $\{j_1, \ldots, j_M\}$.

The assertion that $\bigwedge T$ is a k-algebra homomorphism means that it is k-linear and respects wedge products

$$\left(\bigwedge T\right)(\eta \wedge \omega) = \left(\bigwedge T\right)(\eta) \wedge \left(\bigwedge T\right)(\omega), \quad (11.5)$$

and, in particular,

$$\left(\bigwedge T\right)(v_1 \wedge \ldots \wedge v_M) = T(v_1) \wedge \ldots \wedge T(v_M). \quad (11.6)$$

We also use the notation $\bigwedge^M T : \bigwedge^M V \to \bigwedge^M W$ when we restrict to the Mth exterior power; $\bigwedge^0 T$ is the identity and $\bigwedge^1 T = T$.

Example 11.23 For $T : k^2 \to k^2$ we have

$$\left(\bigwedge T\right)(e_1 \wedge e_2) = (a_{11}e_1 + a_{21}e_2) \wedge (a_{12}e_1 + a_{22}e_2)$$
$$= (a_{11}a_{22} - a_{12}a_{21})e_1 \wedge e_2.$$

For $T : k^2 \to k^3$ we have

$$\left(\bigwedge T\right)(e_1 \wedge e_2) = (a_{11}f_1 + a_{21}f_2 + a_{31}f_3) \wedge (a_{12}f_1 + a_{22}f_2 + a_{32}f_3)$$
$$= (a_{11}a_{22} - a_{12}a_{21})f_1 \wedge f_2$$
$$+ (a_{21}a_{32} - a_{31}a_{22})f_2 \wedge f_3 + (a_{11}a_{32} - a_{12}a_{31})f_1 \wedge f_3.$$

11.4 EXTERIOR ALGEBRA

Proof of Proposition 11.22   In fact, a direct computation shows that Property (11.6) implies that the matrix of $\bigwedge T$ is given by Expression (11.4). However, we shall work in reverse: We verify that the linear transformations $\bigwedge^M T$ defined by Expression (11.4) have Property (11.5).

Since each $\bigwedge^M T$ is linear and the operation $\wedge$ is linear in each factor, we need only check Property 11.5 for $\eta = e_{j_1} \wedge \ldots \wedge e_{j_M}$ and $\omega = e_{j'_1} \wedge \ldots \wedge e_{j'_B}$. By induction on B, it suffices to assume $\omega = e_{j'}$. Recall the matrix identity

$$A_{i_1,\ldots,i_M,i_{M+1};j_1,\ldots,j_M,j'} = \sum_r (-1)^{M+1+r} a_{i_r j'} A_{i_1,\ldots,\hat{i}_r,\ldots,i_{M+1};j_1,\ldots,j_M}$$

obtained by expanding along the j'-column. This yields

$$\left(\bigwedge^{M+1} T\right)(e_{j_1} \wedge \ldots \wedge e_{j_M} \wedge e_{j'})$$

$$= \sum_{i_1<\ldots<i_{M+1}} A_{i_1,\ldots,i_{M+1};j_1,\ldots,j_M,j'} f_{i_1} \wedge \ldots \wedge f_{i_{M+1}}$$

$$= \sum_{i_1<\ldots<i_{M+1};\, r} (-1)^{M+1+r} a_{i_r j'} A_{i_1,\ldots,\hat{i}_r,\ldots,i_{M+1};j_1,\ldots,j_M} f_{i_1} \wedge \ldots \wedge f_{i_r} \wedge \ldots \wedge f_{i_{M+1}}$$

$$= \sum_{i_1<\ldots<i_{M+1};\, r} A_{i_1,\ldots,\hat{i}_r,\ldots,i_{M+1};j_1,\ldots,j_M} f_{i_1} \wedge \ldots \hat{f}_{i_r} \ldots \wedge f_{i_{M+1}} \wedge (a_{i_r j'} f_{i_r})$$

$$= \sum_{i'_1<\ldots<i'_M} A_{i'_1,\ldots,i'_M;j_1,\ldots,j_M} f_{i'_1} \wedge \ldots \wedge f_{i'_M} \wedge \left(\sum_{i'} a_{i' j'} f_{i'}\right)$$

$$= \bigwedge^M T(e_{j_1} \wedge \ldots \wedge e_{j_M}) \wedge T(e_{j'}).$$

In changing indices from $i_1, \ldots, \hat{i}_r, \ldots, i_{M+1}$ to $i'_1, \ldots, i'_M$, it might seem that we are adding extra terms, i.e., the cases where i' equals one of the $i'_1, \ldots, i'_M$. However, in precisely these cases

$$f_{i'_1} \wedge \ldots \wedge f_{i'_M} \wedge f_{i'} = 0. \qquad \square$$

We deduce the following corollaries:

Corollary 11.24   Let $T : V \to V$ be a linear transformation on a vector space of dimension N. Then

$$\left(\bigwedge^N T\right) : \bigwedge^N V \to \bigwedge^N V$$

is multiplication by $\det(T)$.

Corollary 11.25   Let $T : V \to W$ be a linear transformation. Then $\operatorname{rank}(T) < M$ if and only if $\bigwedge^M T = 0$.

Indeed, a matrix A has rank $< M$ exactly when each of its $M \times M$ minors vanish.

Corollary 11.26   Let $v_1, \ldots, v_M \in V$ be linearly independent, and choose

$$v'_1, \ldots, v'_M \in \operatorname{span}\{v_1, \ldots, v_M\}$$

with $v'_i = \sum_{j=1}^{M} a_{ij} v_j$. Then we have

$$v'_1 \wedge \ldots \wedge v'_M = \det(a_{ij}) v_1 \wedge \ldots \wedge v_M.$$

11.4.3 Decomposible elements

Definition 11.27   An element $\omega \in \bigwedge^m V$ is *completely decomposible* if there exist $v_1, \ldots, v_m \in V$ such that $\omega = v_1 \wedge \ldots \wedge v_m$. An element $\omega \in \bigwedge^m V$ is *partially decomposible* if $\omega = v \wedge \eta$ for some $v \in V$ and $\eta \in \bigwedge^{m-1} V$.

Proposition 11.28   Let $\omega \in \bigwedge^m V$.

1. If ω is partially decomposible then $\omega \wedge \omega = 0$.
2. ω is partially decomposible if and only if the linear transformation

$$\phi(\omega) : V \to \bigwedge^{m+1} V$$
$$w \mapsto w \wedge \omega$$

   has nontrivial kernel.
3. If $\{v_1, \ldots, v_M\}$ is a basis for $\operatorname{kernel}(\phi(\omega))$ then

$$\omega = v_1 \wedge \ldots \wedge v_M \wedge \eta, \eta \in \bigwedge^{m-M} V.$$

4. ω is completely decomposible if and only if $\ker(\phi(\omega))$ has dimension m.

Proof   If $\omega = v \wedge \eta$ then

$$\omega \wedge \omega = v \wedge \eta \wedge v \wedge \eta = (-1)^{m-1} v \wedge v \wedge \eta \wedge \eta = 0,$$

which proves the first assertion. It is evident that $v \in \ker(\phi(\omega))$, which is the 'only if' part of the second assertion. Similarly, if $\omega = v_1 \wedge \ldots \wedge v_m$ then $v_1, \ldots, v_m \in \ker(\phi(\omega))$.

For the third assertion, extend $v_1, \ldots, v_M$ to a basis $v_1, \ldots, v_N$ for V. Then we have

$$\omega = \sum_{i_1 < \ldots < i_m} c_{i_1, \ldots, i_m} v_{i_1} \wedge \ldots \wedge v_{i_m}$$

and

$$v_j \wedge \omega = \sum_{i_1 < \ldots < i_m} c_{i_1,\ldots,i_m} v_j \wedge v_{i_1} \wedge \ldots \wedge v_{i_m}$$

$$= \sum_{\substack{i_1 < \ldots < i_m \\ i_r \neq j}} c_{i_1,\ldots,i_m} v_j \wedge v_{i_1} \wedge \ldots \wedge v_{i_m}.$$

This is zero if and only if $c_{i_1,\ldots,i_m} = 0$ for all indices $\{i_1, \ldots, i_m\} \not\ni j$. Similarly,

$$v_1 \wedge \omega = \ldots = v_M \wedge \omega = 0$$

if and only if $c_{i_1,\ldots,i_m} = 0$ for all indices $\{i_1, \ldots, i_m\} \not\supset \{1, \ldots, M\}$. Then we have $\omega = v_1 \wedge \ldots \wedge v_M \wedge \eta$ for

$$\eta = \sum_{j_1,\ldots,j_{M-m}} c'_{j_1,\ldots,j_{M-m}} v_{j_1} \wedge \ldots \wedge v_{j_{M-m}},$$

where

$$c'_{j_1,\ldots,j_{M-m}} = \pm c_I, \quad I = \{1, \ldots, M, j_1, \ldots, j_{m-M}\}.$$

The 'if' parts of the second and fourth assertions follow from this analysis. $\square$

11.5 Grassmannians as projective varieties

The following result realizes the Grassmannian in projective space:

Proposition 11.29 *There is a well-defined map*

$$j : \mathrm{Gr}(M, N) \to \mathbb{P}\left(\bigwedge\nolimits^M k^N\right)$$
$$\mathrm{span}(v_1, \ldots, v_M) \mapsto [v_1 \wedge \ldots \wedge v_M]$$

which is a bijection between elements of the Grassmannian and projective equivalence classes of completely decomposable elements $\omega \in \bigwedge^M k^N$.

Proof We first check that j is well-defined. If $v_1, \ldots, v_M$ and $v'_1, \ldots, v'_M$ are bases for a subspace Λ then

$$[v_1 \wedge \ldots \wedge v_M] = [v'_1 \wedge \ldots \wedge v'_M]$$

by Corollary 11.26.

We can recover $\mathrm{span}(v_1, \ldots, v_M)$ easily from $v_1 \wedge \ldots \wedge v_M$: Proposition 11.28 implies

$$\mathrm{span}(v_1, \ldots, v_M) = \ker(\phi(v_1 \wedge \ldots \wedge v_M)).$$

We conclude that j is injective. $\square$

Proposition 11.30 *The inclusion $j : \mathrm{Gr}(M, N) \hookrightarrow \mathbb{P}(\bigwedge^M k^N)$ realizes the Grassmannian as a closed subset of projective space.*

This is called the *Plücker embedding* in honor of Julius Plücker (1801–1868).

Our argument will also complete the proof that the Grassmannian is an abstract variety (Theorem 11.13.)

Proof For each $S = \{s_1, \ldots, s_M\} \subset \{1, \ldots, N\}$ with complement T, write

$$V_S = \left\{ [c_{i_1 \ldots i_M}] \in \mathbb{P}\left(\bigwedge^M k^N\right) : c_{s_1 \ldots s_M} \neq 0 \right\}$$

for the corresponding distinguished subset of $\mathbb{P}(\bigwedge^M k^N)$. Recall the notation in the discussion of the abstract variety structure on the Graassmannian (cf. Theorem 11.13): let $U_S \simeq \mathbb{A}^{M(N-M)}(k) \subset \mathrm{Gr}(M, N)$ denote the distinguished subset corresponding to subspaces of the form

$$\mathrm{span}\left(e_s + \sum_{t \in T} b_{st} e_t : s \in S\right).$$

Observe that $j^{-1}(V_S) = U_S$: a decomposable element $v_1 \wedge \ldots \wedge v_M$ has nonvanishing coefficient $c_{s_1 \ldots s_M}$ if and only if (after permuting indices) each e_{s_j} appears in v_j.

For notational simplicity, we take $S = \{1, \ldots, M\}$. We can expand out $j(\Lambda(S; B))$ as

$$e_1 \wedge \ldots \wedge e_M + \sum_{1 \leq i \leq M < j \leq N} e_1 \wedge \ldots \wedge e_{i-1} \wedge (e_j b_{i,j}) \wedge e_{i+1} \wedge \ldots \wedge e_M$$
$$+ \sum_{1 \leq i_1 < i_2 \leq M < j_1, j_2 \leq N} e_1 \ldots \wedge (e_{j_1} b_{i_1, j_1}) \wedge \ldots \wedge (e_{j_2} b_{i_2, j_2}) \wedge \ldots e_M + \cdots$$

or in the form

$$\sum_{r=1}^{M} \sum_{1 \leq i_1 < \ldots < i_r \leq M < j_1 < \ldots < j_{M-r} \leq N} p_{i_1, \ldots, i_r; j_1, \ldots, j_{M-r}}(b_{ij}) e_{i_1} \wedge \ldots \wedge e_{i_r} \wedge e_{j_1} \wedge \ldots \wedge e_{j_{M-r}}$$

where $p_{i_1, \ldots, i_r; j_1, \ldots, j_{M-r}}(b_{ij})$ is a polynomial of degree $M - r$ in the b_{ij}, and

$$p_{1, \ldots, \hat{i}, \ldots, M; j} = (-1)^{M-i} b_{ij}. \tag{11.7}$$

Since the b_{ij} are the coordinates on $U_S \simeq \mathbb{A}^{M(N-M)}(k)$, $j : U_S \to V_S$ is a morphism of affine varieties.

Proposition 11.29 implies that j is bijective onto its image. Projection onto the coordinates (11.7) and adjusting for the signs, we obtain a left inverse $V_S \to U_S$ for $j|U_S$. It follows that $j(U_S) \subset V_S$ is closed and $U_S \to j(U_S)$ is an isomorphism of affine varieties. $\square$

11.6 Equations for the Grassmannian

The proof of Proposition 11.30 gives some equations of the Grassmannian:

Example 11.31 Let $M = 2$ and $N = 4$. We want to classify

$$\omega = x_{12} e_1 \wedge e_2 + \cdots + x_{34} e_3 \wedge e_4$$

arising from the Grassmannian. The condition

$$\omega \wedge \omega = 0$$

translates into the quadratic equation

$$x_{12} x_{34} - x_{13} x_{24} + x_{14} x_{23} = 0.$$

The corresponding hypersurface $X \subset \mathbb{P}(\bigwedge^2 k^4)$ contains $\mathrm{Gr}(2, 4)$.

On the other hand, we can compute the image

$$j(U_{12}) \subset \mathbb{P}\left(\bigwedge^2 k^4\right)$$

and its projective closure. We have $j(U_{12}) \subset V_{12}$ where $V_{12} = \{x_{12} \neq 0\}$, and let $y_{13}, y_{14}, y_{23}, y_{24}, y_{34}$ be affine/dehomogenized coordinates for V_{12}. The induced morphism of affine varieties $j|U_{12} \to V_{12}$ takes

$$(e_1 + b_{13} e_3 + b_{14} e_4) \wedge (e_2 + b_{23} e_3 + b_{24} e_4)$$

to

$$(b_{23}, b_{24}, -b_{13}, -b_{14}, b_{13} b_{24} - b_{14} b_{23})$$

with graph

$$y_{13} = b_{23},\ y_{14} = b_{24},\ y_{23} = -b_{13},\ y_{24} = -b_{14},\ y_{34} = b_{13} b_{24} - b_{14} b_{23}.$$

Eliminating the variables b_{ij} gives

$$y_{34} = -y_{23} y_{14} + y_{24} y_{13}$$

and homogenizing gives the projective closure

$$x_{12} x_{34} + x_{23} x_{14} - x_{13} x_{24} = 0.$$

This equation can be put in a more general context. Given

$$\omega = \sum_{i_1 < \ldots < i_M} x_{i_1 \ldots i_M} e_{i_1} \wedge \ldots \wedge e_{i_M},$$

if $[\omega] \in j(\mathrm{Gr}(M, N))$ then ω is partially decomposible and $\omega \wedge \omega = 0$ (Proposition 11.28). This translates into *quadratic equations* in the $x_{i_1 \ldots i_M}$. In the special case $M = 2$, we can write these explicitly: For each set of four indices $i_1 < i_2 < i_3 < i_4$ we have

$$x_{i_1 i_2} x_{i_3 i_4} - x_{i_1 i_3} x_{i_2 i_4} + x_{i_1 i_4} x_{i_2 i_3} = 0.$$

For $M > 2$ there exist partially decomposible elements which are not completely decomposible, so these equations are insufficient to cut out the Grassmannian. However, we do have the following:

Proposition 11.32 (Rough draft of the equations for Gr(M, N)) *An element $[\omega] \in \mathbb{P}(\bigwedge^M k^N)$ is completely decomposible if and only if each of the $(N - M + 1) \times (N - M + 1)$ minors of the matrix of*

$$\phi(\omega) : k^N \to \bigwedge^{M+1} k^N$$
$$w \mapsto w \wedge \omega$$

vanish.

Proof By Proposition 11.28, ω is completely decomposible exactly when $\phi(\omega)$ has rank $\leq N - M$. If $\omega = v_1 \wedge \ldots \wedge v_M \neq 0$, then $(\sum_{j > M} c_j v_j) \wedge \omega \neq 0$ unless each $c_j = 0$, so the rank $< N - M$ only when $\omega = 0$. The $(N - M + 1) \times (N - M + 1)$ minors of a matrix B vanish if and only if rank $(B) < N - M + 1$. □

Remark 11.33 This is only a rough draft! Even in the case $M = 2, N = 4$ we do not get generators for the homogeneous ideal of the Grassmannian. Here we obtain the 3×3 minors of a 4×4 matrix

$$\begin{array}{c c} & \begin{array}{cccc} e_1 & e_2 & e_3 & e_4 \end{array} \\ \begin{array}{c} e_2 \wedge e_3 \wedge e_4 \\ e_1 \wedge e_3 \wedge e_4 \\ e_1 \wedge e_2 \wedge e_4 \\ e_1 \wedge e_2 \wedge e_3 \end{array} & \left(\begin{array}{cccc} 0 & x_{34} & -x_{24} & x_{23} \\ x_{34} & 0 & -x_{14} & x_{13} \\ x_{24} & -x_{14} & 0 & x_{12} \\ x_{23} & -x_{13} & x_{12} & 0 \end{array} \right). \end{array}$$

There are altogether six nonzero equations

$$x_{ij}(x_{12} x_{34} - x_{13} x_{24} + x_{14} x_{23}) = 0.$$

11.6.1 Plücker relations

We give, without proof, the complete set of homogeneous equations for the Grassmannian. These are known as the *Plücker relations*.

For any vector space V over k, the *dual space* V^* consists of the linear transformations $V \to k$. Given a subspace $W \subset V$, we define

$$W^\perp = \{f \in V^* : f(w) = 0 \text{ for each } w \in W\} \subset V^*.$$

11.6 EQUATIONS FOR THE GRASSMANNIAN

Each linear transformation $T: V \to W$ induces a transpose $T^*: W^* \to V^*$ with formula

$$(T^*g)(v) = g(T(v)), \quad g \in W^*, v \in V.$$

Let V be a vector space of dimension N with basis $\{e_1, \ldots, e_N\}$ for V; let $e^1, \ldots, e^N$ denote the dual basis of V^* so that $e^i(e_j) = \delta_{ij}$. Our choice of basis induces an isomorphism

$$\Delta: \bigwedge^N V \xrightarrow{\sim} k$$
$$e_1 \wedge \ldots \wedge e_N \mapsto 1.$$

The wedge-product

$$\bigwedge^M V \times \bigwedge^{N-M} V \to \bigwedge^N V \xrightarrow{\Delta} k$$

is a nondegenerate pairing. Concretely, we have dual basis elements

$$e_{i_1} \wedge \ldots \wedge e_{i_M} \Leftrightarrow \text{sign}(\sigma) e_{j_1} \wedge \ldots \wedge e_{j_{N-M}}$$

where σ is the permutation

$$\{i_1, \ldots, i_M, j_1, \ldots, j_{N-M}\} \xrightarrow{\sigma} \{1, \ldots, N\}.$$

Example 11.34   The dual basis elements for $\bigwedge^1 k^4$ and $\bigwedge^3 k^4$ are

$$\begin{array}{c|c} e_1 & e_2 \wedge e_3 \wedge e_4 \\ e_2 & -e_1 \wedge e_3 \wedge e_4 \\ e_3 & e_1 \wedge e_2 \wedge e_4 \\ e_4 & -e_1 \wedge e_2 \wedge e_3 \end{array}.$$

Our duality induces an isomorphism

$$\gamma: \bigwedge^{N-M} V \xrightarrow{\sim} \bigwedge^M V^* \tag{11.8}$$

given by

$$\gamma(e_{j_1} \wedge \ldots e_{j_{N-M}}) = \text{sign}(\sigma) e^{i_1} \wedge \ldots \wedge e^{i_M}.$$

γ depends on the choice of the isomorphism Δ but not on the precise choice of basis. It is therefore uniquely determined up to scalar multiplication.

Observe that ω is completely decomposable if and only if $\gamma(\omega)$ is completely decomposable: if $\omega = v_1 \wedge \ldots \wedge v_M$ and $v^{M+1}, \ldots, v^N \in V^*$ are a basis for $\text{Span}\{v_1, \ldots, v_M\}^\perp$ then $\gamma(\omega) = c v^{M+1} \wedge \ldots \wedge v^N$ for some scalar $c \in k$.

Consider the linear transformation

$$\psi(\omega) : \bigwedge V^* \to \bigwedge^{N-M+1} V^*$$
$$f \to f \wedge \gamma(\omega).$$

This has rank $\leq M$ if and only if $\gamma(\omega)$ is completely decomposable. Moreover, using the relationship between ω and $\gamma(\omega)$, we find

$$\ker(\psi(\omega)) = \operatorname{span}\{v^{M+1}, \ldots, v^N\}$$
$$= \ker(\phi(\omega))^\perp.$$

Consider the transpose maps

$$\phi(\omega)^* : \bigwedge^{M+1} V^* \to V^*, \quad \psi(\omega)^* : \bigwedge^{N-M+1} V \to V.$$

For a linear transformation $T : W_1 \to W_2$ with transpose $T^* : W_2^* \to W_1^*$, we have

$$\operatorname{image}(T^*) = \ker(T)^\perp.$$

Thus we find

$$\operatorname{image}(\phi(\omega)^*) = \ker(\phi(\omega))^\perp = \ker(\psi(\omega)) = \operatorname{image}(\psi(\omega)^*)^\perp.$$

This means that for each $\alpha \in \bigwedge^{M+1} V^*$ and $\beta \in \bigwedge^{N-M+1} V$

$$\Xi_{\alpha,\beta}(\omega) := \phi(\omega)^t(\alpha)[\psi(\omega)^t(\beta)] = 0;$$

this is a quadratic polynomial in the coordinates of ω.

Theorem 11.35 *[22, ch. 7] The Plücker relations generate the homogeneous ideal of the Plücker embedding of the Grassmannian $j : \operatorname{Gr}(M, N) \hookrightarrow \mathbb{P}(\bigwedge^M k^N)$, i.e.,*

$$J(\operatorname{Gr}(M, N)) = \left\langle \Xi_{\alpha,\beta}(\omega) : \alpha \in \bigwedge^{M+1} V^*, \beta \in \bigwedge^{N-M+1} V \right\rangle.$$

11.7 Exercises

11.1 Consider the plane curve

$$X = \{(x_0, x_1, x_2) : x_0^3 x_1 + x_1^3 x_2 + x_2^3 x_0\} \subset \mathbb{P}^2(\mathbb{C}).$$

(a) Show that X is smooth.
(b) Compute the dual curve $\check{X} \subset \check{\mathbb{P}}^2(\mathbb{C})$.
(c) Find at least one singular point of $\check{X}$.

11.7 EXERCISES

11.2 Enumerate the singularities of the Cayley cubic surface

$$\{[w, x, y, z] : wxy + xyz + yzw + zwx = 0\} \subset \mathbb{P}^3(\mathbb{C})$$

and write down an equation for its dual. Describe the dual to the hypersurface

$$\{[w, x, y, z] : wxy - z^3 = 0\} \subset \mathbb{P}^3(\mathbb{C}).$$

11.3 Consider the plane curve

$$X = \{(x_0, x_1, x_2) : x_0^2 x_1^2 + x_1^2 x_2^2 + x_2^2 x_0^2\} \subset \mathbb{P}^2(\mathbb{C}).$$

Show that X is not smooth. If you are ambitious, work out the following:
(a) Set

$$J = \langle \partial f/\partial x_0, \partial f/\partial x_1, \partial f/\partial x_2 \rangle.$$

Compute the dehomogenization I_i of J with respect to each of the variables x_i. Find a Gröbner basis for each I_i with respect to a graded order. Compute the rehomogenization of each I_i, $J_i \subset \mathbb{C}[x_0, x_1, x_2]$, and the intersection $J' := J_0 \cap J_1 \cap J_2$.
(b) Determine the singular points of X.
(c) Show that $J \neq J'$ and compute primary decompositions of J' and J.

11.4 Consider the complex projective curve

$$C = \{(x_0, x_1, x_2) : x_2 x_1^2 = x_0^2 x_2 + x_0^3\} \subset \mathbb{P}^2(\mathbb{C}).$$

(a) Determine whether C is smooth. If it is not smooth, find its singularities.
(b) Compute the tangent line to the curve at the point $[-1, 0, 1]$.
(c) Decide whether this tangent line is an inflectional tangent.

11.5 Prove the statement

$$\Lambda(S; B) = \Lambda(S; C) \quad \Rightarrow \quad B = C$$

from Lemma 11.14.

11.6 Let $\mathbb{G}(M - 1, N - 1)$ denote the set of all linear subspaces $\Lambda \simeq \mathbb{P}^{M-1} \subset \mathbb{P}^{N-1}$. Show there is an identification

$$\mathbb{G}(M - 1, N - 1) = \mathrm{Gr}(M, N).$$

11.7 Verify directly that the gluing maps for $\mathrm{Gr}(2, 4)$ satisfy the closed-graph condition.

11.8 Show that the Plücker relations for the Grassmannian

$$\mathrm{Gr}(2, N) \subset \mathbb{P}\left(\bigwedge^2 k^N\right) \simeq \mathbb{P}^{\binom{N}{2}-1}$$

are equivalent to

$$x_{ij}x_{kl} - x_{ik}x_{jl} + x_{il}x_{jk} = 0, \quad \{i, j, k, l\} \subset \{1, 2, \ldots, N\}.$$

11.9 Fix a two-dimensional subspace $\Lambda \subset \mathbb{C}^4$. Consider the set

$$X_\Lambda = \{\Lambda' \in \mathrm{Gr}(2, 4) : \Lambda \cap \Lambda' \neq 0\}.$$

Show this is a closed subset of $\mathrm{Gr}(2, 4) \subset \mathbb{P}^5(\mathbb{C})$ and find explicit homogeneous equations for X_Λ in the special case $\Lambda = \mathrm{span}(e_1, e_2)$. This is an example of a *Schubert variety*.

11.10 Consider the lines $\ell_1, \ell_2, \ell_3, \ell_4 \subset \mathbb{P}^3(\mathbb{C})$ with equations

$$\ell_1 = \{x_0 = x_1 = 0\} \qquad \ell_2 = \{x_2 = x_3 = 0\}$$
$$\ell_3 = \{x_0 - x_2 = x_1 - x_3 = 0\}, \quad \ell_4 = \{x_0 - x_3 = x_1 + x_2 = 0\}.$$

How many lines $\Lambda \subset \mathbb{P}^3(\mathbb{C})$ intersect all four? Describe the set

$$\{\Lambda \in \mathbb{G}(1, 3) : \Lambda \cap \ell_i \neq \emptyset, i = 1, 2, 3, 4\}.$$

There is a well-developed theory for enumerating the linear subspaces meeting prescribed configurations of linear subspaces. This is called *Schubert calculus* after the enumerative geometer Hermann Schubert (1848–1911).

11.11 Let $T : k^N \to k^N$ be an invertible linear transformation. For each M-dimensional subspace $\Lambda \subset k^N$, $T(\Lambda)$ is also an M-dimensional subspace. Show this induces an automorphism of the Grassmannian

$$T : \mathrm{Gr}(M, N) \to \mathrm{Gr}(M, N).$$

11.12 Consider the map $S : \bigwedge^2 k^4 \to \bigwedge^2 k^4$ given by

$$S(x_{12}e_1 \wedge e_2 + x_{13}e_1 \wedge e_3 + x_{14}e_1 \wedge e_4 + x_{23}e_2 \wedge e_3 + x_{24}e_2 \wedge e_4 + x_{34}e_3 \wedge e_4)$$
$$= x_{34}e_1 \wedge e_2 + x_{24}e_1 \wedge e_3 + x_{23}e_1 \wedge e_4 + x_{14}e_2 \wedge e_3 + x_{13}e_2 \wedge e_4 + x_{12}e_3 \wedge e_4.$$

(a) Show that S is an invertible linear transformation and thus yields an automorphism

$$S : \mathbb{P}\left(\bigwedge^2 k^4\right) \to \mathbb{P}\left(\bigwedge^2 k^4\right).$$

(b) Show that $S(\mathrm{Gr}(2, 4)) = \mathrm{Gr}(2, 4)$ and that S restricts to an automorphism

$$S : \mathrm{Gr}(2, 4) \to \mathrm{Gr}(2, 4).$$

(c) Can this automorphism be realized as one of the automorphisms introduced in Exercise 11.11?

11.7 EXERCISES

11.13 Consider the incidence correspondence

$$I = \{(\ell, \Lambda) : \ell \subset \Lambda \subset k^4, \ell \in \mathbb{P}^3, \Lambda \in \mathrm{Gr}(2,4)\} \subset \mathbb{P}^3(k) \times \mathrm{Gr}(2,4).$$

(a) Show that I is an abstract variety.
(b) Find equations for

$$I \subset \mathbb{P}^3(k) \times \mathbb{P}^5(k) = \mathbb{P}(k^4) \times \mathbb{P}\left(\bigwedge^2 k^4\right).$$

Hint: Use and prove the following fact from linear algebra: If $\omega = v_1 \wedge v_2 \in \bigwedge^2 k^4$ is decomposable and $w \in k^4$, then $w \in \mathrm{Span}\{v_1, v_2\}$ if and only if $\omega \wedge w = 0$.

11.14 (a) Decide whether

$$\omega = e_1 \wedge e_2 \wedge e_3 + e_2 \wedge e_3 \wedge e_4 + e_3 \wedge e_4 \wedge e_1 + e_4 \wedge e_1 \wedge e_2 \in \bigwedge^3 k^4$$

is partially decomposible. If so, find v and η so that $\omega = v \wedge \eta$.
(b) Show that every element $\omega \in \bigwedge^3 k^4$ is partially decomposible.
(c) Prove or disprove: every element $\omega \in \bigwedge^3 k^5$ is partially decomposible.
(d) *Challenge:* Find equations for the locus of partially decomposible elements in $\bigwedge^3 k^6$. *Hint:* Consider the induced map

$$k^6 \to \bigwedge^4 k^6$$
$$w \mapsto w \wedge \omega.$$

11.15 Consider the subset $Z \subset \mathrm{Gr}(2,4) \times \mathrm{Gr}(2,4)$ defined by

$$Z = \{(\Lambda_1, \Lambda_2) : \Lambda_1, \Lambda_2 \subset k^4, \Lambda_1 \cap \Lambda_2 \neq 0\}.$$

(a) Show that Z is closed in $\mathrm{Gr}(2,4) \times \mathrm{Gr}(2,4)$.
(b) Find bihomogeneous equations for Z in $\mathbb{P}^5(k) \times \mathbb{P}^5(k)$.

11.16 Fix integers M_1, M_2, N with $0 < M_1 < M_2 < N$ and consider the incidence $F(M_1, M_2) \subset \mathrm{Gr}(M_1, N) \times \mathrm{Gr}(M_2, N)$ given by

$$F(M_1, M_2) = \{(\Lambda_1, \Lambda_2) : \Lambda_1 \subset \Lambda_2 \subset k^N\}.$$

(a) Show that $F(M_1, M_2)$ is a closed subset of $\mathrm{Gr}(M_1, N) \times \mathrm{Gr}(M_2, N)$.
(b) Fix $M_1 = 1, M_2 = 2, N = 4$. Write down equations for

$$F(1,2) \subset \mathrm{Gr}(1,4) \times \mathrm{Gr}(2,4) = \mathbb{P}^3(k) \times \mathrm{Gr}(2,4) \subset \mathbb{P}^3(k) \times \mathbb{P}^5(k)$$

where $\mathrm{Gr}(2,4) \subset \mathbb{P}^5(k)$ is the Plücker embedding.

(c) For $\Lambda_1 \in \mathrm{Gr}(M_1, N)$, consider the set

$$W(\Lambda_1) := \{\Lambda_2 : \Lambda_2 \supset \Lambda_1\} \subset \mathrm{Gr}(M_2, N).$$

Show that W is closed. *Hint:* Observe that $W(\Lambda_1) = \pi_2(\pi_1^{-1}(\Lambda_1))$ where $\pi_j : F(M_1, M_2) \to \mathrm{Gr}(M_j, N)$ is the projection.

The varieties $F(M_1, M_2)$ are called two-step *flag varieties*.

12 Hilbert polynomials and the Bezout Theorem

Hilbert polynomials are the main tool for classifying projective varieties. Many invariants of topological and geometric interest, like the genus of a Riemann surface, are encoded in their coefficients. Here we will carefully analyze the Hilbert polynomials of ideals defining finite sets over algebraically closed fields. The Bezout Theorem is the main application.

Hilbert polynomials are defined in terms of the Hilbert function, but the precise relationship between the Hilbert function and the Hilbert polynomial is extremely subtle and continues to be the object of current research. The interpolation problems considered in Chapter 1 involve measuring the discrepancy between these two invariants.

12.1 Hilbert functions defined

While our main focus is homogeneous ideals in polynomial rings, Hilbert functions and polynomials are used in a much broader context:

Definition 12.1   A *graded ring* R is a ring admitting a direct-sum decomposition

$$R = \oplus_{t \in \mathbb{Z}} R_t$$

compatible with multiplication, i.e., $R_{t_1} R_{t_2} \subset R_{t_1+t_2}$ for all $t_1, t_2 \in \mathbb{Z}$. The decomposition is called a *grading* of R and the summands R_t are called its *graded pieces*; elements $F \in R_t$ are *homogeneous* of degree t.

Let k be a field. A *graded k-algebra* is a k-algebra R with a grading compatible with the algebra structure, i.e., for any $c \in k$ and $F \in R_t$ we have $cF \in R_t$.

Observe that the constants in a graded k-algebra necessarily have degree zero.

Example 12.2

1. $S = k[x_0, \ldots, x_n]$ is a graded ring with graded pieces

$$S_t = k[x_0, \ldots, x_n]_t = \text{homogeneous forms of degree } t.$$

2. Let $w_0, \ldots, w_n$ be positive integers. Then $S = k[x_0, \ldots, x_n]$ is graded with graded pieces
$$S_t = \mathrm{span}\bigl(x_0^{\alpha_0} \ldots x_n^{\alpha_n} : w_0\alpha_0 + \cdots + w_n\alpha_n = t\bigr).$$
This is called a *weighted polynomial ring*.

3. Let $J \subset S = k[x_0, \ldots, x_n]$ be a homogeneous ideal. The quotient ring $R = S/J$ is graded with graded pieces
$$R_t = \mathrm{image}(k[x_0, \ldots, x_n]_t \to S/J).$$
Here is a proof: For each t, the inclusion $k[x_0, \ldots, x_n]_t \subset k[x_0, \ldots, x_n]$ induces an inclusion $R_t \subset R$. These together give a surjective homomorphism $\oplus_{t \geq 0} R_t \to R$. We claim this is injective. Suppose we have homogeneous $F_j \in k[x_0, \ldots, x_n]_j$, $j = 0, \ldots, r$, such that $F_0 + \cdots + F_r \equiv 0$ in R, i.e., $F_0 + \cdots + F_r \in J$. By Exercise 9.1, each $F_j \in J$ and thus $F_j \equiv 0$ in R_j.

4. For any projective variety $X \subset \mathbb{P}^n(k)$ the ring
$$R(X) = k[x_0, \ldots, x_n]/J(X)$$
is graded; it is called the *graded coordinate ring* of X.

Definition 12.3 Let R be a graded k-algebra with $\dim_k R_t < \infty$ for each t. The *Hilbert function* $\mathrm{HF}_R : \mathbb{Z} \to \mathbb{Z}$ is defined
$$\mathrm{HF}_R(t) = \dim_k R_t.$$

We compute Hilbert functions in some important examples:

Example 12.4

1. For $S = k[x_0, \ldots, x_n]$ with the standard grading, we have
$$\mathrm{HF}_S(t) = \binom{t+n}{t}.$$

2. If $F \in k[x_0, \ldots, x_n]$ is homogeneous of degree d and $R = k[x_0, \ldots, x_n]/\langle F \rangle$ then
$$\mathrm{HF}_R(t) = \binom{t+n}{n} - \binom{t-d+n}{n}$$
for $t \geq d$. Indeed, the elements of $\langle F \rangle$ of degree t are of the form FG where G is an arbitrary homogeneous polynomial of degree $t - d$.

3. If $a = [a_0, \ldots, a_n] \in \mathbb{P}^n(k)$ and $R = k[x_0, \ldots, x_n]/J(a)$ then $\mathrm{HF}_R(t) = 1$ for $t \geq 0$.

Hilbert functions are invariant under projectivities:

12.1 HILBERT FUNCTIONS DEFINED

Proposition 12.5 *Let $J \subset k[x_0, \ldots, x_n]$ be homogeneous, $\phi : \mathbb{P}^n(k) \to \mathbb{P}^n(k)$ a projectivity, and $J' = \phi^* J$. If $R = k[x_0, \ldots, x_n]/J$ and $R' = k[x_0, \ldots, x_n]/J'$ then $\mathrm{HF}_R(t) = \mathrm{HF}_{R'}(t)$.*

Proof   The coordinate functions of ϕ take the form

$$\phi_i(x_0, \ldots, x_n) = \sum_{j=0}^{n} a_{ij} x_j$$

where $A = (a_{ij})$ is an $(n+1) \times (n+1)$ invertible matrix. The corresponding homomorphism

$$\phi^* : k[x_0, \ldots, x_n] \to k[x_0, \ldots, x_n]$$

restricts to invertible linear transformations

$$(\phi^*)_t : k[x_0, \ldots, x_n]_t \to k[x_0, \ldots, x_n]_t$$

$$x_0^{\alpha_0} \ldots x_n^{\alpha_n} \mapsto \left(\sum_{j=0}^{n} a_{0j} x_j\right)^{\alpha_0} \ldots \left(\sum_{j=0}^{n} a_{nj} x_j\right)^{\alpha_n}$$

for each $t \geq 0$. It follows that

$$\dim_k J'_t = \dim_k (\phi^*)_t J_t = \dim_k J_t$$

for each t, and $\mathrm{HF}_R(t) = \mathrm{HF}_{R'}(t)$. $\square$

Proposition 12.6 *Let $J, J' \subset k[x_0, \ldots, x_n]$ be graded ideals with intersection $J'' = J \cap J'$, and write*

$$R = k[x_0, \ldots, x_n]/J, \; R' = k[x_0, \ldots, x_n]/J', \; R'' = k[x_0, \ldots, x_n]/J''.$$

Then we have

$$\mathrm{HF}_R(t) + \mathrm{HF}_{R'}(t) \geq \mathrm{HF}_{R''}(t).$$

Equality holds for $t \gg 0$ if and only if $J + J'$ is irrelevant.

Proof   Let J_t, J'_t, and J''_t denote the degree-t graded pieces of the corresponding ideals, e.g., $J_t = J \cap k[x_0, \ldots, x_n]_t$. For each $t \geq 0$, we have a surjection

$$J_t \oplus J'_t \twoheadrightarrow (J + J')_t$$

with kernel J''_t. Thus we find

$$\dim J_t + \dim J'_t = \dim (J + J')_t + \dim J''_t$$

and

$$\dim R_t + \dim R'_t = \dim k[x_0, \ldots, x_n]_t / (J + J')_t + \dim R''_t.$$

This yields the identity of Hilbert functions

$$\mathrm{HF}_R(t) + \mathrm{HF}_{R'}(t) = \mathrm{HF}_{R''}(t) + \dim k[x_0, \ldots, x_n]_t/(J + J')_t,$$

and our first inequality follows. $J + J'$ is irrelevant if and only if $(J + J')_N = k[x_0, \ldots, x_n]_N$ for $N \gg 0$; equality holds for precisely these values of t. □

Definition 12.7   The *Hilbert function* of a projective variety X is defined

$$\mathrm{HF}_X(t) = \mathrm{HF}_{R(X)}(t),$$

where $R(X)$ is its graded coordinate ring.

Proposition 12.8   *If $X_1, \ldots, X_r \subset \mathbb{P}^n(k)$ are projective varieties then*

$$\mathrm{HF}_{\cup_j X_j}(t) \leq \sum_{j=1}^r \mathrm{HF}_{X_j}(t).$$

Equality holds for $t \gg 0$ provided $J(X_i) + J(X_j)$ irrelevant for each $i \neq j$.

Proof   By induction, it suffices to address the $r = 2$ case. Just as for affine varieties, we have

$$J(X_1 \cup X_2) = J(X_1) \cap J(X_2).$$

Proposition 12.6 then gives the result. □

An application of the Projective Nullstellensatz (Theorem 9.25) yields

Corollary 12.9   *Let k be algebraically closed. Suppose $X_1, \ldots, X_r \subset \mathbb{P}^n(k)$ are projective varieties which are pairwise disjoint. Then for $t \gg 0$*

$$\mathrm{HF}_{\cup_j X_j}(t) = \sum_{j=1}^r \mathrm{HF}_{X_j}(t).$$

In the special case where each X_j is a point we obtain

Corollary 12.10   *Let $S \subset \mathbb{P}^n(k)$ be finite. Then $\mathrm{HF}_S(t) = |S|$ for $t \gg 0$.*

The following case is crucial for the Bezout Theorem:

Proposition 12.11   *Let $F, G \in S = k[x_0, \ldots, x_n]$ be homogeneous of degree d and e without common factors. The quotient ring $R = S/\langle F, G \rangle$ has Hilbert function*

$$\mathrm{HF}_R(t) = \dim S_t - \dim S_{t-d} - \dim S_{t-e} + \dim S_{t-d-e}$$
$$= \binom{t+n}{n} - \binom{t+n-d}{n} - \binom{t+n-e}{n} + \binom{t+n-d-e}{n}$$
provided $t \geq d + e$.

If $n = 2$ then $\mathrm{HF}_R(t) = de$ for $t \gg 0$.

Proof Let $q : S \to R$ be the quotient homomorphism and $q(t) : S_t \to R_t$ the induced linear transformation. Recall (Exercise 2.15) that $AF + BG = 0$ for $A, B \in k[x_0, \ldots, x_n]$ if and only if $A = CG$ and $B = -CF$ for some $C \in k[x_0, \ldots, x_n]$. We have a series of linear transformations (cf. §5.4)

$$0 \to S_{t-d-e} \xrightarrow{\delta_1(t)} S_{t-d} \oplus S_{t-e} \xrightarrow{\delta_0(t)} S_t \xrightarrow{q(t)} R_t \to 0$$
$$C \mapsto (CG, -CF)$$
$$(A, B) \mapsto AF + BG$$

where $\ker(\delta_0(t)) = \operatorname{image}(\delta_1(t))$, $\ker(q(t)) = \operatorname{image}(\delta_0(t))$, $\delta_1(t)$ is injective, and $q(t)$ is surjective. Applying the rank-nullity theorem successively, we obtain

$$\dim R_t = \dim S_t - \dim S_{t-d} - \dim S_{t-e} + \dim S_{t-d-e}.$$

The formula $\dim S_r = \binom{r+n}{n}$ is valid for $r \geq 0$ and yields

$$\operatorname{HF}_R(t) = \binom{t+n}{n} - \binom{t+n-d}{n} - \binom{t+n-e}{n} + \binom{t+n-d-e}{n}$$

for $t \geq d + e$. We also have

$$\dim S_r = \frac{(r+n)(r+n-1)\ldots(r+1)}{n!}$$

for $r \geq -n$; when $n = 2$, we obtain $\operatorname{HF}_R(t) = de$ for all $t \geq d + e - 2$. $\square$

12.2 Hilbert polynomials and algorithms

Proposition 12.12 *Let $J \subset S = k[x_0, \ldots, x_n]$ be a homogeneous ideal, $>$ a monomial order, and $\operatorname{LT}(J)$ the ideal of leading terms of J. Then $\operatorname{HF}_{S/J}(t) = \operatorname{HF}_{S/\operatorname{LT}(J)}(t)$.*

This reduces the computation of Hilbert functions to the case of monomial ideals.

Proof This follows immediately from the existence of normal forms (Theorem 2.16): The monomials not in $\operatorname{LT}(J)$ form a basis for S/J as a vector space. In particular, the monomials of degree t not in $\operatorname{LT}(J)$ form a basis for $(S/J)_t = S_t/J_t$.
$\square$

From now on, we use binomial notation to designate polynomials with rational coefficients: For each integer $r \geq 0$, we write

$$\binom{t}{r} = \begin{cases} t(t-1)\ldots(t-r+1)/r! & \text{if } r > 0 \\ 1 & \text{if } r = 0, \end{cases}$$

which is a polynomial of degree r in $\mathbb{Q}[t]$. We allow t to assume both positive and negative values.

Proposition 12.13 *Let $f : \mathbb{Z} \to \mathbb{Z}$ be a function such that*

$$(\Delta f)(t) := f(t+1) - f(t)$$

is a polynomial of degree $d - 1$. Then f is a polynomial of degree d and can be written in the form

$$f(t) = \sum_{j=0}^{d} a_j \binom{t}{d-j}, \quad a_j \in \mathbb{Z}. \tag{12.1}$$

In particular, any polynomial $P(t) \in \mathbb{Q}[t]$ of degree d with $P(\mathbb{Z}) \subset \mathbb{Z}$ takes this form.

Proof The key to the proof is the identity

$$\binom{t+1}{r} - \binom{t}{r} = \binom{t}{r-1},$$

which we leave to the reader.

The argument uses induction on degree. The base case $d = 1$ is straightforward: If $(\Delta f)(t)$ is a constant a_0 then

$$f(t) = a_0 t + a_1 = a_0 \binom{t}{1} + a_1 \binom{t}{0}.$$

The inductive hypothesis implies that

$$(\Delta f)(t) = \sum_{j=0}^{d-1} a_j \binom{t}{d-1-j}$$

for some $a_j \in \mathbb{Z}$. Set

$$Q(t) = \sum_{j=0}^{d-1} a_j \binom{t}{d-j}$$

so that $(\Delta Q) = (\Delta f)$ by our identity above. It follows that $f = Q + a_d$ for some constant $a_d \in \mathbb{Z}$ and f takes the desired form. $\square$

Theorem 12.14 (Existence of Hilbert polynomial) *Let $J \subset S = k[x_0, \ldots, x_n]$ be a homogeneous ideal and $R = k[x_0, \ldots, x_n]/J$. Then there exists a polynomial $\mathrm{HP}_R(t) \in \mathbb{Q}[t]$ such that*

$$\mathrm{HF}_R(t) = \mathrm{HP}_R(t), \ t \gg 0.$$

This is called the **Hilbert polynomial** *and takes the form (12.1).*

12.2 HILBERT POLYNOMIALS AND ALGORITHMS

Proof If $J \subset S$ is an ideal, write

$$\mathrm{HF}_J(t) = \dim J_t = \binom{t+n}{n} - \mathrm{HF}_{S/J}(t).$$

Of course, $\mathrm{HF}_{S/J}(t)$ is a polynomial function for $t \gg 0$ if and only if $\mathrm{HF}_J(t)$ is a polynomial function for $t \gg 0$.

We may assume J is a monomial ideal by Proposition 12.12. The argument is by induction on n. The base case $n = 0$ is straightforward: if $J = \langle x_0^N \rangle$ then $\mathrm{HF}_J(t) = 1$ for $t \geq N$. For the inductive step, assume each monomial ideal in $S' = k[x_0, \ldots, x_{n-1}]$ has a Hilbert polynomial.

Define auxilliary ideals

$$J[m] = \{ f \in S' : f x_n^m \in J \}$$

so that

$$J[0] \subset J[1] \subset J[2] \subset \ldots.$$

This terminates at some $J[\infty]$ so that $J[m] = J[\infty]$ for $m \geq M$. Assume that $\mathrm{HF}_{J[\infty]}(t) = \mathrm{HP}_{J[\infty]}(t)$ whenever $t \geq N$.

We can write each element of J_t as a sum of terms $f_m x_n^m$ where $f_m \in J[m]_{t-m}$. Hence we have

$$J_t \simeq \oplus_{m=0}^t J[m]_{t-m} x_n^m$$

and, for $t \gg 0$,

$$\mathrm{HF}_J(t) = \sum_{m=0}^t \mathrm{HF}_{J[m]}(t-m)$$

$$= \sum_{m=0}^{M-1} \mathrm{HF}_{J[m]}(t-m) + \sum_{m=M}^t \mathrm{HF}_{J[m]}(t-m)$$

$$= \sum_{m=0}^{M-1} \mathrm{HF}_{J[m]}(t-m) + \sum_{m=M}^t \mathrm{HF}_{J[\infty]}(t-m)$$

$$= \sum_{m=0}^{M-1} \mathrm{HF}_{J[m]}(t-m) + \sum_{m=M}^{t-N} \mathrm{HF}_{J[\infty]}(t-m) + \sum_{m=t-N+1}^t \mathrm{HF}_{J[\infty]}(t-m)$$

(substituting $\mathrm{HF}_{J[\infty]}(s) = \mathrm{HP}_{J[\infty]}(s)$ for $s \geq N$)

$$= \sum_{m=0}^{M-1} \mathrm{HF}_{J[m]}(t-m) + \sum_{m=M}^{t-N} \mathrm{HP}_{J[\infty]}(t-m) + \sum_{m=t-N+1}^t \mathrm{HF}_{J[\infty]}(t-m)$$

(substituting $\mathrm{HF}_{J[m]}(s) = \mathrm{HP}_{J[m]}(s)$ for $m = 0, \ldots, M-1, s \gg 0$)

$$= \sum_{m=0}^{M-1} \mathrm{HP}_{J[m]}(t-m) + \sum_{m=M}^{t-N} \mathrm{HP}_{J[\infty]}(t-m) + \sum_{m=t-N+1}^t \mathrm{HF}_{J[\infty]}(t-m)$$

$$= \sum_{m=0}^{M-1} \mathrm{HP}_{J[m]}(t-m) + \sum_{j=N}^{t-M} \mathrm{HP}_{J[\infty]}(j) + \sum_{j=0}^{N-1} \mathrm{HF}_{J[\infty]}(j).$$

The first part is a finite sum of polynomials in t, hence is a polynomial in t. The third part is constant in t. As for the second part, write

$$f(t) := \sum_{j=N}^{t-M} \mathrm{HP}_{J[\infty]}(j)$$

and observe that

$$(\Delta f)(t) = f(t+1) - f(t) = \mathrm{HP}_{J[\infty]}(t+1-M)$$

is a polynomial in t. Proposition 12.13 implies f is also a polynomial in t. We conclude that $\mathrm{HF}_J(t)$ is a polynomial in t for $t \gg 0$. □

Definition 12.15 Let $X \subset \mathbb{P}^n(k)$ be a projective variety. The *Hilbert polynomial* of X is defined

$$\mathrm{HP}_X(t) = \mathrm{HP}_{R(X)}$$

where $R(X) = k[x_0, \ldots, x_n]/J(X)$.

In light of our previous results on Hilbert functions, we have

1. $\mathrm{HP}_X(t) = \binom{t+n}{n} - \binom{t-d+n}{n}$ when $J(X) = \langle F \rangle$ where $F \in k[x_0, \ldots, x_n]_d$;
2. $\mathrm{HP}_X(t)$ is invariant under projectivities;
3. $\mathrm{HP}_X(t) = \deg(F)\deg(G)$ when $J(X) = \langle F, G \rangle \subset k[x_0, x_1, x_2]$, where F and G are homogeneous with no common factors;
4. $\mathrm{HP}_S(t) = |S|$ for S finite;
5. $\mathrm{HP}_{\cup_j X_j}(t) = \sum_{j=1}^r \mathrm{HP}_{X_j}(t)$ when the X_j are pairwise disjoint and k is algebraically closed.

Example 12.16 Consider the ideal of the twisted cubic curve $C \subset \mathbb{P}^3(k)$

$$J(C) = \langle x_0 x_2 - x_1^2, x_0 x_3 - x_1 x_2, x_1 x_3 - x_2^2 \rangle,$$

which has leading term ideal (with respect to pure lexicographic order)

$$J = \langle x_0 x_2, x_0 x_3, x_1 x_3 \rangle.$$

We have $J[0] = \langle x_0 x_2 \rangle$ hence

$$\mathrm{HF}_{J[0]}(t) = \mathrm{HP}_{J[0]}(t) = \binom{t}{2}.$$

Furthermore,

$$J[m] = \langle x_0, x_1 \rangle = J[\infty]$$

for all $m \geq 1$ and

$$\mathrm{HF}_{J[\infty]}(t) = \mathrm{HP}_{J[\infty]}(t) = \binom{t+2}{2} - 1,$$

12.3 INTERSECTION MULTIPLICITIES

because $J[\infty]_t$ contains all monomials of degree t besides x_2^t. Thus we have

$$\mathrm{HF}_J(t) = \sum_{j=0}^{t} \mathrm{HF}_{J[t-j]}(j)$$

$$= \binom{t}{2} + \sum_{j=0}^{t-1} \left(\binom{j+2}{2} - 1\right)$$

$$= \sum_{j=0}^{t} \binom{j+2}{2} - t + \binom{t}{2} - \binom{t+2}{2}$$

$$= \binom{t+3}{3} - (3t+1)$$

and $\mathrm{HP}_C(t) = 3t+1$.

Proposition 12.17 If $J \subset S = k[x_0, \ldots, x_n]$ is a homogeneous ideal with saturation $\tilde{J}$ then $\mathrm{HP}_{S/J}(t) = \mathrm{HP}_{S/\tilde{J}}(t)$.

Proof Write $\mathfrak{m} = \langle x_0, \ldots, x_n \rangle$ so that

$$\tilde{J} = \{F \in k[x_0, \ldots, x_n] : F\mathfrak{m}^N \subset J \text{ for some N}\}.$$

The inclusion $J \subset \tilde{J}$ is obvious, so it suffices to show $\tilde{J}_t \subset J_t$ for $t \gg 0$. If $\tilde{J}$ is generated by homogeneous $F_1, \ldots, F_r$ then there exists an N such that $\mathfrak{m}^N F_i \subset J$ for $i = 1, \ldots, r$. For $t \geq \max_i\{\deg(F_i)\} + N$ we have $\tilde{J}_t \subset J_t$. $\square$

12.3 Intersection multiplicities

How do we count the number of points where two varieties meet? We want this number to be constant even as the varieties vary in families. Such a method satisfies the *continuity principle*.

Example 12.18

1. Consider the complex plane curves $C_t = \{y = x^2 - t\}$ and $D = \{y = 0\}$. The intersection

$$C_t \cap D = \{(\pm\sqrt{t}, 0)\}$$

is two distinct points for $t \neq 0$ and one point for $t = 0$. This reflects the fact that C_0 is tangent to D at the origin.

2. Consider the complex plane curves $C_t = \{y = x^3 - tx\}$ and $D = \{y = 0\}$. The intersection

$$C_t \cap D = \{(0,0), (\pm\sqrt{t}, 0)\}$$

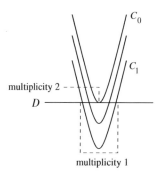

Figure 12.1 Intersections of families of plane curves.

is three distinct points for $t \neq 0$ and one point for $t = 0$. Note that D is an inflectional tangent to C_0 at the origin.

We now define the multiplicity: let $p = (a_1, \ldots, a_n) \in \mathbb{A}^n(k)$ and $\mathfrak{m}_p = \langle x_1 - a_1, \ldots, x_n - a_n \rangle$ the corresponding maximal ideal. Let $Q \subset R := k[x_1, \ldots, x_n]$ be an ideal with

$$\mathfrak{m}_p^N \subset Q \subset \mathfrak{m}_p$$

for some N. Such ideals are precisely the primary ideals in R with associated prime $\mathfrak{m}_p$ (see Exercise 8.4). The induced quotient map $R/\mathfrak{m}_p^N \twoheadrightarrow R/Q$ is surjective and

$$\dim_k R/\mathfrak{m}_p^N = \dim_k k[x_1, \ldots, x_n]/\mathfrak{m}_p^N = \binom{n+N-1}{n},$$

so $\dim_k R/Q < \infty$. We define the *multiplicity* of Q at p by

$$\text{mult}(Q, p) = \dim_k R/Q.$$

We extend this to more general classes of ideals $I \subset k[x_1, \ldots, x_n]$. Assume that $\mathfrak{m}_p$ is a minimal associated prime of I. Then the *multiplicity* of I at p is defined

$$\text{mult}(I, p) = \text{mult}(Q, p),$$

where Q is the primary component of I corresponding to $\mathfrak{m}_p$, which is unique by Theorem 8.32. If $\mathfrak{m}_p \not\supset I$ then we define $\text{mult}(I, p) = 0$.

Remark 12.19 Assume that k is an algebraically closed field. Then $\mathfrak{m}_p$ is a minimal associated prime of I if and only if p is an irreducible component of $V(I)$ (by Corollary 8.29).

Definition 12.20 Let k be algebraically closed, $V_1, \ldots, V_n \subset \mathbb{A}^n(k)$ hypersurfaces with $I(V_i) = \langle f_i \rangle$, $i = 1, \ldots, n$. Suppose that p is an irreducible

12.3 INTERSECTION MULTIPLICITIES

component of $V_1 \cap \ldots \cap V_n$. The *multiplicity* of $V_1 \cap \ldots \cap V_n$ along p is defined as $\mathrm{mult}(\langle f_1, \ldots, f_n \rangle, p)$.

Example 12.21 We return to Example 12.18, computing the multiplicities of $C_t \cap D$ along each of the points of intersection:

In the first example

$$\begin{aligned} I_t &= \langle y, -y + x^2 - t \rangle = \langle y, -t + x^2 \rangle \\ &= \langle y, x - \sqrt{t} \rangle \cap \langle y, x + \sqrt{t} \rangle \\ &\quad \text{primary decomposition if } t \neq 0 \\ &= \langle y, x^2 \rangle \\ &\quad \text{primary decomposition if } t = 0. \end{aligned}$$

In the first instance each primary component has codimension 1, so the two points of intersection have multiplicity 1. In the second instance we have just one primary component with codimension 2, so there is a single intersection point of multiplicity 2.

In the second example

$$\begin{aligned} I_t &= \langle y, y - x^3 + tx \rangle = \langle y, -x^3 + tx \rangle \\ &= \langle y, x \rangle \cap \langle y, x + \sqrt{t} \rangle \cap \langle y, x - \sqrt{t} \rangle \\ &\quad \text{primary decomposition if } t \neq 0 \\ &= \langle y, x^3 \rangle \\ &\quad \text{primary decomposition if } t = 0. \end{aligned}$$

In the first instance each of the three intersection points has multiplicity 1; in the second, there is one point with multiplicity 3.

Example 12.22 Consider the ideal $I = \langle yx, (x-2)^2 x \rangle$ with primary decomposition

$$I = \langle x \rangle \cap \langle y, (x-2)^2 \rangle,$$

thus

$$V(I) = \{y - \text{axis}\} \cup \{(2, 0)\}.$$

The second component is associated with $\mathfrak{m}_p$ for $p = (2, 0)$ and $\mathrm{mult}(I, p) = 2$.

12.3.1 Methods for computing multiplicities

We describe an algorithm for computing intersection multiplicities. Let $I \subset k[x_1, \ldots, x_n]$ be an ideal, $p \in \mathbb{A}^n(k)$ a point, and $\mathfrak{m}_p$ the corresponding maximal ideal. Assume that $\mathfrak{m}_p$ is a minimal prime of I. Fix an irredundant primary decomposition

$$I = Q \cap Q_1 \cap \ldots \cap Q_l,$$

where is Q the primary component associated to $\mathfrak{m}_p$.

1. Localize to Q. Find some
$$f \in Q_1 \cap \ldots \cap Q_l$$
with $f \notin Q$, so that
$$Q = \langle uf - 1 \rangle + I.$$
We can interpret this geometrically when k is algebraically closed: throw out the irreducible components of $V(I)$ other than p.

2. Compute a Gröbner basis for Q: since $k[x_1, \ldots, x_n]/Q$ is finite dimensional, the monomials *not* appearing in $\mathrm{LT}(Q)$ form a basis for the quotient (Theorem 2.16). The number of these monomials equals $\mathrm{mult}(I, p)$.

Example 12.23   Consider the ideal
$$I = \langle y, y - x^2 + x^3 \rangle$$
so that $V(I) = \{(0,0), (1,0)\}$. To compute $\mathrm{mult}(I, (0,0))$ we first extract the primary component
$$Q_{(0,0)} = \langle u(x-1) - 1 \rangle + I = \langle u(x-1), y, y - x^2 + x^3 \rangle,$$
which has Gröbner basis
$$\{y, x^2, u + x + 1\}.$$
The monomials not in $\mathrm{LT}(Q_{(0,0)})$ are $\{1, x\}$ so $\mathrm{mult}(I, (0,0)) = 2$.

The primary component
$$Q_{(1,0)} = \langle ux - 1, y, y - x^2 + x^3 \rangle$$
has Gröbner basis
$$\{u - 1, x - 1, y\}.$$
The only monomial not in $\mathrm{LT}(Q_{(1,0)})$ is $\{1\}$ so $\mathrm{mult}(I, (1,0)) = 1$.

12.3.2 An interpolation result

Theorem 12.24   *Let $I \subset k[y_1, \ldots, y_n]$ be an ideal whose associated primes are all of the form $\mathfrak{m}_p$ for some $p \in \mathbb{A}^n(k)$. Then we have*
$$\dim k[y_1, \ldots, y_n]/I = \sum_{p \in V(I)} \mathrm{mult}(I, p).$$

Proof   We fix some notation. Choose an irredundant primary decomposition
$$I = Q_1 \cap \ldots \cap Q_s$$

with associated primes $\mathfrak{m}_{p_1}, \ldots, \mathfrak{m}_{p_s}$. We have the linear transformation

$$\Pi : k[y_1, \ldots, y_n] \to \oplus_{j=1}^s k[y_1, \ldots, y_n]/Q_j$$
$$f \to (f \pmod{Q_1}, \ldots, f \pmod{Q_s}).$$

It suffices to show Π is surjective: since $\ker \Pi = I$, the definition of the multiplicity and the rank-nullity theorem yield our result.

Recall that for large N each $Q_j \supset \mathfrak{m}_{p_j}^N$, so the quotients

$$k[y_1, \ldots, y_n]/\mathfrak{m}_{p_j}^N \to k[y_1, \ldots, y_n]/Q_j$$

are surjective. Thus Π is surjective provided

$$\Psi : k[y_1, \ldots, y_n] \to \oplus_{j=1}^s k[y_1, \ldots, y_n]/\mathfrak{m}_{p_j}^N$$
$$f \to \left(f \pmod{\mathfrak{m}_{p_1}^N}, \ldots, f \pmod{\mathfrak{m}_{p_s}^N}\right)$$

is surjective. This means there exists a polynomial with prescribed Taylor series of order N at the points $p_1, \ldots, p_s$ (at least over a field of characteristic zero where Taylor series make sense). Note that the $N = 1$ case, i.e., finding a polynomial with prescribed values on a finite set, was addressed in Exercise 3.6.

The proof is by induction on the number of points s. The base case $s = 1$ is straightforward, since after translation we may assume p_1 is the origin. For the inductive step, consider the polynomials mapping to zero in

$$k[y_1, \ldots, y_n]/\mathfrak{m}_{p_j}^N$$

for $j = 1, \ldots, s-1$, which form an ideal $\tilde{I}$. It suffices to show that the induced map

$$\Psi_s : \tilde{I} \to k[y_1, \ldots, y_n]/\mathfrak{m}_{p_s}^N$$

is surjective. The image of Ψ_s is an ideal, so it suffices to check it contains a unit, i.e., an element that does not vanish at p_s. Let $\ell_i, i = 1, \ldots, s-1$, be linear forms with $\ell_i(p_i) = 0$ by $\ell_i(p_s) \neq 0$. The polynomial

$$P = \prod_{j=1}^{s-1} \ell_j^N \in \tilde{I}$$

but $P(p_s) \neq 0$. □

12.4 Bezout Theorem

Our first task is to analyze ideals with constant Hilbert polynomial:

Proposition 12.25 Let $J \subset k[x_0, \ldots, x_n]$ be a homogeneous ideal and $R = k[x_0, \ldots, x_n]/J$. If $\operatorname{HP}_R(t)$ has degree zero then $X(J)$ is finite.

Proof Suppose $X(J)$ is infinite. Then one of the distinguished open sets $U_i \subset \mathbb{P}^n(k)$ (say U_0) contains an infinite number of points in $X(J)$. Let $I \subset k[y_1, \ldots, y_n]$ be the dehomogenization of J with respect to x_0, which is contained in $I(U_0 \cap X(J))$. We therefore have surjections

$$k[x_0, \ldots, x_n]/J \xrightarrow{\mu_0} k[y_1, \ldots, y_n]/I \to k[U_0 \cap X(J)].$$

For each t, write

$$W_t = \text{image}(R_t = k[x_0, \ldots, x_n]_t/J_t \to k[U_0 \cap X(J)]),$$

i.e., the functions on $U_0 \cap X(J)$ that can be realized as polynomials of degree $\leq t$. We have $\dim W_t \leq \dim R_t$.

By Exercise 3.6, $\dim_k k[U_0 \cap X(J)] = |U_0 \cap X(J)| = \infty$ and W_t is unbounded for $t \gg 0$. On the other hand, $\dim R_t$ is bounded because $HP_R(t)$ is constant, a contradiction. $\square$

Over algebraically closed fields, we can sharpen this:

Proposition 12.26 *Suppose k is algebraically closed, $J \subset k[x_0, \ldots, x_n]$ is a homogeneous ideal with $R = k[x_0, \ldots, x_n]/J$, and $HP_R(t)$ is constant and nonzero. Then the minimal associated primes of J are the ideals $J(p)$ for $p \in X(J) \subset \mathbb{P}^n(k)$.*

Proof Irrelevant ideals yield trivial Hilbert polynomials (Proposition 12.17), so our hypothesis guarantees J is not irrelevant. The Projective Nullstellensatz (Theorem 9.25) then guarantees that $X(J)$ is nonempty.

Lemma 12.27 *If $J \subset k[x_0, \ldots, x_n]$ is homogeneous then each associated prime of J is homogeneous.*

We can put this into geometric terms using Corollary 8.29. Recall that $V(J) \subset \mathbb{A}^{n+1}(k)$ is the cone over the projective variety $X(J)$. Irreducible components of $V(J)$ correspond to the cones over irreducible components of $X(J)$.

Proof of lemma By Theorem 8.22, each associated prime of J can be written $P = \sqrt{J:f}$ for some $f \in k[x_0, \ldots, x_n]$. Write $f = F_0 + \cdots + F_d$ as a sum of homogeneous pieces and $K_i = \sqrt{J:F_i}$. Since J is homogeneous,

$$J:f = J:F_0 \cap J:F_1 \cap \ldots \cap J:F_d$$

and Exercise 7.18 gives

$$\sqrt{J:f} = \sqrt{J:F_0} \cap \sqrt{J:F_1} \cap \ldots \cap \sqrt{J:F_d}.$$

Since prime ideals are irreducible (Proposition 8.4), $P = \sqrt{J:F_i}$ for some i, and thus is homogeneous. $\square$

12.4 BEZOUT THEOREM

Each irreducible component of $X(J)$ is a point $p \in \mathbb{P}^n(k)$ (by Proposition 12.25). Thus each irreducible component of $V(J)$ is the line $\ell \subset \mathbb{A}^{n+1}(k)$ parametrized by that point. The description of the minimal associated primes of J follows. □

Corollary 12.28 *Retain the assumptions of Proposition 12.26 and assume in addition that J is saturated. Then the associated primes of J are the ideals $J(p)$ for $p \in X(J)$.*

Proof The only possible embedded prime is $\mathfrak{m} = \langle x_0, \ldots, x_n \rangle$, which would correspond to an irrelevant primary component. In the saturated case, these do not occur (see Exercise 10.2). □

Our next task is to tie Hilbert polynomials and multiplicities together. We start with a fun fact from linear algebra:

Lemma 12.29 *Assume k is infinite and let $S \subset \mathbb{P}^n(k)$ be a finite set. Then there exists a linear $L \in k[x_0, \ldots, x_n]_1$ such that $L(p) \neq 0$ for each $p \in S$.*

Proof First, we construct an infinite collection of points $p_1, p_2, \ldots, p_N, p_{N+1}, \ldots$ in $\mathbb{P}^n(k)$ such that any $n+1$ of the points are in linear general position. Given distinct $\alpha_1, \alpha_2, \ldots \in k$ set

$$p_i = [1, \alpha_i, \alpha_i^2, \ldots, \alpha_i^n] \in \mathbb{P}^n(k).$$

Recall the formula for determinant of the *Vandermonde matrix*

$$\det \begin{pmatrix} 1 & u_0 & u_0^2 & \cdots & u_0^n \\ 1 & u_1 & u_1^2 & \cdots & u_1^n \\ \vdots & \vdots & \vdots & \vdots & \vdots \\ 1 & u_{n-1} & u_{n-1}^2 & \cdots & u_{n-1}^n \\ 1 & u_n & u_n^2 & \cdots & u_n^n \end{pmatrix} = \prod_{0 \leq i < j \leq n} (u_j - u_i).$$

In particular, the determinant is nonzero unless $u_i = u_j$ for some $i \neq j$. Taking $u_0, \ldots, u_n$ to be any of $n+1$ distinct values of the α_i, we get the desired linear independence. In geometric terms, the images of any collection of distinct points in $\mathbb{P}^1(k)$ under the Veronese embedding

$$\nu_n : \mathbb{P}^1(k) \to \mathbb{P}^n(k)$$

are in linear general position.

There exists then a collection of $n+1$ points in $\mathbb{P}^n(k)$ in linear general position, but not contained in S. There exists a projectivity taking these to the coordinate vectors

$$e_0 = [1, 0, \ldots, 0], e_1 = [0, 1, 0, \ldots, 0], \ldots, e_n = [0, \ldots, 0, 1].$$

We may then assume that S does not include any of these points.

We finish by induction on n. Suppose $n = 1$ and $S = \{(a_1, b_1), \ldots, (a_N, b_N)\}$. If $t \in k$ is distinct from each b_j/a_j then $L = x_1 - tx_0$ has the desired property. For the inductive step, project from $e_n = [0, \ldots, 0, 1]$

$$\pi_{e_n} : \mathbb{P}^n(k) \dashrightarrow \mathbb{P}^{n-1}(k)$$
$$[x_0, \ldots, x_n] \mapsto [x_0, \ldots, x_{n-1}],$$

which is well-defined along S. By induction, there exists $L \in k[x_0, \ldots, x_{n-1}]_1$ that is nonzero at each point of $\pi_{e_n}(S)$. Regarding $L \in k[x_0, \ldots, x_n]$, we get a form non-vanishing at each point of S. $\square$

Our next task is to explain how Hilbert polynomials and multiplicities are related.

Proposition 12.30 *Suppose k is algebraically closed, $J \subset k[x_0, \ldots, x_n]$ is a homogeneous ideal with $R = k[x_0, \ldots, x_n]/J$, and $\mathrm{HP}_R(t)$ is constant. Assume in addition that $X(J) \subset U_0$. If $I \subset k[y_1, \ldots, y_n]$ denotes the dehomogenization of J then*

$$\sum_{p \in X(J) = V(I)} \mathrm{mult}(I, p) = \mathrm{HP}_R(t).$$

Proof We may replace J by its saturation without changing either the Hilbert polynomial or the dehomogenization (see Proposition 12.17). From now on, we assume that J is saturated.

Proposition 12.25 implies that $X(J)$ finite and consequently $V(I)$ is finite. The minimal associated primes of I are all of the form $\mathfrak{m}_p$, where $p \in V(I)$ (see Corollary 8.29). Our interpolation result (Theorem 12.24) implies

$$\dim_k k[y_1, \ldots, y_n]/I = \sum_{p \in I} \mathrm{mult}(I, p).$$

Again, consider the dehomogenizations

$$R := k[x_0, \ldots, x_n]/J \xrightarrow{\mu_0} k[y_1, \ldots, y_n]/I$$

and the images of the induced

$$R_t := k[x_0, \ldots, x_n]_t/J_t \to k[y_1, \ldots, y_n]/I$$

for each t. We claim that these are isomorphisms for $t \gg 0$, which implies our result.

Surjectivity is not too difficult. The quotient $k[y_1, \ldots, y_n]/I$ has finite dimension, with a basis consisting of monomials of bounded degree. To prove injectivity, suppose that F is homogeneous of degree d and dehomogenizes to a polynomial $f \in I$. This is the dehomogenization of some homogeneous $F' \in J$, with $F' = x_0^e F$ for some e.

To show that F is also in J, we establish that the multiplication map

$$R_t \to R_{t-1}$$
$$F \mapsto x_0 F$$

is injective.

Corollary 12.28 gives the primary decomposition of J. If $X(J) = \{p_1, \ldots, p_s\}$ then

$$J = Q_1 \cap \ldots \cap Q_s,$$

where Q_i is associated to $P_i := J(p_i)$ for $i = 1, \ldots, s$. However, $x_0(p_i) \neq 0$ for each i by assumption, so $x_0 \notin P_i$. The only zero divisors in R are in the images of the associated primes of J (Corollary 8.24); thus x_0 is not a zero divisor in R. □

Theorem 12.31 (Bezout Theorem) *Let k be an algebraically closed field, $F, G \in k[x_0, x_1, x_2]$ homogeneous polynomials without common factors, and $J = \langle F, G \rangle$. There exist coordinates z_0, z_1, z_2 such that $z_0(p) \neq 0$ for each $p \in X(J)$. If I is the dehomogenization of J with respect to z_0 then*

$$\sum_{p \in X(J)} \mathrm{mult}(I, p) = \deg(F) \deg(G).$$

Proof Lemma 12.29 guarantees the existence of coordinates with the desired property. Then the theorem is just the combination of Propositions 12.11 and 12.30. □

Corollary 12.32 *Two plane curves of degrees d and e with no common components meet in de points, when counted with multiplicities.*

Étienne Bézout (1733–1783) wrote the *Théorie générale des équations algébraiques* in 1779. An English translation [3] was published in 2006.

12.4.1 Higher-dimensional generalizations

Theorem 12.33 *Let $F_1, \ldots, F_n$ be homogeneous in $S = k[x_0, x_1, \ldots, x_n]$, $J = \langle F_1, \ldots, F_n \rangle$, and $R = S/J$. Suppose either of the following equivalent conditions holds*

1. *for each $j = 2, \ldots, n$, F_j is not a zero divisor (mod $\langle F_1, \ldots, F_{j-1} \rangle$);*
2. *$X(J)$ has a finite number of points over an algebraically closed field.*

Then we have

$$\mathrm{HP}_R(t) = \deg(F_1) \ldots \deg(F_n).$$

Idea of proof Let d_j denote the degree of F_j. The computation of $\mathrm{HP}_R(t)$ requires knowledge of the *Koszul complex*

$$0 \to K_n \xrightarrow{\delta_n(t)} K_{n-1} \ldots K_r \xrightarrow{\delta_r(t)} K_{r-1} \ldots K_0 \xrightarrow{\delta_0} R_t \to 0.$$

Here the terms are

$$K_0 = S_t$$
$$K_1 = \sum_j S_{t-d_j}$$
$$K_2 = \sum_{j(1)<j(2)} S_{t-d_{j(1)}-d_{j(2)}}$$
$$\vdots \quad \vdots$$
$$K_r = \sum_{j(1)<j(2)<\ldots<j(r)} S_{t-d_{j(1)}-\cdots-d_{j(r)}}$$
$$\vdots \quad \vdots$$
$$K_n = S_{t-d_1-\cdots-d_n}$$

and the maps $\delta_r(t)$ are direct sums of multiplication maps

$$S_{t-d_{j(1)}-\cdots-d_{j(r)}} \to S_{t-d_{j(1)}-\cdots-d_{j(s-1)}-d_{j(s+1)}-\cdots-d_{j(r)}}$$
$$G \mapsto (-1)^s F_{j(s)} G.$$

The signs are chosen so that image $\delta_r(t) \subset \ker \delta_{r-1}(t)$ for each r; see §5.4 for a special case.

Under our assumptions, the only syzygies among the F_j are the 'obvious' ones, e.g., $F_i(F_j) - F_j(F_i) = 0$. In this context, the Koszul complex is exact, i.e., the kernel of each map equals the image of the one before. We then have

$$\dim R_t = \sum_{r=0}^n (-1)^r \sum_{j(1)<\ldots<j(r)} \dim S_{t-d_{j(1)}-\cdots-d_{j(r)}}$$
$$= \sum_{r=0}^n (-1)^r \sum_{j(1)<\ldots<j(r)} \binom{t - d_{j(1)} - \cdots - d_{j(r)} + n}{n}$$
$$= d_1 \ldots d_n,$$

where the last step is a formal combinatorial identity. □

This granted, our proof of Theorem 12.31 yields the following.

Theorem 12.34 (Higher-dimensional Bezout Theorem) *Let k be an algebraically closed field, $F_1, \ldots, F_n \in k[x_0, \ldots, x_n]$ homogeneous polynomials, and let $J = \langle F_1, \ldots, F_n \rangle$. Assume that $X(J)$ is finite and $x_0(p) \neq 0$ for each $p \in X(J)$. If I is the dehomogenization of J with respect to x_0 then*

$$\sum_{p \in X(J)} \mathrm{mult}(I, p) = \prod_{i=1}^n \deg(F_i).$$

12.4.2 An application to inflectional tangents

Let $C \subset \mathbb{P}^2(\mathbb{C})$ be a smooth plane curve over the complex numbers with homogeneous equation $F \in \mathbb{C}[x, y, z]$ of degree d.

A line $\ell \subset \mathbb{P}^2$ is *tangent* to C at p if the multiplicity of $\ell \cap C$ at p is greater than or equal to 2. It is an *inflectional tangent* to C at p if the multiplicity is greater than or equal to 3. In our discussion of the Plücker formulas in §11.2.1, we observed that the number of inflectional tangents to C equals the number of cusps in the dual curve $\check{C}$.

How do we count the number of inflectional tangents? Recall that the dual curve is the image of the morphism

$$\phi : C \to \check{\mathbb{P}}^2$$
$$[x, y, z] \to [\partial F/\partial x, \partial F/\partial y, \partial F/\partial z].$$

The inflectional tangents are precisely the critical points of this map. The differential of ϕ has nontrivial kernel when the *Hessian*

$$H(F) = \det \begin{pmatrix} \frac{\partial^2 F}{\partial x^2} & \frac{\partial^2 F}{\partial x \partial y} & \frac{\partial^2 F}{\partial x \partial z} \\ \frac{\partial^2 F}{\partial x \partial y} & \frac{\partial^2 F}{\partial y^2} & \frac{\partial^2 F}{\partial y \partial z} \\ \frac{\partial^2 F}{\partial x \partial z} & \frac{\partial^2 F}{\partial y \partial z} & \frac{\partial^2 F}{\partial z^2} \end{pmatrix} = 0.$$

This is a polynomial of degree $3(d - 2)$. We conclude therefore that

$$\#\{\text{inflectional tangents of } C\} = \#\{\text{critical points of } \phi : C \to \check{\mathbb{P}}^2\}$$
$$= \#\{\text{solutions of } F = H(F) = 0\}$$
$$= 3d(d-2),$$

where the last equality is the Bezout Theorem. We obtain the following table:

deg(F)	#{inflectional tangents}
2	0
3	9
4	24
5	45

The Bezout Theorem counts intersection points with multiplicity. The points where C intersects $\{H(F) = 0\}$ with multiplicity > 1 include inflectional tangent ℓ meeting C with multiplicity ≥ 4.

For more information on multiplicities and plane curves, see [11].

12.5 Interpolation problems revisited

We apply some of these ideas to the interpolation problem first considered in Chapter 1. Before stating the General Interpolation Problem, we need a definition:

HILBERT POLYNOMIALS AND BEZOUT

Definition 12.35 Let $f \in k[y_1, \ldots, y_n]$ and $p = (b_1, \ldots, b_n) \in \mathbb{A}^m(k)$. The polynomial f has *multiplicity* $\geq m$ *at* p (or *vanishes to order* m *at* p) if

$$f \in \mathfrak{m}_p^m = \langle y_1 - b_1, \ldots, y_n - b_n \rangle^m.$$

When k has characteristic zero, this means the the Taylor series for f at p only has terms of degree $\geq m$.

Problem 12.36 (General Interpolation Problem) Fix interpolation data

$$\mathcal{S} = (p_1, m_1; \ldots; p_N, m_N)$$

consisting of distinct points $p_1, \ldots, p_N \in \mathbb{A}^n(k)$ and positive integers $m_1, \ldots, m_N$. What is the dimension of the vector space of polynomials of degree $\leq d$ vanishing at each p_i to order m_i?

Let $P_{n,d}$ denote the polynomials of degree $\leq d$ and

$$I_d(\mathcal{S}) = P_{n,d} \cap \left(\cap_{j=1}^N \mathfrak{m}_{p_j}^{m_j} \right)$$

denote the polynomials satisfying the interpolation data.

Definition 12.37 The number of conditions imposed by

$$\mathcal{S} = (p_1, m_1; \ldots; p_N, m_N)$$

on polynomials of degree $\leq d$ is defined as

$$C_d(\mathcal{S}) := \dim P_{n,d} - \dim I_d(\mathcal{S}).$$

$\mathcal{S}$ is said to *impose independent conditions on* $P_{n,d}$ if

$$C_d(\mathcal{S}) = \sum_{j=1}^N \binom{n + m_j - 1}{n}.$$

It *fails to impose independent conditions* otherwise.

The expected number of conditions is given by naively counting the Taylor coefficients we set equal to zero!

Fix N and $m_1, \ldots, m_N$. It may not be possible to choose $p_1, \ldots, p_N$ such that the data $\mathcal{S}$ impose independent conditions on $P_{n,d}$. There are counterexamples even when the expected number of conditions is less than the dimension:

$$\sum_{j=1}^N \binom{n + m_j - 1}{n} \leq \binom{n + d}{n}.$$

Example 12.38

1. Let $p_1 = (0,0)$, $p_2 = (1,0) \in \mathbb{A}^2(k)$ and consider the conditions imposed by $(p_1, 2; p_2, 2)$ on quadrics. Since x_2^2 vanishes to order 2 at p_1 and p_2, we have

$$C_2(\mathcal{S}) \leq 6 - 1 = 5.$$

On the other hand,

$$2\binom{2+2-1}{2} = 6.$$

2. Let $p_1, \ldots, p_5 \in \mathbb{A}^2(k)$ and consider the conditions imposed by

$$(p_1, 2; p_2, 2; p_3, 2; p_4, 2; p_5, 2)$$

on quartics ($d = 4$). There is a unique (up to scalar) nonzero $f \in P_{2,2}$ such that f vanishes at each of the p_j. Hence $f^2 \in P_{2,4}$ vanishes to order 2 at each of the p_j and $C_4(\mathcal{S}) \leq 15 - 1 = 14$. However, the expected number of conditions is

$$5\binom{2+2-1}{2} = 15.$$

We recast these problems using Hilbert functions and polynomials. Realize $\mathbb{A}^n(k)$ as the distinguished affine open $U_0 \subset \mathbb{P}^n(k)$ and consider $p_1, \ldots, p_N \in \mathbb{P}^n(k)$. We write $S = k[x_0, \ldots, x_n]$. Dehomogenization with respect to x_0

$$\mu_0 : k[x_0, \ldots, x_n] \to k[y_1, \ldots, y_m]$$

identifies $J(p_i)$ with $\mathfrak{m}_{p_i}$ and S_d with $P_{n,d}$. Homogeneous forms of degree d with multiplicity m_j at each p_j (cf. Exercise 10.8) are identified with $I_d(\mathcal{S})$. Writing

$$J(\mathcal{S}) = \cap_{j=1}^N J(p_i)^{m_i}, \quad R(\mathcal{S}) = S/J(\mathcal{S}),$$

we have an isomorphism

$$\mu_0 : R(\mathcal{S})_d \xrightarrow{\sim} P_{n,d}/I_d(\mathcal{S})$$

and thus the equality

$$\mathrm{HF}_{R(\mathcal{S})}(d) = C_d(\mathcal{S}).$$

Proposition 12.39 *Let $\mathcal{S} = (p_1, m_1; \ldots; p_N, m_n)$ be a collection of distinct points in $\mathbb{P}^n(k)$ and positive integers $m_1, \ldots, m_N$. Write*

$$J(\mathcal{S}) = \cap_{j=1}^N J(p_i)^{m_i}, \quad R(\mathcal{S}) = S/J(\mathcal{S}).$$

Then we have

$$\mathrm{HP}_{R(\mathcal{S})}(t) = \sum_{j=1}^{N} \binom{n+m_j-1}{n}.$$

Proof   Let $p = [0, \ldots, 0, 1] \in \mathbb{P}^n(k)$ so that $J(p) = \langle x_0, \ldots, x_{n-1} \rangle$ and the monomials

$$\{x_0^{\alpha_0} \ldots x_{n-1}^{\alpha_{n-1}} x_n^{\alpha_n} : \alpha_0 + \cdots + \alpha_{n-1} < m\}$$

form a basis for $S/J(p)^m$. It follows that

$$\mathrm{HF}_{S/J(p)^m}(t) = \binom{n+m-1}{n}$$

for $t \geq m$. By Proposition 12.5, this analysis applies to each $p \in \mathbb{P}^n(k)$; Proposition 12.6 the gives the result. $\square$

Corollary 12.40   *Consider interpolation data $\mathcal{S} = (p_1, m_1; \ldots; p_N, m_N)$ in $\mathbb{P}^n(k)$ and the corresponding graded ring $R(\mathcal{S})$. $\mathcal{S}$ imposes independent conditions on polynomials of degree $\leq d$ if and only if $\mathrm{HP}_{R(\mathcal{S})}(d) = \mathrm{HF}_{R(\mathcal{S})}(d)$. Thus $\mathcal{S}$ imposes independent conditions on polynomials of degree $\leq t$ provided $t \gg 0$.*

The last assertion is the defining property of the Hilbert polynomial (Theorem 12.14). Thus our General Interpolation Problem reduces to the following.

Problem 12.41 (Projective Interpolation Problem)   Fix interpolation data

$$\mathcal{S} = (p_1, m_1; \ldots; p_N, m_N)$$

consisting of distinct points $p_1, \ldots, p_N \in \mathbb{P}^n(k)$ and positive integers $m_1, \ldots, m_N$. If we write

$$J(\mathcal{S}) = \cap_{j=1}^{N} J(p_i)^{m_i}, \quad R(\mathcal{S}) = k[x_0, \ldots, x_n]/J(\mathcal{S}),$$

for which t is

$$\mathrm{HF}_{R(\mathcal{S})}(t) = \mathrm{HP}_{R(\mathcal{S})}(t)?$$

Lemma 12.29 implies we can always choose coordinates such that $p_1, \ldots, p_N \in U_0 \simeq \mathbb{A}^n(k)$, so the Projective Interpolation Problem reduces to the affine case.

Even for generic points in the plane, the General Interpolation Problem is still open! There are precise conjectures of Hirschowitz [21] and Harbourne [15, 16] predicting which interpolation data fail to impose independent conditions on polynomials of degree $\leq d$. A good survey can be found in [31].

12.6 Classification of projective varieties

The importance of the Hilbert polynomial is that most important invariants of projective varieties can be defined in terms of it.

Definition 12.42   Let X be a projective variety.

1. The *dimension* $\dim(X)$ is defined as the degree of HP_X.
2. The *degree* of $\deg(X)$ defined as the normalized leading term of HP_X:

$$\mathrm{HP}_X(t) = \frac{\deg(X)}{\dim(X)!} t^{\dim(X)} + \text{lower-order terms.}$$

In our discussion of Hilbert polynomials, we showed that each can be expressed in the form

$$\mathrm{HP}_X(t) = \sum_{i=0}^{\dim(X)} a_i \binom{t}{\dim(X) - i},$$

with the a_i integers. The coefficient $a_0 = \deg(X)$.

These definitions can be related to existing notions of 'dimension' and 'degree', when these can be easily formulated. A finite set X has dimension zero and degree $|X|$. Projective space $\mathbb{P}^n(k)$ has dimension n and degree 1. As for hypersurfaces, the following holds.

Proposition 12.43   Let $X \subset \mathbb{P}^n(k)$ be a hypersurface with $J(X) = \langle F \rangle$, with F a polynomial of degree d. Then $\dim(X) = n - 1$ and $\deg(X) = d$.

Proof   We have already computed the Hilbert polynomial

$$\begin{aligned}
\mathrm{HP}_X(t) &= \binom{t+n}{n} - \binom{t-d+n}{n} \\
&= \frac{(t+n)(t+n-1)\ldots(t+1)}{n!} - \frac{(t-d+n)\ldots(t+1-d)}{n!} \\
&= t^{n-1} \frac{n + (n-1) + \cdots + 1 - (n-d) - (n-1-d) - \cdots - (1-d)}{n!} \\
&\quad + \text{lower-order terms} \\
&= t^{n-1} \frac{nd}{n!} + \text{lower-order terms.}
\end{aligned}$$

In particular, the Hilbert polynomial has degree $n - 1$ and $a_0 = d$. □

It takes some work to prove in general that the Hilbert-polynomial definition of dimension agrees with the transcendence-base definition given in Chapter 7. See [9, ch. 8] for a discussion of the various notions of dimension.

Not only the leading term of the Hilbert polynomial gets a special name. The constant term is also significant:

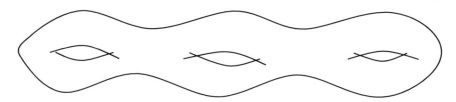

Figure 12.2 A curve of genus 3.

Definition 12.44 The *arithmetic genus* of an irreducible projective variety X is defined as

$$p_a(X) = (-1)^{\dim(X)}(\mathrm{HP}_X(0) - 1).$$

Example 12.45

1. The curve $X \subset \mathbb{P}^3(k)$ with equations

$$\{x_0 x_2 - x_1^2,\ x_0 x_3 - x_1 x_2,\ x_2^2 - x_1 x_3\}$$

   has $\deg(X) = 3$, $p_a(X) = 0$.
2. Let $X \subset \mathbb{P}^2(k)$ be a plane curve of degree d. Then we have

$$\mathrm{HP}_X(t) = \binom{t+2}{2} - \binom{t-d+2}{2}$$
$$= dt + \left[1 - \frac{(d-1)(d-2)}{2}\right]$$

so that $p_a(X) = (d-1)(d-2)/2$.

The geometric meaning of the arithmetic genus is a bit elusive in general. However, for curves it has a very nice geometric interpretation.

Theorem 12.46 *Let $X \subset \mathbb{P}^n(\mathbb{C})$ be a smooth irreducible complex projective curve. Then $X(\mathbb{C})$ is an oriented compact Riemann surface of genus $p_a(X)$.*

The proof, which uses the Riemann–Roch Theorem and significant analysis, would take us too far afield. We refer the interested reader to a book on complex Riemann surfaces, e.g., [30, p. 192].

Example 12.47 Consider the plane curve $X \subset \mathbb{P}^2(\mathbb{C})$ given by the equations $x_0^4 + x_1^4 = x_2^4$. This is smooth (and thus irreducible by the Bezout Theorem!) with genus

$$\frac{3 \cdot 2}{2} = 3.$$

The complex points $X(\mathbb{C})$ are displayed in Figure 12.2.

We sketch briefly how projective varieties are classified using Hilbert polynomials. Fix a polynomial $f(t) \in \mathbb{Q}[t]$ with $f(\mathbb{Z}) \subset \mathbb{Z}$, and consider all varieties $X \subset \mathbb{P}^n$ with $\mathrm{HP}_X(t) = f(t)$. Choose $M \gg 0$ such that $\mathrm{HP}_X(M) = f(M)$; while it appears that M depends on X, it is possible to choose a uniform value for all the varieties with Hilbert polynomial f. Let

$$\mathrm{Gr} := \mathrm{Gr}(\binom{n+M}{M} - f(M), k[x_0,\ldots,x_n]_M)$$

denote the Grassmannian of codimension-$f(M)$ subspaces of the space of polynomials of degree M. The homogeneous polynomials of degree M vanishing on X define a point

$$[X]_M := J(X)_M \in \mathrm{Gr}.$$

For $M \gg 0$, the set of all projective varieties with Hilbert polynomial f are parametrized by a locus

$$\mathcal{H}ilb_{f(t)} \subset \mathrm{Gr}$$

known as the *Hilbert scheme*. For details on the construction and discussion of $\mathcal{H}ilb_{f(t)}$ as a projective variety see [32, ch. 14].

Example 12.48 Consider all plane curves $X \subset \mathbb{P}^2(k)$ of degree d. These have Hilbert polynomial

$$f(t) = \binom{t+2}{2} - \binom{t-d+2}{2}$$

so that $f(d) = \binom{2+d}{d} - 1$. If X is defined by $F \in k[x_0, x_1, x_2]_d$ then

$$[X]_d = [F] \in \mathrm{Gr}(1, k[x_0, x_1, x_2]_d) = \mathbb{P}^{\binom{d+2}{2}-1}.$$

A similar analysis applies to arbitrary hypersurfaces of degree d in $\mathbb{P}^n$.

12.7 Exercises

12.1 Consider the ideal

$$J = \langle x_0 x_1, x_2^2 \rangle \subset k[x_0, x_1, x_2].$$

Compute the Hilbert polynomial $\mathrm{HP}_J(t)$.

12.2 Let $f = x^3 + y^3$ and $g = x^4 + y^5$ and write

$$I = \langle f, g \rangle, \quad p = (0,0).$$

Compute the multiplicity $\mathrm{mult}(I, p)$.

12.3 Consider the interpolation data $S = ((0,0), 2; (0,1), 2; (1,0), 2)$ for $\mathbb{A}^2(\mathbb{R})$. Show that this imposes independent conditions on polynomials in $P_{2,3}$.

12.4 (a) Let $p_1, \ldots, p_7$ be any distinct points in $\mathbb{A}^2(\mathbb{R})$. Show that there exists a nonzero cubic polynomial vanishing at p_1 to order 2 and at $p_2, \ldots, p_7$ to order 1.

(b) Compute the expected number of conditions $C_6(S)$ imposed on $P_{2,6}$ by the interpolation data

$$S = (p_1, 4; p_2, 2; p_3, 2; p_4, 2; p_5, 2; p_6, 2; p_7, 2)$$

for generic points $p_1, \ldots, p_7 \in \mathbb{A}^2(\mathbb{R})$.

(c) Show that S fails to impose independent conditions on $P_{2,6}$.

12.5 Compute Hilbert polynomials for the following varieties:

(a) the quadratic Veronese varieties

$$\nu(2)(\mathbb{P}^n(k)) \subset \mathbb{P}^{\binom{n+2}{2}-1}(k);$$

(b) the Segre variety

$$X = \mathbb{P}^m(k) \times \mathbb{P}^m(k) \subset \mathbb{P}^{nm+n+m}(k);$$

(c) the Grassmannian

$$\mathrm{Gr}(2,4) \subset \mathbb{P}^5(k).$$

Compute the degree and dimension from the Hilbert polynomial.

12.6 Let $X = \{p_1, p_2, p_3, p_4\} \subset \mathbb{P}^2(\mathbb{Q})$ be a collection of four distinct points. List all possible Hilbert functions $\mathrm{HF}_X(t)$.

12.7 Let $S = k[x_0, x_1]$ be a weighted polynomial ring, where x_0 has weight w_0 and x_1 has weight w_1.

(a) Assume that $w_0 = 2$ and $w_1 = 3$. For each $t \geq 0$, express $t = 6q + r$ where $0 \leq r \leq 5$. Show that

$$\mathrm{HF}_S(t) = \begin{cases} q+1 & \text{if } r \neq 1 \\ q & \text{if } r = 1. \end{cases}$$

Explain why this does not contradict Theorem 12.14.

(b) Let M denote the least common multiple of w_0 and w_1 and pick an integer R with $0 \leq R \leq M - 1$. Show that $\mathrm{HF}_S(Mq + R)$ is a degree-1 polynomial function in q for $q \gg 0$.

12.8 Show that any smooth plane curve is irreducible. *Hint:* The irreducible components of a plane curve are themselves plane curves! Check that the following complex plane curves $C \subset \mathbb{P}^2(\mathbb{C})$ are irreducible:

$$x_0^5 + x_1^5 + x_2^5 = 0$$
$$x_0^3 + x_1^3 + x_2^3 = \lambda x_0 x_1 x_2, \quad \lambda^3 \neq 27.$$

12.9 Find all points of intersection of the following pairs of plane curves $\{F = 0\}$, $\{G = 0\} \subset \mathbb{P}^2(\mathbb{C})$, and the multiplicities of the intersections.
   (a) $F = x_0^2 + x_1^2 + x_2^2$ and $G = x_0 x_2 - x_1^2$;
   (b) $F = x_0 x_2^2 - x_1^3$ and $G = x_0 x_2^2 - x_0 x_1^2 - x_1^3$;

12.10 Compute the inflectional tangents of the curve $C \subset \mathbb{P}^2(\mathbb{C})$ given by
$$x_0^3 + x_1^3 + x_2^3 = 2 x_0 x_1 x_2.$$

12.11 Consider the complex projective curves
$$C = \{(x_0, x_1, x_2) : x_0^2 + x_1^2 = x_2^2\} \subset \mathbb{P}^2(\mathbb{C})$$
$$D = \{(x_0, x_1, x_2) : x_0^2 - x_1^2 = x_2^2\} \subset \mathbb{P}^2(\mathbb{C}).$$

Describe the intersection $C \cap D$ and compute the multiplicity of each point. Make sure you prove that you have found every point of the intersection!

12.12 Let V be an affine variety with $p \in V$ and $\mathfrak{m}_p \subset k[V]$ the corresponding maximal ideal. Consider the graded ring
$$R = \oplus_{t \geq 0} R_t = k[V]/\mathfrak{m}_p \oplus \mathfrak{m}_p/\mathfrak{m}_p^2 \oplus \mathfrak{m}_p^2/\mathfrak{m}_p^3 \cdots \oplus \mathfrak{m}_p^t/\mathfrak{m}_p^{t+1} \oplus \cdots,$$

i.e., $R_0 = k$ and $R_t = \mathfrak{m}_p^t/\mathfrak{m}_p^{t+1}$ for $t > 0$. This is called the *graded ring associated* to $k[V]$ and $\mathfrak{m}_p$.

(a) Let $x_0, \ldots, x_n$ be generators of $R_1 = \mathfrak{m}_p/\mathfrak{m}_p^2$ as a vector space over k. Show that $x_0, \ldots, x_n$ generate R as a k-algebra. *Hint:* Check that the monomials x^α with $|\alpha| = t$ span R_t.

(b) Show that the kernel $J = \ker(k[x_0, \ldots, x_n] \twoheadrightarrow R)$ is homogeneous. The corresponding affine variety
$$V(J) \subset \mathbb{A}^{n+1}(k)$$
is the *tangent cone* to V at p and the projective variety
$$X(J) \subset \mathbb{P}^n(k)$$
is called the *projective tangent cone*.

(c) Suppose that $0 = p \in V \subset \mathbb{A}^m(k)$ and $I(V) = \langle f \rangle$ for some $f \in k[y_1, \ldots, y_m]$ with graded pieces
$$f = F_M + F_{M+1} + \cdots + F_d, \quad F_M \neq 0.$$
Show that the associated graded ring is $k[y_1, \ldots, y_m]/\langle F_M \rangle$.

(d) Draw the graph of $V = \{(y_1, y_2) : y_2^2 = y_1^2 + y_1^3\} \subset \mathbb{A}^2(\mathbb{R})$ and its tangent cone. Do the same for $V' = \{y_2^2 = y_1^3 + y_1^4\} \subset \mathbb{A}^2(\mathbb{R})$.

12.13 Find an explicit formula for the arithmetic genus of a degree d hypersurface $X \subset \mathbb{P}^3(\mathbb{C})$.

12.14 Extract equations for $\mathcal{H}ilb_{f(t)}$ for the following classes of varieties:

(a) Lines $\ell \in \mathbb{P}^3$ with $f(t) = t + 1$. *Hint:* Use $\mathrm{Gr}(2, k[x_0, x_1, x_2, x_3]_1)$.

(b) Pairs of points $X = \{p_1, p_2\} \subset \mathbb{P}^2$ with $f(t) = 2$. *Hint:* Use the Grassmannian $\mathrm{Gr}(4, k[x_0, x_1, x_2]_2)$.

Is $\mathcal{H}ilb_{f(t)}$ closed in the Grassmannian? If not, describe the points in the closure.

12.15 *Challenge:* Each bihomogeneous $F \in \mathbb{C}[x_0, x_1, y_0, y_1]$ defines a curve in $\mathbb{P}^1(\mathbb{C}) \times \mathbb{P}^1(\mathbb{C})$. State and prove a Bezout Theorem for a pair of bihomogeneous forms F, G without common factors.

Appendix A Notions from abstract algebra

This appendix is a brief resumé of the abstract algebra used in this book. Sketch proofs are sometimes included to highlight the key ideas. We encourage the reader to consult a general text in abstract algebra for detailed arguments and discussion. Michael Artin's book [1] contains all that we require (and much more besides).

A.1 Rings and homomorphisms

A *commutative ring* R is a set with two operations, addition

$$R \times R \to R$$
$$(a, b) \mapsto a + b$$

and multiplication

$$R \times R \to R$$
$$(a, b) \mapsto ab$$

satisfying the following conditions:

1. R is an abelian group under addition:
   - addition is associative, i.e., $(a + b) + c = a + (b + c)$ for each $a, b, c \in R$;
   - addition is commutative, i.e., $a + b = b + a$ for each $a, b \in R$;
   - there exists $0 \in R$ such that $0 + a = a$ for each $a \in R$;
   - there exist additive inverses, i.e., for each $a \in R$ there exists an element $b \in R$ such that $a + b = 0$.

2. Multiplication on R satisfies the following:
   - $(ab)c = a(bc)$ for each $a, b, c \in R$;
   - $ab = ba$ for each $a, b \in R$;
   - there exists $1 \in R$ such that $1a = a$ for each $a \in R$.

3. The addition and multiplication operations are compatible, i.e., for all $a, b, c \in R$ we have $(a + b)c = ac + bc$.

235

We will not consider rings with non-commutative multiplication, so we frequently shorten 'commutative ring' to 'ring'.

Example A.1 The integers form a ring $\mathbb{Z}$ under the standard operations of addition and multiplication.

If $N > 1$ is an integer then congruence classes a (mod N) form a ring $\mathbb{Z}/N\mathbb{Z}$ under the operations of addition and multiplication modulo N.

A *homomorphism of rings* is a function $\phi : R \to S$ satisfying the following conditions

- $\phi(a+b) = \phi(a) + \phi(b)$ and $\phi(ab) = \phi(a)\phi(b)$ for all $a, b \in R$;
- $\phi(1) = 1$.

A *domain* is a ring R with no nontrivial zero-divisors, i.e., if $ab = 0$ then $a = 0$ or $b = 0$. A *field* k is a domain such that the nonzero elements $k^* := k \setminus \{0\}$ form a group under multiplication, i.e., for each $a \neq 0 \in k$ there exists $b \in k$ with $ab = 1$. Every domain R has a *field of fractions*

$$K := \left\{ \frac{r}{s} : r, s \in R, s \neq 0 \right\},$$

where we identify r_1/s_1 and r_2/s_2 when $r_1 s_2 = r_2 s_1$. The operations are addition and multiplication of fractions

$$\frac{r_1}{s_1} + \frac{r_2}{s_2} = \frac{(r_1 s_2 + r_2 s_1)}{s_1 s_2}, \quad \left(\frac{r_1}{s_1}\right)\left(\frac{r_2}{s_2}\right) = \frac{r_1 r_2}{s_1 s_2}.$$

We realize $R \subset K$ as the fractions $r/1$.

Example A.2 The real numbers $\mathbb{R}$ and complex numbers $\mathbb{C}$ are fields under the standard operations of addition and multiplication. The rational numbers

$$\mathbb{Q} = \left\{ \frac{r}{s} : r, s \in \mathbb{Z}, s \neq 0 \right\}$$

are the field of fractions of the integers $\mathbb{Z}$.

Let k be a field. A *k-algebra* is a ring R along with a ring homomorphism $k \to R$. By Exercise A.4, this homomorphism is injective provided $R \neq 0$, so we may regard k as a subset of R. A *homomorphism of k-algebras* $\phi : R \to S$ is a ring homomorphism such that $\phi(ca) = c\phi(a)$ for each $a \in R$ and $c \in k$.

A.2 Constructing new rings from old

Let R be a ring. An *ideal* is a nonempty subset $I \subset R$ satisfying the following properties:

- for any $f_1, f_2 \in I$ we have $f_1 + f_2 \in I$;
- for any $f \in I$ and $r \in R$ we have $fr \in I$.

Given a finite set

$$\{f_1, f_2, \ldots, f_r\} \subset R,$$

the ideal generated by this set is denoted $\langle f_1, f_2, \ldots, f_r \rangle$ and consists of all the sums $f_1 h_1 + f_2 h_2 + \cdots + f_r h_r$ where the $h_j \in R$. (Showing this defines an ideal is an exercise.) More generally, for any subset $\{f_i\}_{i \in I} \subset R$, we can consider the ideal $\langle f_i \rangle_{i \in I} \subset R$ consisting of all finite sums $f_{i_1} h_{i_1} + \cdots + f_{i_r} h_{i_r}$ with $i_j \in I$ and $h_{i_j} \in R$.

Let R be a ring and I an ideal of R. Two elements $a, b \in R$ are *congruent modulo I*,

$$a \equiv b \pmod{I},$$

if $a - b \in I$. As the notation suggests, this is an equivalence relation on R. Addition and multiplication are compatible with congruence: if $a_1 \equiv a_2 \pmod{I}$ and $b_1 \equiv b_2 \pmod{I}$ then $a_1 + b_1 \equiv a_2 + b_2 \pmod{I}$ and $a_1 b_1 \equiv a_2 b_2 \pmod{I}$. In particular, addition and multiplication are well-defined on congruence classes. The *quotient ring* R/I is the set of congruence classes modulo I under these operations. We have natural *quotient homomorphism*

$$R \twoheadrightarrow R/I$$
$$a \mapsto a \pmod{I}.$$

Example A.3 Let $R = \mathbb{Z}$, N a positive integer, and $I = \langle N \rangle$. Then $\mathbb{Z}/\langle N \rangle = \mathbb{Z}/N\mathbb{Z}$ is the corresponding quotient ring.

Given a ring R, the *polynomials with coefficients in R* is the set of finite formal sums

$$R[x] = \{p_0 + p_1 x + \cdots + p_d x^d : p_0, p_1, \ldots, p_d \in R\};$$

the largest power x^i appearing with a nonzero coefficient is the *degree* of the polynomial. This is a ring under the operations of addition

$$(p_0 + p_1 x + \cdots + p_d x^d) + (q_0 + q_1 x + \cdots + q_e x^e)$$
$$= (p_0 + q_0) + (p_1 + q_1)x + (p_2 + q_2)x^2 + \cdots$$

and multiplication

$$(p_0 + p_1 x + \cdots + p_d x^d)(q_0 + q_1 x + \cdots + q_e x^e) =$$
$$p_0 q_0 + (p_1 q_0 + p_0 q_1)x + \cdots + (p_i q_0 + p_{i-1} q_1 + \cdots + p_0 q_i)x^i + \cdots + p_d q_e x^{d+e}.$$

We use the shorthand

$$R[x_1, \ldots, x_n] = R[x_1][x_2]\ldots[x_n];$$

this does not depend on the order of the variables.

We can combine these constructions, e.g.,
$$R = \mathbb{Z}[x]/\langle 3, x^2 + 1 \rangle.$$

A.3 Modules

When we develop linear algebra over general rings, modules play the rôle of vector spaces: an *R-module M* is an abelian group M (written additively) equipped with an action
$$R \times M \to M$$
$$(r, m) \mapsto rm$$

satisfying the following properties:

- $r(m_1 + m_2) = rm_1 + rm_2$ for each $r \in R$ and $m_1, m_2 \in M$;
- $(r_1 + r_2)m = r_1 m + r_2 m$ for each $r_1, r_2 \in R$ and $m \in M$;
- $(r_1 r_2)m = r_1(r_2 m)$ for each $r_1, r_2 \in R$ and $m \in M$.

Given R-modules M and N, a *homomorphism of R-modules* or *R-linear homomophism* is a function $\phi : M \to N$ satisfying the following:

- $\phi(m_1 + m_2) = \phi(m_1) + \phi(m_2)$, for all $m_1, m_2 \in M$;
- $\phi(rm) = r\phi(m)$, for each $r \in R$ and $m \in M$.

Example A.4

(a) Every additive abelian group M is a $\mathbb{Z}$-module under the action
$$(r, m) \mapsto rm, \quad r \in \mathbb{Z}, m \in M.$$

(b) R is itself an R-module with action
$$(r, s) \mapsto rs, \quad r, s \in R.$$

(c) if M_1 and M_2 are R-modules then so is the *direct sum*
$$M_1 \oplus M_2 = \{(m_1, m_2) : m_1 \in M_1, m_2 \in M_2\},$$
where addition is taken componentwise and R acts by 'scalar multiplication'
$$r(m_1, m_2) = (rm_1, rm_2).$$

(d) For each $n > 0$ we have the R-module
$$R^n = \{(c_1, \ldots, c_n) : c_i \in R\} = \underbrace{R \oplus \ldots \oplus R}_{n \text{ times}}.$$

We will use the standard notation
$$e_1 = (1, 0, \ldots, 0), \ldots, e_n = (0, \ldots, 0, 1).$$

Given an R-module M, an R-submodule $N \subset M$ is a subgroup such that, for each $r \in R$ and $n \in N$, we have $rn \in N$. Two elements $m_1, m_2 \in M$ are *congruent modulo N*,

$$m_1 \equiv m_2 \pmod{N},$$

if $m_1 - m_2 \in N$. The R-action respects congruence classes: if $m_1 \equiv m_2 \pmod{N}$ then $rm_1 \equiv rm_2 \pmod{N}$ for each $r \in R$. The resulting set of equivalence classes is denoted M/N.

Proposition A.5 *If M is an R-module and $N \subset M$ an R-submodule, then M/N naturally inherits an R-module structure, known as the quotient module structure.*

Example A.6 A subset $I \subset R$ is a submodule if and only if it is an ideal. The resulting quotient ring R/I is also naturally an R-module.

An R-module M is *finitely generated* if there exists a finite set of elements $m_1, \ldots, m_n \in M$ such that every $m \in M$ can be expressed

$$m = r_1 m_1 + \cdots + r_n m_n, \quad r_1, \ldots, r_n \in R.$$

This is equivalent to the existence of a surjective R-linear map $\phi : R^n \to M$.

A.4 Prime and maximal ideals

Let R be a ring. An ideal $\mathfrak{m} \subsetneq R$ is *maximal* if there exists no ideal I with $\mathfrak{m} \subsetneq I \subsetneq R$. An ideal $P \subset R$ is *prime* if, for any $a, b \in R$ with $ab \in P$, either $a \in P$ or $b \in P$.

These notions are useful in constructing rings with prescribed properties.

Proposition A.7 *Consider an ideal $I \subset R$. I is maximal if and only if R/I is a field. I is prime if and only if R/I is a domain. In particular, every maximal ideal is prime.*

Proof If I is maximal then, for any $x \in R \setminus I$, we have $I + \langle x \rangle = R$. Thus there exist $y \in R$ and $r \in I$ with $xy + r = 1$, and $xy \equiv 1 \pmod{I}$. Conversely, if R/I is a field then for any $x \in R \setminus I$ we have $xy \equiv 1 \pmod{I}$ for some $y \in R$, and any ideal $J \supsetneq I$ contains 1.

The remaining assertions are left as an exercise. □

Example A.8 If p is a prime number then the ideal $\langle p \rangle \subset \mathbb{Z}$ is maximal. Indeed, given $a \in \mathbb{Z}$ not divisible by p, we show that $\langle p, a \rangle = \mathbb{Z}$. Consider the powers

$$\{a, a^2, a^3, \ldots, a^p\} \subset \mathbb{Z}/p\mathbb{Z}.$$

None of the a^i is a zero divisor in $\mathbb{Z}/p\mathbb{Z}$: if $a^i b \equiv 0 \pmod{p}$ then $p | a^i b$, but since p does not divide a we must have $p|b$, i.e., $b \equiv 0 \pmod{p}$. Now $\mathbb{Z}/p\mathbb{Z}$ has $p - 1$

nonzero elements, so we find that $a^i \equiv a^j$ (mod p) for some $1 \le i < j \le p$. Since $a^i(1 - a^{j-i}) \equiv 0$ (mod p), we must have $1 - a^{j-i} \equiv 0$ (mod p), i.e., $aa^{j-i-1} \equiv 1$ (mod p) and $aa^{j-i-1} = 1 + np$ for some integer n. This implies that $1 \in \langle a, p \rangle$ and thus $\langle p, a \rangle = \mathbb{Z}$.

Proposition A.7 guarantees $\mathbb{Z}/p\mathbb{Z}$ is a field, called the *finite field with p elements*.

A.5 Factorization of polynomials

A domain R is a *principal ideal domain* (PID) if every ideal I is principal, i.e., there exists an $f \in R$ such that $I = \langle f \rangle$.

Theorem A.9 *Let k denote a field. Then $k[x]$ is a principal ideal domain.*

Proof The key ingredient is following systematic formulation of polynomial long division, which can be found in any abstract algebra text:

Algorithm A.10 (Euclidean Algorithm) *Let k denote a field, $f \ne 0$ a polynomial in $k[x]$, and $g \in k[x]$ a second polynomial. Then there exist unique $q, r \in k[x]$ such that $g = qf + r$ and $\deg(r) < \deg(f)$ (where $\deg(0) = -\infty$). We say q is the quotient of g by f, and r is the remainder.*

Let $I \subset k[x]$ be an ideal. Let $0 \ne f \in I$ have minimal degree among nonzero elements of I. Given another element $g \in I$ we apply the division algorithm to find q and r such that $g = fq + r$ and $\deg(r) < \deg(f)$. Note that $r = g - fq \in I$, so by our assumption on f we have $r = 0$, and g is a multiple of I. We conclude that $I = \langle f \rangle$. $\square$

The *units* of a ring R are the elements with multiplicative inverses

$$R^* = \{u \in R : \text{ there exists } v \in R \text{ with } uv = 1\};$$

this forms a group under multiplication. Suppose $a \in R$ is neither a zero-divisor nor a unit; a is *irreducible* if, for any factorization $a = bc$ with $b, c \in R$, either $b \in R^*$ or $c \in R^*$. A domain R is a *unique factorization domain* (UFD) if the following hold.

- Each $a \in R$ can be written as a product of irreducibles

$$a = p_1 \ldots p_r, \quad p_1, \ldots, p_r \in R.$$

- This factorization is unique in the following sense. Suppose we have another factorization $a = q_1 \ldots q_s$. Then $s = r$ and after permuting $p_1, \ldots, p_r$ we can find units $u_1, \ldots, u_r \in R^*$ such that $p_i = u_i q_i, i = 1, \ldots, r$.

Proposition A.11 *If f is an irreducible element of a unique factorization domain R then $\langle f \rangle$ is prime. If f is irreducible in a principal ideal domain then $\langle f \rangle$ is maximal.*

A.5 FACTORIZATION OF POLYNOMIALS

Proof Suppose R is a UFD. Given $ab \in \langle f \rangle$ then $ab = hf$ for some $h \in R$, and f must be an irreducible factor of either a or b. Now suppose R is a PID. Given an ideal $I \subset R$ with $\langle f \rangle \subsetneq I$, we can write $I = \langle g \rangle$ for some $g \in R$. We have $f = gh$ for some $h \in R$; $h \notin R^*$ because $\langle f \rangle \neq \langle g \rangle$. Since f is irreducible, $g \in R^*$ and $I = R$. $\square$

Proposition A.12 (Gauss' Lemma) *Let R be a unique factorization domain with fraction field L. Let $g, h \in R[x]$ and assume that the coefficients of g have no common irreducible factor. If $g|h$ in $L[x]$ then $g|h$ in $R[x]$.*

Proof Given a polynomial $f = f_d x^d + \cdots + f_0 \in R[x]$, we define

$$\text{content}(f) = \gcd(f_0, \ldots, f_d),$$

the greatest common divisor of the coefficents, which is well-defined up to multiplication by a unit. A polynomial has *content 1* if its coefficients have no common irreducible factor.

We shall establish the formula

$$\text{content}(fg) = \text{content}(f) \cdot \text{content}(g)$$

for $f, g \in R[x]$. Dividing through f and g by their contents, it suffices to prove this when f and g have content 1. Suppose $p \in R$ is irreducible dividing all the coefficients of fg. If f and g have degrees d and e respectively, we obtain

$$p \mid f_d g_e$$
$$p \mid f_d g_{e-1} + f_{d-1} g_e$$
$$p \mid f_d g_{e-2} + f_{d-1} g_{e-1} + f_{d-2} g_e$$
$$\vdots$$

Suppose p does not divide f_d. The first expression shows it divides g_e. The second expression shows it divides $f_d g_{e-1}$, so it divides g_{e-1}. The third expression shows it divides $f_d g_{e-2}$, so it divides g_{e-2}. Continuing, we conclude p divides each g_i, hence $p|\text{content}(g)$.

We can divide the coefficients of h by their common factors to obtain a polynomial of content 1. Without loss of generality, we may assume that h has content 1. Suppose we have $h = \hat{f} g$ with $\hat{f} \in L[x]$. Clear denominators, i.e., choose $r \in R$ such that $f := r\hat{f} \in R[x]$ with content 1. We have $rh = fg$ so our claim implies rh has content 1. This is only possible if $r \in R^*$, in which case $\hat{f} = r^{-1} f \in R[x]$. $\square$

Corollary A.13 *Let R be a UFD with fraction field L and $f \in R[x]$ irreducible and nonconstant. Then f is irreducible in $L[x]$.*

The following fundamental result about unique factorization can be found in most abstract algebra textbooks:

Theorem A.14 *If R is a UFD then $R[x]$ is a UFD. In particular, if k is a field then $k[x_1, \ldots, x_n]$ has unique factorization.*

Sketch Proof Suppose we want to write $h \in R[x]$ as a product of irreducibles. We first express $h = r\hat{h}$ where $\hat{h}$ has content 1 and $r \in R$, and factor r as a product of irreducibles over R. Then we factor

$$\hat{h} = p_1 \cdots p_r$$

where p_j is irreducible over the fraction field L of R. Successively applying Gauss' Lemma, we can choose the p_j to be polynomials of content 1 in $R[x]$. □

A.6 Field extensions

A *field extension* L/k is a nontrivial homomorphism of fields

$$k \hookrightarrow L;$$

Exercise A.4 guarantees these are injective. The extension is *finite* if L is finite-dimensional as a vector space over k.

Here is the main source of finite extensions: Suppose $f \in k[x]$ is an irreducible, so that $\langle f \rangle$ is maximal by Proposition A.11. Then $L = k[x]/\langle f(x) \rangle$ is a field and the induced homomorphism

$$k \hookrightarrow k[x] \twoheadrightarrow L$$

is nonvanishing as f is nonconstant; we have $\dim_k L = \deg f$ and $k = L$ if and only if $\deg f = 1$.

Definition A.15 Given a field extension L/k, an element $z \in L$ is *algebraic over k* if there exists a nonzero polynomial $f \in k[x]$ with $f(z) = 0$. The extension is *algebraic* if each element $z \in L$ is algebraic over k.

Given a field extension L/k and elements $z_1, \ldots, z_N \in L$, let $k(z_1, \ldots, z_N)$ denote the smallest subfield of L containing k and $z_1, \ldots, z_N$.

Proposition A.16 *Let L/k be a field extension.*

1. *If L/k is finite then it is algebraic.*
2. *An element $z \in L$ is algebraic over k if and only if $k(z)/k$ is finite.*
3. *The collection of all elements of L algebraic over k forms a field.*
4. *If M/L and L/k are algebraic extensions then M/k is algebraic.*

A.6 FIELD EXTENSIONS

Proof Suppose L/k is finite and take $z \in L$. Consider $k(z) \subset L$, a finite-dimensional vector space over k. For d sufficiently large, the set $\{1, z, \ldots, z^d\}$ is linearly dependent, i.e.,

$$c_d z^d + c_{d-1} z^{d-1} + \cdots + c_0 = 0.$$

We take $f = c_d x^d + \cdots + c_0 \in k[x]$.

Consider the k-algebra homomorphism

$$\mathrm{ev}(z) : k[x] \to k[z]$$
$$x \mapsto z.$$

Since $k[x]$ is a PID, $\ker(\mathrm{ev}(z)) = \langle f \rangle$ for some $f \in k[x]$. Now $k[z]$ is an integral domain because it sits inside L, so f is either irreducible or zero. Of course, f is irreducible precisely when z is algebraic, in which case $\dim_k k[z] = \deg f$. (We call f the *irreducible polynomial of z over k*.) Furthermore, $\langle f \rangle$ is maximal by Proposition A.11 so

$$k[z] \simeq k[x] \langle f(x) \rangle$$

is a field. It follows that $k[z] = k(z)$ and $\dim_k k(z) = \deg f$. On the other hand, $f = 0$ when z is not algebraic, in which case $\dim_k k[z] = \infty$.

Suppose z_1 and z_2 are nonzero and algebraic over k; *a fortiori*, z_2 is algebraic over $k(z_1)$. By our previous analysis, $k(z_1)$ is finite-dimensional over k and $k(z_1, z_2)$ is finite-dimensional over $k(z_1)$. It follows that $k(z_1, z_2)$ is finite-dimensional over k. We have

$$k \subset k(z_1 + z_2), k(z_1 - z_2), k(z_1 z_2), k(z_1/z_2) \subset k(z_1, z_2),$$

so all the intermediate fields are finite extensions of k. This implies that $z_1 + z_2$, $z_1 - z_2$, $z_1 z_2$, z_1/z_2 are all algebraic over k, so the algebraic elements form a field.

We prove the last assertion. Given $z \in M$, there exists a nonzero polynomial

$$f(x) = c_d x^d + \cdots + c_0 \in L[x]$$

with $f(z) = 0$. This means that $k(z, c_0, \ldots, c_d)$ is finite-dimensional over $k(c_0, \ldots, c_d)$. However, each c_i is algebraic over L and so $k(c_i)$ is finite over k. Iterating the argument of the last paragraph, we find that $k(c_0, \ldots, c_d)$ and hence $k(c_0, \ldots, c_d, z)$ is finite over k. But then $k(z) \subset k(c_0, \ldots, c_d, z)$ is finite over k, and z is algebraic over k. $\square$

A field k is *algebraically closed* if it has no nontrivial finite extensions; equivalently, any nonconstant polynomial in $k[x]$ has a root in k (see Exercise A.14). Standard texts in complex analysis and abstract algebra prove the following theorem.

Theorem A.17 (Fundamental Theorem of Algebra) *The complex numbers are algebraically closed.*

More generally, every field k admits an extension L/k which is algebraically closed [1, p. 528].

A.7 Exercises

A.1 Let R be a ring.
   (a) Show that there is a unique element $e \in R$ with $ea = a$ for each $a \in R$.
   (b) Show that if $0 = 1$ then $R = 0$.

A.2 Show that $\mathbb{Z}/N\mathbb{Z}$ is not a domain if $N > 1$ is a composite number. Show that $\mathbb{R}[x]/\langle x^2 \rangle$ is not a domain.

A.3 Let R be a domain with $|R| < \infty$. Show that R is a field.

A.4 Show that any ring homomorphism $k \to R$ from a field to a nonzero ring is injective.

A.5 Let $\phi : R \to S$ be a ring homomorphism. Show that the kernel

$$\ker(\phi) = \{a \in R : \phi(a) = 0\} \subset R$$

is an ideal. Show that the image

$$\mathrm{image}(\phi) = \{b \in S : b = \phi(a) \text{ for some } a \in R\} \subset S$$

is a ring.

A.6 Show that the intersection of two ideals $I, J \subset R$ is an ideal. Show by example that the union of two ideals need not be an ideal.

A.7 Show that any quotient of a finitely generated module is finitely generated.

A.8 Finish the proof of Proposition A.7.

A.9 Show that the ring

$$R = \mathbb{Z}[x]/\langle 3, x^2 + 1 \rangle$$

is a field with nine elements.

A.10 (a) Let k be a ring. Show there exists a unique nonzero ring homomorphism $j : \mathbb{Z} \to k$.
   (b) If k is a field, show that $j(\mathbb{Z})$ is a domain. Prove that $\ker(j) = \langle 0 \rangle$ or $\ker(j) = \langle p \rangle$ for some prime p.
   In the first case, we say that k is a *field of characteristic zero*. In the second case, k is a *field of characteristic p*.

A.11 Let k be a field. Show that $k[x]^* = k^*$.

A.12 Prove Corollary A.13.

A.13 Let k be a field, $f \in k[x]$ a nonzero polynomial, and $\alpha \in k$.
   (a) Show that $x - \alpha$ is irreducible in $k[x]$.
   (b) Show that $f(\alpha) = 0$ if and only if $x - \alpha$ divides $f(x)$.
   (c) Show that if $f(\alpha_1) = \cdots = f(\alpha_m) = 0$ for distinct $\alpha_1, \ldots, \alpha_m \in k$ then $(x - \alpha_1) \cdots (x - \alpha_m)$ divides f and $m \leq \deg(f)$.

A.14 Let k be a field. Show that the following statements are equivalent:
   (1) k is algebraically closed;

(2) every nonconstant polynomial $g \in k[x]$ has a root in k;
(3) every irreducible in $k[x]$ has degree 1;
(4) every nonconstant polynomial $g \in k[x]$ factors

$$g = c \prod_{i=1}^{d}(x - \alpha_i), \quad c, \alpha_1, \ldots, \alpha_d \in k.$$

Bibliography

[1] Michael Artin. *Algebra*. Englewood Cliffs, NJ: Prentice Hall Inc., 1991.

[2] Hyman Bass. A nontriangular action of $\mathbf{G}_a$ on $\mathbf{A}^3$. *J. Pure Appl. Algebra*, **33**:3(1984),1–5.

[3] Etienne Bézout. *General Theory of Algebraic Equations*. Princeton, NJ: Princeton University Press, 2006. [Translated from the 1779 French original by Eric Feron.]

[4] W. Dale Brownawell. Bounds for the degrees in the Nullstellensatz. *Ann. Math. (2)*, **126**:3(1987), 577–591.

[5] Bruno Buchberger. Ein Algorithmus zum Auffinden der Basiselemente des Restklassenrings nach einem nulldimensionalen Polynomideal. PhD thesis, Universität Innsbruck, 1966.

[6] Arthur Cayley. On the theory of elimination. *Cambridge and Dublin Math. Journal*, **3**(1848),116–120. [Reprinted in Vol. 1 of the *Collected Papers*, Cambridge University Press, 1889; also reprinted in [12].]

[7] C. Christensen, G. Sundaram, A. Sathaye, and C. Bajaj, eds. *Algebra, Arithmetic and Geometry with Applications*. Berlin: Springer-Verlag, 2004. [Papers from Shreeram S. Abhyankar's 70th birthday conference held at Purdue University, West Lafayette, IN, July 20–26, 2000.]

[8] David Cox, John Little, and Donal O'Shea. *Ideals, Varieties, and Algorithms: An Introduction to Computational Algebraic Geometry and Commutative Algebra*. Undergraduate Texts in Mathematics. New York: Springer-Verlag, second edition, 1997.

[9] David Eisenbud. *Commutative Algebra: With a View Toward Algebraic Geometry*. Graduate Texts in Mathematics, 150. New York: Springer-Verlag, 1995.

[10] David Eisenbud, Craig Huneke, and Wolmer Vasconcelos. Direct methods for primary decomposition. *Invent. Math.*, **110**:2(1992), 207–235.

[11] William Fulton. *Algebraic Curves: An Introduction to Algebraic Geometry*. Advanced Book Program. Redwood City, CA: Addison-Wesley, 1989. [Notes written with the collaboration of Richard Weiss, Reprint of 1969 original.]

[12] I. M. Gel′fand, M. M. Kapranov, and A. V. Zelevinsky. *Discriminants, Resultants, and Multidimensional Determinants*. Mathematics: Theory & Applications. Boston, MA: Birkhäuser Boston Inc., 1994.

[13] Hermann Grassmann. *A New Branch of Mathematics*. Chicago, IL: Open Court Publishing Co., 1995. [The *Ausdehnungslehre* of 1844 and other works, translated from the German and with a note by Lloyd C. Kannenberg, With a foreword by Albert C. Lewis.]

[14] Phillip Griffiths and Joseph Harris. *Principles of Algebraic Geometry*. New York: Wiley-Interscience, 1978. [Pure and applied mathematics.]

[15] Brian Harbourne. The geometry of rational surfaces and Hilbert functions of points in the plane. In *Proceedings of the 1984 Vancouver Conference in Algebraic Geometry*, Vol. 6, ed. A. V. Geramita and P. Russell. Providence, RI: American Mathematical Society, 1986, pp. 95–111.

[16] Brian Harbourne. Points in good position in $\mathbf{P}^2$. In *Zero-Dimensional Schemes: Proceedings of the International Conference Held in Ravello, Italy, June 8–13, 1992*, ed. F. Orecchia and L. Chiantini. Berlin: de Gruyter, 1994, pp. 213–229.

[17] Joe Harris. *Algebraic Geometry: A First Course*. Graduate Texts in Mathematics, 133. New York: Springer-Verlag, 1992.

[18] Joe Harris, Barry Mazur, and Rahul Pandharipande. Hypersurfaces of low degree. *Duke Math. J.*, **95**:1(1998), 125–160.

[19] Robin Hartshorne. *Algebraic Geometry*. Graduate Texts in Mathematics, 52. New York: Springer-Verlag, 1977.

[20] Brendan Hassett. Some rational cubic fourfolds. *J. Algebraic Geom.*, **8**:1(1999), 103–114.

[21] André Hirschowitz. Une conjecture pour la cohomologie des diviseurs sur les surfaces rationnelles génériques. *J. Reine Angew. Math.*, 397(1989), 208–213.

[22] W. V. D. Hodge and D. Pedoe. *Methods of Algebraic Geometry*, Vol. I. Cambridge University Press, 1947.

[23] Amit Khetan. The resultant of an unmixed bivariate system. *J. Symbolic Comput.*, **36**:3-4(2003), 425–442. [International Symposium on Symbolic and Algebraic Computation (ISSAC'2002) (Lille).]

[24] János Kollár. Sharp effective Nullstellensatz. *J. Am. Math. Soc.*, **1**:4(1988), 963–975.

[25] János Kollár. Low degree polynomial equations: arithmetic, geometry and topology. In *European Congress of Mathematics*, Vol. 1 *(Budapest, 1996)*, ed. A. Balog, G. O. H. Katona, A. Recski and D. Szász. Progress in Mathematics, 168. Basel: Birkhäuser, 1998, pp. 255–288.

[26] János Kollár. Unirationality of cubic hypersurfaces. *J. Inst. Math. Jussieu*, **1**:3(2002), 467–476.

[27] Serge Lang. *Algebra*, second edition. Advanced Book Program, Reading, MA: Addison-Wesley, 1984.

[28] F. S. Macaulay. On some formula in elimination. *Proc. Lond. Math. Soc.*, **3**(1902), 3–27.

[29] Hideyuki Matsumura. *Commutative ring theory*, Cambridge Studies in Advanced Mathematics, 8, second edition. Cambridge University Press, 1989. [Translated from the Japanese by M. Reid.]

[30] Rick Miranda. *Algebraic Curves and Riemann Surfaces*, Graduate Studies in Mathematics, 5. Providence, RI: American Mathematical Society, 1995.

[31] Rick Miranda. Linear systems of plane curves. *Not. Am. Math. Soc.*, **46**:2(1999), 192–201.

[32] David Mumford. *Lectures on Curves on an Algebraic Surface*. [With a section by G. M. Bergman.] Annals of Mathematics Studies, 59. Princeton, NJ: Princeton University Press, 1966.

[33] Emmy Noether. Idealtheorie in Ringbereichen. *Math. Ann.*, **83**(1921), 24–66.

[34] Kapil H. Paranjape and V. Srinivas. Unirationality of the general complete intersection of small multidegree. In *Flips and Abundance for Algebraic Threefolds*. Paris: Société Mathématique de France, 1992, pages 241–248. [Papers from the Second Summer Seminar on Algebraic Geometry held at the University of Utah, Salt Lake City, Utah, August 1991, Astérisque No. 211 (1992).]

[35] G. C. M. Ruitenburg. Invariant ideals of polynomial algebras with multiplicity free group action. *Compositio Math.*, **71**:2(1989), 181–227.

[36] Frank Olaf Schreyer. Die Berechnung von Syzygien mit dem verallgemeinerten Weierstrass'schen Divisionssatz, 1980. Diploma thesis.

[37] Igor R. Shafarevich. *Basic Algebraic Geometry 1: Varieties in Projective Space*, second edition. Berlin: Springer-Verlag, 1994. [Translated from the 1988 Russian edition and with notes by Miles Reid.]

[38] Igor R. Shafarevich. *Basic Algebraic Geometry 2: Schemes and Complex Manifolds*, second edition. Berlin: Springer-Verlag, 1994. [Translated from the 1988 Russian edition by Miles Reid.]

[39] B. L. van der Waerden. *Moderne Algebra, Unter Benutzung von Vorlesungen von E. Artin und E. Noether I,II*. Die Grundlehren der mathematischen Wissenschaften in Einzeldarstellungen mit besonderer Berücksichtigung der Anwendungsgebiete Bd.33, 44. Berlin: J. Springer, 1930/1931. [1937, 1940 (second edition); Berlin-Göttingen-Heidelberg: Springer-Verlag, 1950, 1955 (third edition); 1955, 1959 (fourth edition); 1960 (fifth edition of vol. I); 1964 (sixth edition of vol. I); 1967 (seventh edition of vol. I); 1967 (fifth edition of vol. II); 1971 (eighth edition of vol. I). English translations: Frederick Ungar Publishing Co., New York, 1949 (second edition); Frederick Ungar Publishing Co., New York, 1970 (seventh edition of volume I and fifth edition of volume II); Springer-Verlag, New York, 1991; Springer-Verlag, New York, 2003.

Index

k-algebra, 235

Abhyankar, Shreeram S., 57, 86
affine linear subspace, 70
affine open covering, 160
affine open subset, 48
affine space, 1
algebraic extension, 241
algebraic group, 56
algebraically closed field, 242
algebraically independent elements, 105
arithmetic algebraic geometry, 38
arithmetic genus, 229
Artinian ring, 32
ascending chain condition, 20, 31, 91
associated graded ring, 233
associated prime, 120
    embedded, 124
    minimal, 124
automorphism
    of affine line, 55
    of affine space, 43
    of an affine variety, 43

Bézout Theorem, 223
    higher-dimensional, 224
Bézout, Étienne, 223
base field, 1
bihomogeneous form, 178
birational map, 94
birational morphism, 47
birational varieties, 94
Buchberger, Bruno, 29
Buchberger Algorithm, 24
    Buchberger Criterion, 23

cardioid, 68
Cayley cubic surface, 178, 203
Cayley, Arthur, 86
characteristic, 243

classical adjoint, 83
closed graph condition, 157
closed graph property, 160
closure
    projective, 140
    Zariski, 38
cone
    over a projective variety, 177
    over a variety, 68
consistent system of linear equations, 100
content, 240
continuous function, 39
coordinate ring, 41
Cramer's rule, 56, 83, 173
cusp singularity, 186

degree of projective variety, 229
dehomogenization, 79, 138
Descartes Circle Theorem, 70
descending chain condition, 32
determinant, 73
    of a complex, 86
diagonal, 54, 167
dimension, 115
    of projective variety, 228
    of variety, 111
Diophantine geometry, 38
direct sum
    of modules, 237
    of rings, 99
discriminant, 78
distinguished open subset, 135
division algorithm, 13
domain, 91, 235
dominant morphism, 92
dominant rational map, 51, 93
dual projective plane, 164
dual projective space, 181
dual space, 200
dual variety, 185

Eisenbud, David, xii
elimination projective, 172
elimination order, 60
Elimination Theorem, 60
Emmy, Noether, 20
equivalence of morphisms, 41
Euclidean Algorithm, 239
Euler's Formula, 184
evaluation homomorphism, 103
extension
    of a morphism, 41
    of a rational map, 46
exterior algebra, 194
exterior power, 191

faithfully flat neighborhood, 84
Fermat's Last Theorem, 37
field, 235
field extension, 51, 241
finite field, 239
    difficulties over, 33, 41
finitely generated algebra, 108
finitely generated field extension, 104
finitely generated ideal, 19
finitely generated module, 28, 238
finitely generated module of syzygies, 27
flag variety, 206
flat module, 98
fractions
    field of, 235
    ring of, 45
Fulton, William, xii
function field, 92, 104

Gauss Lemma, 107, 240
general linear group, 56
general position
    linear, 65, 221
genus, 230
gluing maps, 159, 160
Gröbner basis, 15
    reduced, 30
Gröbner, Wolfgang, 29
graded algebra, 207
graded commutative, 191
graded coordinate ring, 208
graded lexicographic order, 13
graded monomial order, 138, 170
graded reverse lexicographic order, 13
    efficiency of, 30
graded ring, 207
graph
    of a morphism, 58
    of a rational map, 64
Grassmann, Herman Günter, 187
Grassmannian, 187

Harbourne/Hirschowitz Conjecture, 228
Harris, Joe, xii
Hessian, 225
Hilbert Basis Theorem, 19, 91
    generalized, 27
Hilbert function, 208
Hilbert polynomial, 212, 214
Hilbert scheme, 231
Hilbert, David, 19
homogeneous ideal, 138
homogeneous pieces, 137
homogeneous polynomial, 1, 170
homogenization, 79, 138
    relative to a set of variables, 170
homomorphism
    of k-algebras, 235
    of rings, 235
hyperplane at infinity, 142
hypersurface, 4, 115, 182
    irreducible, 110
    projective, 140

ideal, 235
    of polynomials vanishing on a set, 33
ideal membership, 4
ideal product, 35
ideal sum, 35
implicitization, 1, 2
    generalized, 57
incidence correspondence, 164, 182
independent conditions, 6, 226
indeterminacy
    ideal, 50, 128
    locus, 44, 49, 50, 146
inflectional tangent, 186, 225
integral element, 106
interpolation, 218
    generalized, 226
    simple, 5
intersection
    of ideals algorithm, 95
    of monomial ideals, 100
    of varieties, 36
intersection multiplicity, 217
irreducible component, 90
irreducible element, 239
irreducible ideal, 117
irreducible polynomial, 89, 92, 242
irreducible variety, 90
irredundant decomposition
    into irreducible components, 90
    into irreducible ideals, 117
    into primary components, 126
irrelevant ideal, 144, 209
isomorphism of affine varieties, 43

INDEX

join, 68

Koszul complex, 86, 224

Lang, Serge, xii
leading monomial, 13
leading term, 13
   ideal of, 14, 97
least common multiple, 22
lemniscate, 69
lexicographic order
   inefficiency of, 30
line at infinity, 165
linear morphism to projective space, 148
linear subspace of projective space, 149, 203
localization, 50

maximal ideal, 238
module, 237
monomial, 11
   ideal, 14
   order, 11
morphism
   defined over k, 2, 42
   of abstract varieties, 161
   of affine spaces, 1, 39
   of affine varieties, 42
   projection, 36, 58
multiple root, 78
multiplicative subset, 50
multiplicity
   of a polynomial at a point, 179, 225
   of an ideal at a point, 216

node singularity, 186
Noetherian ring, 20, 117
nontrivial zero, 78
normal form, 16, 97
Nullstellensatz, 38, 92
   effective, 102
   Hilbert, 102
   projective, 144

O'Shea, Donal, xii
order of vanishing at a point, 179, 225

parallel lines, 164
parametrization
   rational, 51
   regular, 4
Plücker embedding, 198
Plücker formulae, 186, 225
Plücker relations, 200
Plücker, Julius, 198
polynomial map

of projective spaces, 145
polynomial ring, 236
primary decomposition, 94, 121
primary ideal, 119
prime avoidance lemma, 129
prime ideal, 91, 238
principal ideal domain, 74, 239
product order, 60
product variety, 36
projection from a subspace, 149
projective space, 134
projectivity, 149, 209
projectivization of a vector space, 137
pull-back, 40
pure lexicographic order, 12

quotient ideal, 118
quotient ring, 236

radical
   ideal, 101
   of monomial ideals, 114
rational map
   of abstract varieties, 161
   of affine spaces, 43
   of affine varieties, 46
rational normal curve, 152
rational variety, 52
reduced row echelon form, 188
reducible ideal, 117
reducible variety, 90
refinement, 161
resultant, 115
   Sylvester, 75
Riemann surface, 230
ring, commutative, 234

saturation, 171
Schreyer, Frank Olaf, 29
Schubert calculus, 204
Schubert variety, 204
Schubert, Hermann, 204
scroll, 67
secant
   subspace, 66
   variety, 66
Segre imbedding, 154
Segre, Corrado, 154
sign of a permutation, 192
similar matrices, 133
singular point, 182
smooth hypersurface, 185
smooth point, 182
span, affine, 65
Steiner Roman surface, 165, 166
Steiner, Jakob, 166

submodule, 238
Sylvester, James Joseph, 75
symbolic power, 131
syzygy, 26, 85

tangent cone, 233
tangent space
  affine, 182
  projective, 183
topology, abstract, 38
transcendence
  basis, 105
  degree, 105

union
  disjoint, 99
  of varieties, 36
unique factorization domain, 89, 107, 239
uniqueness of primary decomposition, 123, 126
unirational hypersurfaces of small degree, 53
unirational variety, 51
units, group of, 239

van der Waerden, Bartel Leendert, 86
Vandermonde matrix, 221
variety
  abstract, 160
  affine, 34
  projective, 140
Veronese embedding, 150
  quadratic, 152
Veronese, Giuseppe, 150

weakly irredundant primary decomposition, 122
wedge product, 191
weighted polynomial ring, 208

Zariski topology, 39
Zariski, Oscar, 39

Made in the USA
Lexington, KY
17 January 2014